THERMAL ENERGY STORAGE

ON ENERGY SYSTEMS AND TECHNOLOGY

A series devoted to the publication of courses and educational seminars given at the
Joint Research Centre, Ispra Establishment, as part of its education and training program.
Published for the Commission of the European Communities,
Directorate-General Information Market and Innovation.

Already published:

ENERGY STORAGE AND TRANSPORTATION
PROSPECTS FOR NEW TECHNOLOGIES

Edited by G. BEGHI
1981, x + 497 pp.

Additional volumes in preparation.

THERMAL ENERGY STORAGE

*Lectures of a Course held at the Joint Research Centre,
Ispra, Italy, June 1-5, 1981*

Edited by

G. BEGHI

Joint Research Centre, Ispra, Italy

D. REIDEL PUBLISHING COMPANY
Dordrecht : Holland / Boston : U.S.A.
London : England

Library of Congress Cataloging in Publication Data
Main entry under title:

Thermal energy storage.

　　　Includes index.
　　　1.　Heat storage–Addresses, essays, lectures.　I.　Beghi, G.
II.　Commission of the European Communities. Joint Research Centre.
TJ260.T483　　　621.402'8　　　　　　　　82–5330
ISBN-13: 978-94-009-7845-4　　　　e-ISBN-13: 978-94-009-7843-0
DOI: 10.1007/978-94-009-7843-0

Commission of the European Communities　　　　　　　Joint Research Centre Ispra (Varese), Italy

Publication arrangements by
Commission of the European Communities
Directorate-General Information Market and Innovation, Luxembourg

EUR 7955 EN
Copyright © 1982, ECSC, EEC, EAEC, Brussels and Luxembourg
Softcover reprint of the hardcover 1st edition 1982

Published by D. Reidel Publishing Company
P.O. Box 17, 3300 AA Dordrecht, Holland

Sold and distributed in the U.S.A. and Canada
by Kluwer Boston Inc.,
190 Old Derby Street, Hingham, MA 02043, U.S.A.

In all other countries, sold and distributed
by Kluwer Academic Publishers Group,
P.O. Box 322, 3300 AH Dordrecht, Holland

D. Reidel Publishing Company is a member of the Kluwer Group

TABLE OF CONTENTS

INTRODUCTION TO THE ISPRA COURSE

G. Beghi
Commission of the European Communities
Joint Research Centre - Ispra Establishment
21020 Ispra (Va) - Italy

I have the pleasure of welcoming you, on behalf of the Joint Research Centre of the Commission of the European Community.

Since it was established, our organization has been active in the field of energy with research and development work, and different actions for dissemination of information. In recent years increasing attention has been devoted, within the programmes of the Ispra Establishment, to the so-called alternative or renewable energies.

It is well known that energy systems include primary energies and secondary energies; these are the carriers which make energy available to the users in the form in which it will be used (e.g. light, heat, movement, etc.). According to the development in the availability of different primary energies, the energy systems are changing; considering the renewable energies, transformation into energy carriers is more and more necessary, and through increasingly expensive processes. The problem of matching the availability (energy production) with the needs (energy use) is linked to discontinuity, out-of-phase and distance. The importance of storage, in this context, is growing.

Between the primary energy and the end use we may have different forms of energy:
- mechanical
- electrical
- chemical
- thermal.

The system (see Fig. 1) is completed by storage, (i) in different forms, (ii) at different levels of temperature, if thermal, (iii) with different sizes, i.e. daily or seasonal. The introduction of storage, in the right form, may help obtain a better use of a system based on the use of different primary energies.

Storage is particularly important, and may be critical for thermal energy: this is the particular subject of the present Ispra Course, which hopes to show the problems, the technological solutions and the future prospects.

As well as its role in energy systems, thermal energy storage is important as part of

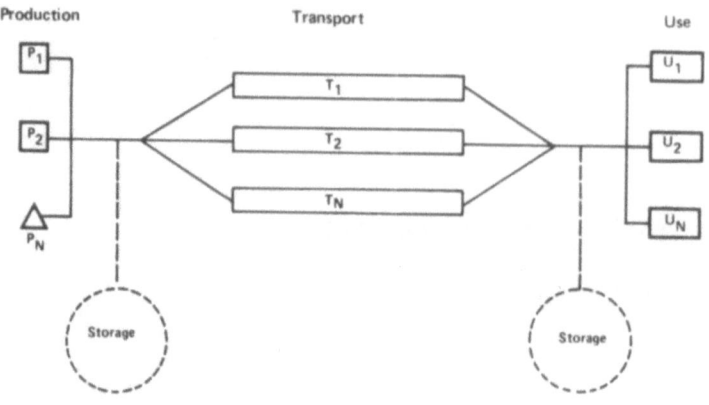

Production   Transport   Use

Energy system, including storage, from primary energy (production) to the end use. Interposition of storage can be helpful in solving the problem of fluctuating sources coupled with variable demands.

Fig. 1

energy conservation: energy saving through the recovery of industrial process heat, otherwise wasted, should be mentioned.

Research may contribute to an extended and efficient introduction of thermal energy storage into present and future energy systems. Simple and reliable technologies, which are economic and competitive are needed; large size realizations are useful to demonstrate feasibility and economic viability.

We hope that the present Ispra Course, with your participation, may make a small but useful contribution towards making thermal energy storage an important factor in future energy systems.

ENERGY STORAGE AND THE ENERGY SYSTEMS

N. Kurti

Department of Engineering Science,
University of Oxford,
U.K.

ABSTRACT

Storage does not save energy but forms nevertheless an important part of energy strategies. It enhances the use of the renewable energy sources and permits the more efficient use of high capital cost conversion plant.

After surveying the main objectives of energy consumption the paper lists the energy sources provided by nature, the forms in which energy is used by society and the storage periodicity needed for various combinations of source and use, in particular for electricity production. The paper concludes with a review and comparison of various storage methods.

INTRODUCTION

Preoccupation with energy questions has become fashionable recently - perhaps a decade ago. For instance, in a representative world-wide sample of periodicals 75% of those having the word "Energy" in their title were started after 1970. Among the remaining 25% pre-1970 "energy" periodicals there are quite a few which embraced that magic word only in the last few years. Thus "Actualités, Combustibles, Energie" had originally the title "Bulletin Synoptique de Documentation Thermique" and, similarly, "Energy World" and the "Journal of the Institute of Energy" started life as the "Journal of the Institute of Fuel". Why is there this more than threefold increase in 10 years?

Let me begin by telling you a true story. Early in the

morning of Friday, 5th December 1952, dense fog enveloped the whole of London and continued for four whole days, until Tuesday morning, 9th December. This in itself was not a very unusual occurrence. The proverbial London fog had come to be regarded almost as one of the touristic sights of London like the British Museum, Buckingham Palace or Westminster Abbey. What made this incident particularly noteworthy was that it was accompanied by a very marked temperature inversion: the air near the ground remained cold while the upper layers were warm. As a result, convective mixing in the atmosphere was greatly reduced and the noxious gases (e.g. SO_2) and the soot emitted by open domestic fireplaces, factories, power stations remained near the ground. This had tragic consequences and, as Fig. 1 shows, during the week immediately following the incidence of the fog the death rate in London was almost <u>three times</u> as high as in previous years (1).

One might well ask why I should start the opening lecture of a course devoted to Thermal Energy Storage with such a sad tale. Although I would be justified in using this as a dramatic example of how bad energy management and inefficient energy utilization can lead to misery and death, my reason is more subtle. It had been known for many decades that the inefficient and archaic methods of combustion were the chief reason of the London fog, of the smoke-laden atmosphere of our cities, of the uniform greyness and blackness of many of our cherished architectural monuments, but nothing was done about it. Anyone interested in disheartening accounts of the efforts of public-spirited people to introduce legislation to curb atmospheric pollution should read the scholarly article by Lord Ashby and Dr. Mary Anderson (2). All those efforts were of no avail. With cheap and plentiful coal available, appeals to the conscience of parliamentarians, of industrialists, of public servants fell on deaf ears and it needed the dramatic impact of a few thousand sudden deaths to jolt politicians into action, and in less than 4 years the Clean Air Act, covering our large cities and many of our towns, was passed by Parliament. As a result the greyness of our buildings has given way to warm gold or glittering white, we can enjoy breathing the air in our cities (at least as far as smoke emission is concerned, motorcar exhausts are another story) and our laundry bills are dramatically reduced.

A similar incident - a relatively small cause followed by a large effect - happened 21 years later, in October 1973. It had been known for decades that the world's energy resources are not inexhaustible and that difficulties might arise by the end of the century, but apart from studies and discussions - many of them under the auspices of the World Energy Conference - few active or significant steps were taken. It needed the occurrence of a relatively minor war in the Middle East in

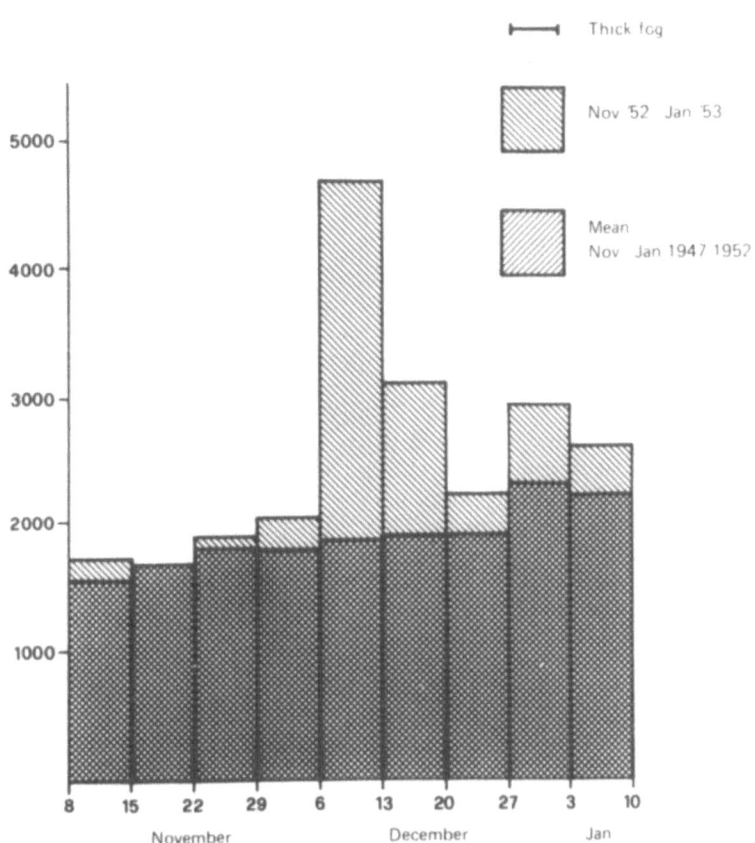

Figure 1. Weekly mortality in LONDON

October 1973 and its immediate effect on oil supplies to jolt people out of their complacency and the previously muted warnings about possible energy shortages turned into a deafening chorus. Perhaps scientists and engineers should always be on the look-out for dramatic illustrations in order to persuade politicians to do the right thing.

WHY STORE ENERGY?

It should be made clear at the outset that storage does not save energy, at least not directly. On the contrary, energy is invariably degraded or lost during storage. In the case of heat storage imperfect thermal insulation can bring about appreciable losses. When storing free energy such as kinetic energy, e.g. in flywheels, friction causes degradation and the resulting low grade heat is lost. Even fuels, be they fossils or nuclear, are not immune, although the effects of slow oxidation in the first case and radioactive decay in the second are imperceptible. And yet there are very good reasons why energy storage should have become such an important part of our energy strategies.

The raison d'être of energy storage has two aspects. First, there is a growing demand for the more widespread use of renew-able energy sources such as tides, waves, wind, solar radiation. But the output of these sources is not governed by mankind, as is the mining of coal or the extraction of oil. It is nature that determines the intensity and the timing of the supply and, while rough predictions are possible, they are seldom sufficiently accurate to be relied on for planning. Clearly if one wants to make full use of the energy that nature provides free and without depleting its stock, any surplus should be collected, stored and used when it is needed.

Second, even if one uses fossil or fissile fuel when, in principle energy production can be adjusted to meet the demand, and no more fuel is used than necessary, storage can play an important role - not in saving energy but in reducing cost. The conversion and ultimate utilization of energy usually employs costly plant, such as power stations, boiler plants, heat distribution networks etc., and it is advantageous to run all high capital cost plant with as big a load factor as possible. However human activities show great diurnal, weekly or seasonal variations and, short of drastically altering our working and living habits, we must aim at storing energy produced or converted during periods of low demand and use the energy so stored to supplement the plant's output when demand is high.

HOW IS ENERGY USED BY SOCIETY?

I shall limit this discussion to the so-called "developed" countries which are responsible for an overwhelming part of the world's energy consumption. The uses can be divided into three large categories, each of them responsible for very roughly one third of the total:

1. The maintenance of individuals in adequate bodily comfort. This is achieved by warming (or cooling) them, by feeding them, by providing them with hot water, light etc. In temperate climates this consumption has rather strong seasonal variations because of the importance of heating. But this is by no means a general rule; thus in the U.S.A. because of the wide use of air-conditioning the seasonal variation is smoothed out. It may even be reversed in some instances: on the East coast of the U.S.A. electricity peaks may occur in the summer, because air-conditioning is run mostly by electricity while heating is done largely by gas or oil or coal.

2. Industry, business, commerce. Here we are governed by our working and living habits. We prefer to work during the day and to take the same days of rest as the other members of the community. This results in strong daily and weekly variations. It is unlikely that people would take kindly to the idea of a work routine which would permit our factories and offices to be used for 24 hours a day and for 7 days a week.

3. In the case of transport weekly and seasonal variations are less marked than in the two previous categories, but short term storage could be important. Thus the success of the electric car or lorry depends on the existence of light and cheap storage batteries or of flywheels. The latter could also be useful if one were to replace, at least partially, the present frictional dissipative braking by a reversible transfer of the vehicle's kinetic energy to a flywheel or, through the intermediary of a dynamo, to a storage battery.

ENERGY SOURCES PROVIDED BY NATURE.

Apart from tidal, nuclear and geothermal energy all our primary energy sources derive ultimately from the sun. However, it is more convenient to consider solar energy that reaches us in the form of radiation separately from "converted" solar energy, especially since, depending on the process, the conversion may take days, years, thousands or millions of years to become completed. Let us now consider one by one the different forms of energy available to mankind and indicate the intermittence of its availability. This in turn will determine the desirability or necessity of storage and also the periodicity and duration required.

For convenience four categories will be used in this discussion:

a) short ($\underline{s}$), i.e. appreciable variations occur in a matter of minutes, say up to an hour.

b) daily ($\underline{d}$). Here the variations occur at intervals of several hours, say up to a day.

c) medium ($\underline{m}$). Variations over periods of days, say up to a month.

d) long ($\underline{l}$). Refers to periods of several months, which are usually seasonal in nature but can extend to years.

Energy sources may also be placed in several categories according to their nature. This classification is somewhat arbitrary - it has its origins in custom and usage and purists may frown on it. It is nevertheless adopted here for convenience.

Kinetic and Potential Energy.

a) Wind, next to waterfalls or rivers (see d) is probably the oldest in use. From the point of view of storage $\underline{s}$, $\underline{d}$, $\underline{m}$ and $\underline{l}$ may all have to be considered. $\underline{s}$ i.e. short term variations in wind intensity, call for storage mainly to smooth out variations. $\underline{d}$ is perhaps less important, but both $\underline{m}$ and $\underline{l}$ need storage in most parts of the world if wind is to be used as the main source of energy.

b) Waves have broadly the characteristics of wind, except that, because of the inertia in the building up of waves, $\underline{s}$ may be neglected.

c) Tides represent the most predictable renewable energy source. The daily variations can be calculated accurately, and so can, though to a lesser extent so far as amplitudes or intensities go, the medium and long term variations.

d) "Hydro" (for lack of a better name) represents another ancient renewable energy source. Waterfalls and fast rivers are the most important examples. The variations here are chiefly seasonal ($\underline{l}$) but medium ($\underline{m}$) must also be reckoned with.

Chemical Energy.

a) Fossil fuels, which include coal, natural gas, peat, oil.

These constitute excellent storage materials in themselves. In most cases storage can be at the source, i.e. in the coal mines or the oil wells, though to make continuous production possible or for other reasons the fuel may have to be moved from the source and stored elsewhere.

b) Wood, unlike the fossil fuels, is a renewable energy source with a periodicity of years or decades, so that storage is always of a long term nature. The importance of wood as an energy source should not be underestimated. Even in a highly developed country such as the U.S. its contribution is not negligible. Thus, a few years ago - it may still be true today - wood provided about the same amount of energy as nuclear power stations.

c) Crops for human and animals consumption or to be used as biomass, i.e. for conversion into other forms of energy. These obviously require long term storage.

Nuclear Fuels.

As in the case of fossil fuels, no conversion is necessary for storage.

Thermal Energy.

Geysers, hot springs, hot rocks. Only exceptionally is storage necessary.

Solar Energy.

This is a highly variable energy source and all types of storage s, d, m and l may be advisable or necessary if one wants to rely entirely on solar radiation to satisfy one's energy needs and to put energy conversion equipment to full use.

IN WHAT FORMS IS ENERGY USED.

Just as in the previous section, the grouping of energy uses into categories is somewhat arbitrary and can lead to over-lap.

1. Electrical. Although this energy form represents, even for the most developed countries, a rather small fraction of the total energy used - 10-15% in the U.K. - it is the most versatile form. Also it should be emphasised, that, unlike chemical and thermal energy, it can be readily converted, with very high efficiency, into kinetic energy. The four pricipal uses are:

a) <u>Transport</u> (electrified railways, trams, trolley-buses).
It is a fluctuating use, the intermittence being mainly <u>s</u>,
(stopping and starting) and <u>d</u> (less urban traffic and, apart
from freight, less long distance traffic at night).
b) <u>Heating</u>. There are marked daily (<u>d</u>) and seasonal (<u>l</u>)
variations.
c) <u>Lighting</u>. Mainly daily (<u>d</u>) variations; seasonal
variations (<u>l</u>) are less important.
d) <u>Industry</u>. As mentioned earlier our working and living
habits cause strong daily (<u>d</u>) and weekly (<u>m</u>) variations
(weekends). However, extensive holidays, which in many
countries are not staggered so that an appreciable part of
industrial production comes to a standstill at the same
time, cause also long term (<u>l</u>) seasonal variations.

2. <u>Kinetic</u> energy is used in transport and in industry and
we have seen above the intermittence of these activities.

3. <u>Chemical</u> energy manifests itself in chemical reactions
and these are used in <u>heating</u> (combustion) and in <u>transport</u>
(e.g. the internal combustion engine) and in <u>industry</u> - both
chemical and manufacturing.

4. <u>Thermal</u> energy, i.e. a situation in which the energy is
supplied to the user directly in the form of heat. District
<u>heating</u>, provision of <u>process steam for industry</u> are covered by
this category.

WHAT STORAGE PERIODICTY IS WANTED FOR DIFFERENT COMBINATIONS
OF ENERGY SOURCE AND ENERGY USE?

This question will be discussed by reference to Fig. 2.
The left-hand side gives the various energy source categories,
indicating in each case the corresponding intermittence as
explained in the previous section. The right-hand side gives
the various uses, again with an indication of intermittence.
Lines originating from each source and typographically identified
lead to the various uses. We shall now discuss one by one the
combinations represented by these lines and try to establish
the storage requirements of the various conversion utilization
modes.
1) Since <u>potential energy</u>, e.g. that of water at the top
of a waterfall or of a heavy weight used for demolishing buildings,
is always converted into kinetic energy before use, the discussion
can be limited to <u>kinetic energy</u> as the primary energy source.
We see lines indicating conversion into electricity and lines
showing direct use of kinetic energy. The division into these
two uses is somewhat artificial, since even when conversion into
electricity takes place the kinetic energy provided by nature is

first converted into another form better adapted to electricity production (translational kinetic energy into rotational kinetic energy) – however, it is a convenient classification. A line which has been omitted is one going from "wind" or "hydro" to "thermal". The excuse for this is that the conversion of natur- ally occuring kinetic energy into heat has been little practised so far. But it has been used on a small scale where a watermill or a windwill provides solely low-grade heat, say at a temperature of about $30^{\circ}C$, to warm hot-houses. Conversion first into electricity would add to the capital cost without increasing the amount of useful heat gained. Considerable heat amplification could be obtained by using the kinetic energy to run a heat pump, but here again increased capital cost might render the method uneconomical. A further advantage of direct conversion into heat is that it lends itself to thermal storage. The highly viscous liquid in which the kinetic energy is dissipated by friction can serve as storage medium.

Let us now consider the lines originating from the four main sources.

a) Wind to Electrical. Storage needs are mainly determined by the intermittence of the energy received. As said earlier s, m, and l are called for, but it seems unlikely that long-term storage of the energy harvested during winter storms for use during windless summer periods would be practical and economical.

Wind to Kinetic, e.g. in sailing boats or windmills. I do not believe that there have been any attempts to store any surplus wind-energy for use when the boat is becalmed. In the case of windmills it may be possible to store surplus wind energy either as kinetic energy (e.g. by flywheels) or, via electricity, electrochemically.

b) Wave energy has moved into the forefront during the last decade. It seems to be considered exclusively for conversion into electricity and moreover for feeding into a large network rather than for local use. Therefore the storage requirements are determined by the needs of the electricity supply industry and not by intermittence of the energy source.

Interesting storage problems might arise if, as has been suggested, factory ships using wave power for propulsion and for industrial processes, (e.g. recovery of chemicals from sea water) became reality.

c) Tides, just as waves, are thought of mainly for electrical generation. From the point of view of storage they have the advantage of predictability which renders their integration into an electricty network even easier than in the case of wave energy.

d) <u>Hydraulic</u> energy can be used, just as wind energy, for direct electricity generation (local or supplying a network) after transformation from translational into rotational kinetic energy. Variations are, as mentioned above, mainly seasonal (<u>1</u>). Shorter term variations can be smoothed out since a hydro-electric power station with its reservoir and dam has its own built-in storage.

Much of what has been said in the case of <u>wind</u> energy, when used directly, also applies to <u>hydraulic</u> energy, (watermills instead of windmills) and in particular the generation of low grade heat is a possible application.

2. <u>Chemical</u> energy and 3. <u>Nuclear</u> energy may be considered together. As shown in Fig. 2 the only difference is that, while the energy stored in chemical bonds can be used directly in industrial processes (petrochemical industry), nuclear energy is always first converted into heat. The distribution between the two end-users, namely electrical and thermal, is arbitrary but convenient.

Both fossil and nuclear fuels are, as said above, excellent storage materials in themselves. Any additional storage, occurring after conversion, is determined by the process in which they are used. We shall discuss the storage needs of electricity supply in a subsequent section. In the case of thermal applications the need for storage other than in the fuel will be less for fossil fuels than for nuclear fuels. The reason for this is that the capital cost of a nuclear "boiler", that is the nuclear reactor and the associated steam raising plant is higher than that of a coal, oil or gas-fired boiler. It is therefore advantageous to run a nuclear reactor continuously and, since heating (domestic, commercial, institutional and the provision of process heat to industry) show strong diurnal (<u>d</u>) and seasonal (<u>1</u>) variations, storage is called for.

4. <u>Geothermal</u> energy is used either for electricity generation or to provide heat directly. The rate at which thermal energy is extracted can usually be controlled and there-fore the need for storage is not great.

5. <u>Solar</u> energy. This, the original source of all the energy, except tidal and nuclear, used by mankind provides at present a very small fraction of human needs - at least in the developed countries. But it would be a mistake to underestimate its contribution as the following simple example shows. Let us consider France, a country which is self-sufficient in food and even has an exportable surplus. Taking 100 W as the human metabolic rate we find that the annual food comsumption of France corresponds to 40-45 TWh (tera-watt hours, 10^{12}Wh). This

-12-

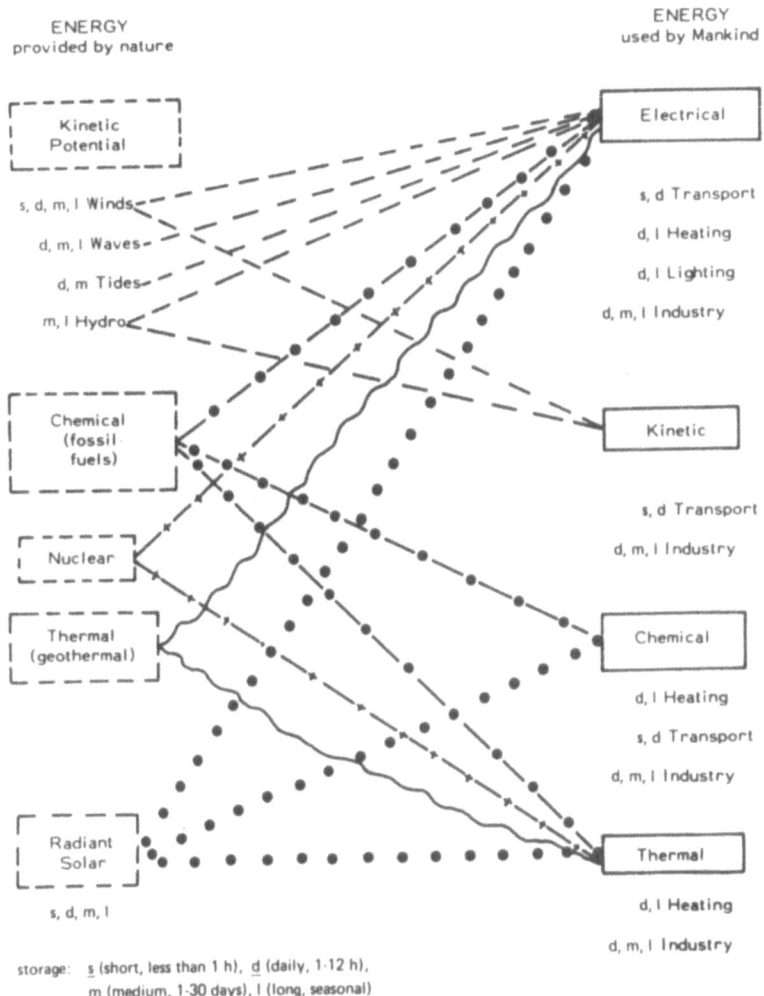

ENERGY
provided by nature

ENERGY
used by Mankind

Kinetic
Potential

s, d, m, l Winds
d, m, l Waves
d, m Tides
m, l Hydro

Chemical
(fossil
fuels)

Nuclear

Thermal
(geothermal)

Radiant
Solar

s, d, m, l

Electrical

s, d Transport
d, l Heating
d, l Lighting
d, m, l Industry

Kinetic

s, d Transport
d, m, l Industry

Chemical

d, l Heating
s, d Transport
d, m, l Industry

Thermal

d, l Heating
d, m, l Industry

storage: s (short, less than 1 h), d (daily, 1-12 h),
m (medium, 1-30 days), l (long, seasonal)

Figure 2.

– 13 –

is also the annual consumption of solar energy in the form of food. The energy content of artificial fertilizers had been left out of the calculation – on the other hand food exports have not been allowed for. 45 TWh equals about one quarter of France's annual electrical energy consumption in 1977.

Unlike the other renewable energy sources, solar energy could be regarded as the sole, or almost sole, energy source for certain regions or communities. Clearly diurnal (d) storage is essential irrespective of the intermittence of energy use. We shall return to this when discussing the role of energy storage in electricity production.

The conversion of solar into chemical energy refers mainly to photochemical reactions, that means agriculture, forestry, biomass production, but also as a future possibility, high temperature chemical reactions. The technique of solar furnaces has probably been sufficiently developed for this to be regarded as feasible. It should be mentioned in this connection that, although it is not indicated in the Fig. 2 nuclear heat might also be used in this fashion, e.g. for producing H_2.

When solar energy is used through the "thermal" route, seasonal (1) storage becomes desirable especially in temperate climates: summer heat to beat the winter cold. An interesting Danish development might be mentioned in this connection. The efficiency or Coefficient of Performance $(COP = \dfrac{\text{useful heat out}}{\text{energy in}})$

of a heat pump is given, under ideal conditions, by $\dfrac{T_h}{T_h - T_c}$
(T_h and T_c being the output and heat source temperature respectively); it depends very strongly on T_c, especially when the temperature difference T_h-T_c is small as is the case usually for domestic heating. A domestic heat pump system has been developed in Denmark which uses the ground under the garden as a heat source. It was found that, provided the heat is collected at a depth of 2-3 m an area adjacent to the house roughly equal four times the floor area of the house will provide the necessary amount of heat, without the ground temperature dropping below 0°C. COP values higher than 3 could be maintained through-out the winter even with output temperatures of 50°C: a clear case of the summer's heat being used in winter.

ENERGY STORAGE IN ELECTRICITY PRODUCTION

As Fig. 2 shows both reasons for energy storage given in WHY STORE ENERGY? apply, since all energy sources can be used and, in fact are used, for electricity generation. Let me recall some basic facts. Electrical energy is a free energy, i.e. its conversion into work or mechanical energy can be achieved with a high efficiency, approaching 100%. Once electrical energy is generated it is desirable, or even imperative, to store it as free energy. This is possible, at least in principle, by one or other of the following methods to be described briefly in the last chapter: potential energy (pumped water storage, also compressed gas storage); kinetic energy (flywheel, kinetic ring); electromagnetic (superconductors); electrochemical (batteries); chemical (electrolysis of H_2).

The conservation of surplus electrical energy as heat is justifiable in certain special cases, e.g. in the so-called night-storage heaters which use electricity when demand is low. This method is used mainly in "all-electric" dwellings or offices and this is a way of smoothing out demand.

There are, however, conditions in which large scale heat storage at power stations is desirable. If the full output of the station is needed for only part of the day, one could reduce the size of the boiler (thereby saving appreciable capital cost), run it continuously, for 24 hours a day, store the heat generated during off-peak hours and add it to the boiler-output during peak hours. It is of course important that storage should occur at as high a temperature as possible.

This brings us to an important, although often overlooked, aspect when discussing electricity generation. When electricity is produced by the "thermal" route the amount of electrical energy or work W_{therm} produced is given in the ideal case by the second law of thermodynamics,

$$W_{therm} = \Delta H (1 - T_c/T_h)$$ where T_h and T_c are the boiler

and condenser temperatures respectively and ΔH is the enthalpy of the chemical reaction used to raise steam, usually the heat of combustion of a fossil fuel. But if instead of burning the fuel irreversibly, one carried out the reaction reversibly, e.g. by an electrochemical route, (fuel-cells), the corresponding electrical energy produced W_{rev} equals, in the ideal case, the free enthalpy (or Gibbs free energy) of the reaction ΔG. Therefore,

$$W_{rev}/W_{therm} = \frac{\Delta G}{\Delta H} \frac{1}{1 - T_c/T_h}$$

Let us now look at some specific cases, always remembering that the above relations assume ideal conditions, i.e. a Carnot cycle and fully reversible fuel-cells. However, since we shall consider _ratios_ only, the error will not be large unless deviation from _ideal_ behaviour are very different in the two cases.

We assume $T_h = 750^\circ K$ $T_c = 300^\circ K$ $T_c/T_h = 0.4$

1) $C + O_2 = CO_2$ $\Delta H = 393.5 \text{ kJgmol}^{-1}$, $\Delta G = 394.2 \text{ kJmol}^{-1}$

$\Delta G/\Delta H = 1$ $W_{rev}/W_{therm} = 1.7$

2) $H_2 + \frac{1}{2}O_2 = H_2O$ $\Delta H = 285 \text{ kJ}$, $\Delta G = 240 \text{ kJ}$

$\Delta G/\Delta H = 0.84$ $W_{rev}/W_{therm} = 1.4$

This result is relevant to energy storage. Surplus electricity could be used to produce H_2 and O_2 by the electrolysis of water. If the H_2 and the O_2 stored were to be used to produce electricity during peak demand it would be 40% more efficient to do it in a fuel-cell than in a thermal power station. However, high capital cost seems to rule out this as a practical possibility, at least for the present.

REVIEW AND COMPARISON OF ENERGY STORAGE METHODS.

1. General Considerations

It is evident from what was said in the preceding chapter that energy storage methods may be divided into two large categories: storage of free energy and storage of heat. While, in the former case, the stored energy can be converted, in principle, without any loss into some other form of energy, in the case of heat storage the conversion into any form of free energy is never complete. A corollary of this is that, since the efficiency of conversion depends on the temperature at which thermal energy is available, it is not sufficient to characterize a thermal energy storage method by the energy density; the temperature of the storage must also be given. The concept of "exergy" which has found increased use in chemical engineering during the last few decades is helpful in this connection. Exergy indicates the amount of work that may be obtained from a system given the temperature of its surroundings, with which it can exchange heat.

This definition points to an important difference between exergy and the traditional thermodynamic functions such as energy, enthalpy, entropy, free energy. The latter functions are intrinsic properties of the system in question; their values can be stated once certain reference conditions, e.g. $273^\circ K$, 1 atm etc. are defined. Exergy is not an intrinsic material property – it depends on the temperature of the surroundings. Thus the thermal energy of 1 kg of H_2O at $40^\circ C$ – with $0^\circ C$ as reference point – is $4.2 \times 40 - 168$ kJ, irrespective of the ambient temperature. But it will have a considerable exergy in winter, while on a hot summer's day its exergy will be small.

Before reviewing the various energy storage methods let us first of all define "thermal energy storage". The following definition, which also covers "thermochemical storage" has been found useful.

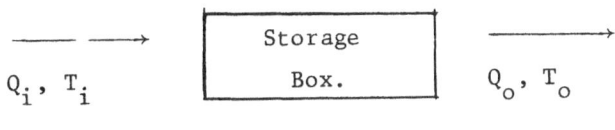

Q_i, T_i   Storage Box.   Q_o, T_o

Fig. 3

The "storage box" of Fig. 3 is completely isolated from the surroundings, i.e. there is no exchange of work or heat or matter. However, it can receive on the left-hand side, at the temperature T_i, a certain heat-input Q_i which is to be stored, and can release, on the right-hand side, a certain heat-output Q_o at the temperature T_o. If at the end of a storage cycle the contents of the box return to their original state then, from the 1st law of thermodynamics $Q_i = Q_o$, while it follows from the 2nd law of thermodynamics that $T_o < T_i$.

This definition covers not only sensible and latent heat storage but also thermochemical processes like the "Eva-Adam" storage method. In this process Q_i enters the box at a temperature of between $850^\circ C$ and $1200^\circ C$ and actuates the endothermic steam-reforming reaction $CH_4 + H_2O \longrightarrow CO + 3H_2$. The resulting gas mixture may be stored at any temperature. The heat Q_o is extracted from the box by actuating catalytically the exothermic reaction $CO + 3H_2 \rightarrow CH_4 + H_2O$ which occurs in the temperature range $350^\circ C - 700^\circ C$.

The expression "Eva-Adam" derives from the acronym EVA of the German description of the rig used for the heat input: Einzelspaltrohr-VersuchsAnlage (single cracking-tube experimental rig). The name for the output-end (catalytic combination of CO and H_2) was prompted by the recognition that every Eve needs an Adam.

2. Energy Storage Methods.

Since this course is concerned with thermal energy storage and detailed description of these will be given in the next few days the only reason for including them in this review is to enable them to be compared with the other methods. It seems that at present more effort is directed at thermal energy storage than at all the other methods taken together. This, at least, is the impression one gets by analysing the programme of the most recent International Energy Storage Conference, held in Brighton, England, 29th April - 1st May 1981 (3). Forty papers were given, and of these twenty-three were devoted to a great variety of heat storage methods, five to electrochemical storage, four to chemical storage, four to compressed air storage, two to kinetic energy storage, one to hydraulic storage, and there was one general paper.

We shall now list in order of increasing storage density the main groups of energy storage methods. The energy density ($\underline{u}$) values are only approximative and for any group of methods they may vary by as much as a factor 2 in either direction. More information can be found in Table 1 and a coloured chart prepared by Dr. F. M. Russell, of the Energy Research Support Unit (ERSU) of the Rutherford-Appleton Laboratory, Chilton, U.K. from whom copies may be obtained. Figures are taken from that chart and reproduced here with Dr. Russell's kind permission.

ENERGY STORAGE METHODS

		Storage Density		
		$kWhm^{-3}$	MJm^{-3}	MJt^{-1}
Pumped Water (100m)	F	0.25	0.90	–
Compressed Air (70 bar)	F	5	18	–
Compressed Air and thermal	F	10	36	–
Electromagnetic (superconductive)	F	10	36	5
Hot Water, Steam	Q	22	80	–
Kinetic Energy (Kinetic Ring)	F	50	180	25
Kinetic Energy (Flywheel)	F	80	290	40
Thermochemichal (Eva-Adam)	Q	60	220	–
Thermal (sand $300^{o}C$-$800^{o}C$)	Q	180	650	–
Thermal (molten salt $200^{o}C$-$500^{o}C$)	Q	250	900	–
Electrochemical (battery)	F	250	900	200
Chemical (liquid H_2)	F/Q	2,500	9,000	120,000
Chemical (hydrocarbons)	F/Q	10,000	36,000	40,000

F = Free Energy   Q = Heat

Table 1.

a) Pumped water storage. The storage density depends of course on the difference of the level of the upper and lower reservoirs. For an elevation of 100 m u = 0.09 MJ m^{-3}.

b) Compressed air storage (70 bar) usually in caverns has u = 18 MJ m^{-3}.

c) Improved performance results from hot compressed air storage, leading to u = 36 MJ m^{-3}.

d) Electromagnetic energy storage. This is based on superconducting alloys and compounds which retain their zero resistivity even at high current densities and when exposed to strong magnetic fields. With superconductive self inductances carrying loss-less, persistent currents one can attain u = 36 MJ m^{-3}. The inductances must be kept at the temperature of liquid He, i.e. about 4°K.

e) Hot water, steam: u = 80 MJ m^{-3}.

f) Kinetic energy storage. Until recently this was virtually synonymous with storage in fly-wheels, to which the energy was transmitted through the axis. A new development, due to F. M. Russell and S. H. Chew (3) consists of a heavy ring, without spokes, energy being trasmitted through rollers on the periphery. For flywheels u = 220 MJ m^{-3} while for the kinetic ring u = 110 MJ m^{-3}.

g) Thermochemical storage in the gas phase (low pressure) e.g. "Eva-Adam" process u = 220 MJ m^{-3}.

h) Thermal energy storage (sand), 300°C - 800°C u = 650 MJ m^{-3}.

i) Thermal energy storage (molten salts) 250°C - 500°C. u = 900 MJ m^{-3}.

j) Electrochemical storage (batteries). Considerable advances have been made in recent years, and values of u = 900 MJ m^{-3} are probably within reach. Particularly interesting is the development of batteries base on TiS_2 or MoS_2 intercalated with Li. They work at normal temperature, have low density and can withstand, without damage, high discharge current. A battery of about 150 mm x 80 mm x 10 mm is reputed to be capable of starting a motor-car engine.

k) Chemical energy storage. As mentioned above H_2 could be a useful medium for storing surplus electrical energy. In liquid form u = 9,000 MJ m^{-3}. For hydorcarbons u = 36,000 MJm^{-3}.

CONCLUSION.

   Energy storage is likely to pay an increasing role in the
energy economy. While, as emphasized in the above discussion,
energy storage does not lead to direct energy savings, it can
result in saving traditional energy resources such as coal,
oil, natural gas through a more intensive use of renewable
but intermittent energy sources. It also seems that for large
scale storage thermal energy has many advantages and it is
hoped that this course will lead, through the exchange of ideas
of many experts to further advances.

REFERENCES.

(1) Logan, W. P. D.: 1953, The Lancet, 264, 336.

(2) Interdisciplinary Science Reviews: 1976, 1, 279;
     1977, 2, 9, 190.

(3) Papers presented at the International Conference on
     Energy Storage, April 29th - May 1st, 1981, Brighton,
     U.K. Vol. 1. Edited by H. S. Stephen and B. Jarvis,
     Published B.H.R.A., I.S.B.N. 0-906085-500.

(4) Ref. 3 pp. 373-384.

SENSIBLE HEAT STORAGE

N. A. Mancini, F. Simone

Istituto di Struttura della Materia, Università
di Catania
Programma Finalizzato Energetica – C.N.R.
Gruppo Nazionale Struttura della Materia – C.N.R.

Presently most thermal storage devices use sensible heat
storage and a good technology is developed for the design of
such systems. Between liquid materials, water appears to be
the most convenient because it is inexpensive and has a high
specific heat. However, above 100°C, the storage tank must be
able to contain water at its vapor pressure and the storage tank
cost rises sharply for temperatures above this point. Organic
oils, molten salts and liquid metals do not exibit the same pressure problems but their use is limited because of their handling,
containment, storage capacities and cost. In Table I[1] storage
characteristics are showed of liquids mainly used.
The difficulties and limitations relative to liquids can be avoided by using solid materials for storing thermal energy as sensible heat. But larger amounts of solids are needed than using
water, due to the fact that solids, in general, exibit a lower
storing capacity than water. The cost of the storage media per
unit energy stored is, however, still acceptable for rocks.
Fig.1 shows storage capacity per unit volume as a function of temperature for commonly used solids.
Fig.2 shows the storage media cost per unit energy stored as a
function of temperature.

Direct contact between the solid storage media and a heat storage
fluid is necessary to minimize the cost of heat exchange in a solid storage medium.

Medium	Fluid Type	Cost ($/kg)	Temperature Range (°C)	Heat Capacity ($Jkg^{-1}K^{-1}$)	Comments
Water	——	0	0 to 374	4190	Pressure vessel required above 100°C.
Caloria HT43	Oil	.30	-10 to 315	2300	Cracking occurs at high temperatures and may form volatile products lowering flash point. May polymerize at high temperatures to increase viscosity. Nonoxidizing environment required at high temperatures.
Therminol 55	Oil	.60	-18 to 315	2400	
Therminol 66	Oil	2.03	-9 to 343	2100	
Hitec	Molten Salt	.59	150 to 590	1550	Long-term stability unknown above 550°C. Stainless steel or other expensive containers probably required above 450°C. Inert atmosphere required at high temperatures. Heated lines required to prevent freezing.
Draw Salt	Molten Salt	.44	250 to 590	1550	
Sodium	Liquid Metal	.90	125 to 760	1300	Stainless steel or suitable alternate containers required. Requires sealed system. Reacts violently with water, oxygen, and other materials.

Tab.I

Fluids such as high pressure helium or oils are generally used for high temperature storage in solids. The heat transfer fluid must also be compatible with the solids and the problems of finding fluids with low vapor pressure, high heat capacity and low cost are similar to those for storage in a liquid.
The use of rocks for thermal storage provides the following advan
tages:
a) rocks are not toxic and non-flammable
b) rocks are inexpensive
c) rocks act as both heat transfer surface and storage medium
d) the heat transfer between air and a rock bed is good, due to the very large heat transfer area, and the effective heat con ductance of the rock pile is low, due to the small area of contact beween the rocks. Then the heat losses from the pile are low.

Air leaving a rock bed is at a temperature nearly equal to the temperature of the hot air entering the pile. This makes it possible to deliver heat from the storage at almost the maximum temperature of operation, indipendently of the amount of energy stored.
e) rock beds may be used for winter humidification, by spraying water over them; and for partial summer cooling by drawing cool summer night air through them and spraying cold water over

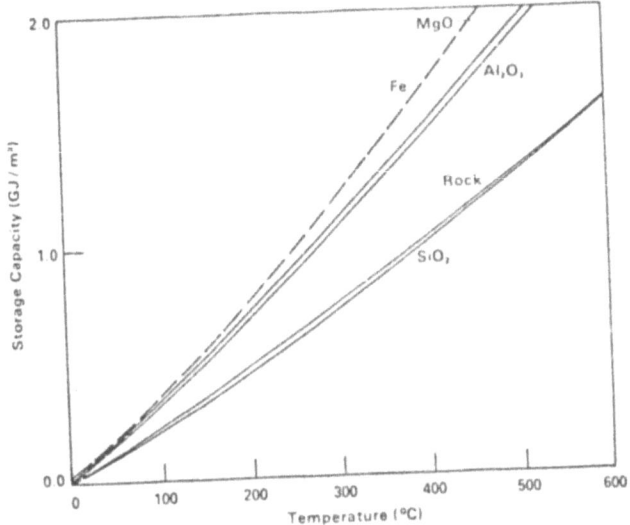

(Fig.　1　)

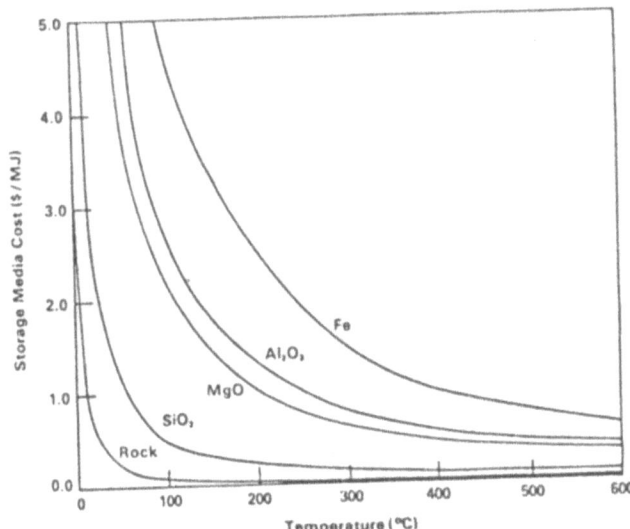

(Fig.　2　)

them. To insure a satisfactory operation of the rock pile, the hot air should be admitted from top and the cold air from the bottom of the pile, similar to a hot water storage tank.

Typically[2] the rock bed is constructed directly on a area of earth which is graded to provide a southward facing sloping surface. The earth is covered with a plastic sheet on which is placed a wooden frame whose height varies from about 20cm to some meters. Typical frame dimensions are of the order of some square meters. The volume inside the frame is filled with rocks which are painted black. The rocks and frame are covered with a sheet of clear plastic. Vent holes are provided in the upper and lower wooden frame members.

As the rocks in the rock bed are heated by sun, the air contained between the plastic sheet becomes hot and moves over and around the rocks and through the vent holes in the upper frame member and into the area to be heated. Cool air is drawn into the rock bed through the vent holes in the lower frame member. The thermal energy stored in the pile rock bed supplies the heat necessary to provide a source of heated air during the night.

Assumptions generally made for mathematical analysis of heat transfer in a rock bed storage systems are the following:

a) A plenum space is assumed at the top and the bottom of the rock bed, so there are no radial differences in pressure drop across the bed.
b) The consequent air flow is assumed to be vertical and uniform across the bed.
c) The temperature gradients in the radial direction are assumed negligible.
d) The bed is assumed uniformly packed having the same apparent density and the same uniform apparent thermal capacity throughout.
e) Thermal gradient within solid particles is neglected.
f) Losses to environment are ignored.
g) Internal heat generation is absent.

Typical dimensions of solid particles are about 15÷50 mm of diameter. The particle size has, in fact, to be chosen small enough to minimize temperature gradients in the particles themselves during charging-discharging processes. Very small dimensions of particles should be excluded to prevent voids being filled, which both increase standing thermal losses and increase the fan energy power needed to drive the air through the bed.

If the one-dimensional two phase model of a packed bed, an energy balance is performed on each of the fluid and solid phases[3], yielding to a coupled set of partial differential equations:

$$\rho_a c_a \left\{ \frac{\partial T_a}{\partial t} + v_a \frac{\partial T_a}{\partial x} \right\} = h_\nu (T_b - T_a) + k_a \frac{\partial^2 T_a}{\partial x^2} \qquad \text{fluid} \qquad (1)$$

$$\rho_b c_b \frac{\partial T_b}{\partial t} = h_\nu (T_a - T_b) + k_b \frac{\partial^2 T_b}{\partial x^2} \qquad \text{solid} \qquad (2)$$

subject to appropriate boundary conditions. In eqs.(1) and (2) the void fraction is considered to be already included in the ρ and k parameters.

$\rho_{a(b)}$ = density of fluid (bed)

$c_{a(b)}$ = specific heat of fluid (bed)

$T_{a(b)}$ = fluid (bed) temperature

v_a = velocity of fluid flow

$k_{a(b)}$ = thermal conductivity of fluid (bed)

h_ν = volumetric heat transfer coefficient.

The equations (1) and (2) cannot be solved exactly in simple analytical forms so that digital computer solutions are usually sought. In the case of an air rock packed bed, it is justified to ignore the thermal capacity and the axial conduction terms in the gas phase eq.(1). If the axial conduction in the bed material is also neglected the eqs. (1) and (2) can be written as follows:

$$\rho_a c_a v_a \frac{\partial T_a}{\partial x} = h_\nu (T_b - T_a) \qquad (1a)$$

$$\rho_b c_b \frac{\partial T_b}{\partial t} = h_\nu (T_a - T_b) \qquad (2a)$$

A single phase conductivity model can be derived formally from the two phase equations by letting h_ν tend to infinity. In this limit, T_a and T_b assume the same value and we get only the following equations from (1) and (2):

$$(\rho_a c_a + \rho_b c_b) \frac{\partial T}{\partial t} + v_a \rho_a c_a \frac{\partial T}{\partial x} = (k_a + k_b) \frac{\partial^2 T}{\partial x^2} \qquad (1b)$$

In the case of an air-gravel system, $\rho_a c_a \ll \rho_b c_b$ and $k_a \ll k_b$, so eq.(1b) can be written as:

$$\frac{\partial T}{\partial t} + v \frac{\partial T}{\partial x} = \alpha \frac{\partial^2 T}{\partial x^2} \tag{1c}$$

where $v = v_a c_a \rho_a / c_b \rho_b$ and $\alpha = k_b / c_b \rho_b$.

Solutions of these equations have been studied[3] for a variety of time dependence of air inlet temperature, in order to analyze the response of storage system in different operation conditions. The response of a packed bed to a sudden change in inlet fluid temperature is shown in fig.3

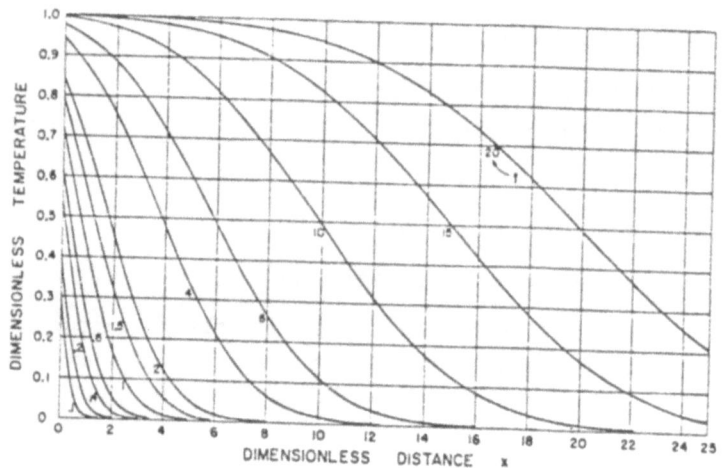

Fig.3

This figure gives the space-time distribution of temperature in the bed, that is the space-time profile of the thermal wave propagating in the system.

Some authors[4] made calculations utilizing the exact solutions for the temperature distributions of the solid and the fluid flowing through a packed bed with arbitrary initial bed temperature and arbitrary inlet fluid temperature.

In particular assuming a daily cycle operating mode and a time dependence of inlet fluid temperature of the following type:

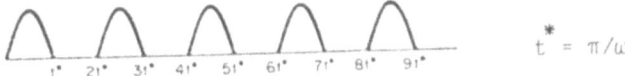

$$\overset{*}{t} = \pi/\omega$$

Fig.4

They obtained the steady periodic temperature distributions of
the bed at the end of the heating and cooling (lower sets of
graphes) half cycles. (See fig.5)

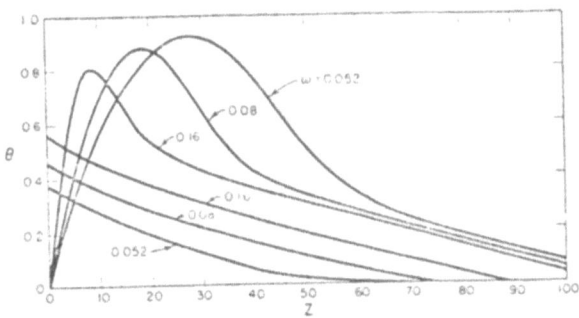

Fig.5

Sensible heat storage systems which use water as a storage medium are essentially of two types[6]:

a) Tanks
b) Solar ponds
Regarding the tanks, if at a pressure, they enable us to store energy up to a temperature of about 200°C. Above such a temperature value the pressure necessary for storing the water in a liquid state is very great (vapor pressure of water at 240°C is
33atm.).

For storage at higher temperatures liquids having a low vapor pressure are used (molten salts, diathermic oils). In
fig.6 a tank for a storage system is shown, it is made up of a
metal container, externally insulated and filled with water.

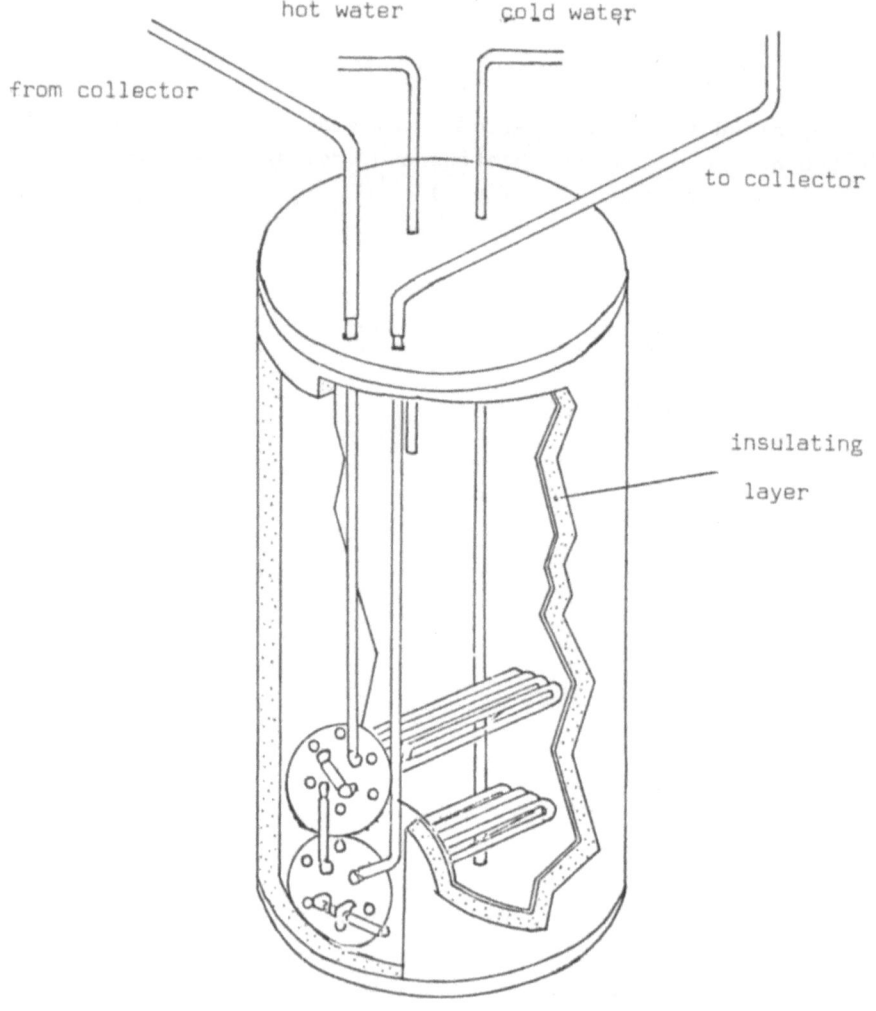

hot water　　　cold water

from collector

to collector

insulating
layer

(Fig. 6)

The heat is supplied to the water by a heat exchanger when the thermovector fluid differs from water or by direct immission of hot water when the thermovector fluid is the water itself.

Its method of use can be demonstrated as follows: the thermovector fluid of the collectors, when it is a temperature T_a higher than the temperature of the water contained in the tank T_w, passes through a coil and supplies the heat to the accumulator. The hot water needed by the consumer is extracted directly from the upper part while the cold water enters from the lower.

The comparison between T_c and T_w is made by a differential thermostat which working on a by-pass or directly on the primary circuit pump allows the circulation of the thermovector fluid in the exchanger only if $T_c > T_w$.

As the thermal conduction coefficient of the water is very low $(0,6 W/m^2 C)$ the transmission of the heat of the heat inside the tank is due essentially to convective motion; the warmer water rises towards the surface, the colder sinks to the bottom. Consequently the temperature of the water inside the tank is not uniform, but at the same time it is not perfectly layered; that is it does not decrease in a continuous way from the bottom to the top of the container. The efficiency of the accumulator becomes very low when the temperature of the water which is in direct contact with the coil is greater than T_c; the transmission of heat from the thermovector fluid to the water contained in the tank is impeded also if inside the accumulator a layer of water exists with a temperature lower than T_c. If, on the contrary, the temperature of the water inside the tank were perfectly layered the hot water needed by the consumer and the water returning to the collectors could be taken respectively from the warmer and the colder zones. Thus it would satisfy the consumer in the best way, and so the collector could work more efficiently at a lower temperature. An experimental study[7] to determine a method of extracting hot water from the tank and adding cold water maintaining the temperature steps inside the tank, was carried out. These authors demonstrated how it is possible to keep the layering in cylindrical tanks also with high flow rates. The layering also increases, when the relationship between the length L and the diameter D of the tank increases, the temperature difference ΔT of the fluid at the entrance and exit also increases with the increase of the diameter of the entrance and exit; it decreases when the rate of flow increases. Different systems have been proposed to obtain layering.

In Holland[8]) an accumulator was made using a tank of $5m^3$. Such an accumulator is equiped with an entrance for the water leading from the collectors, this entrance is made up of a plastic flexible tube with the same density as the water. Varying the temperature of the hot water coming from the collectors, the free end of the plastic tube moves up and down inside the accumulator because of the difference in density between the fluid which is contained and that which surrounds it. Such an end will remain still only when it meets a layer of water with the same density and with the same temperature of water coming from the collectors. In this way the immission of the water in the tank does not disturb the layering.

Again in Holland, in Enschede, where the first system was constructed improvements have been made which consist of supplying the system with two flexible tubes at the entrance, one for the return of the collector and one the return of the heating system.

Using this last entrance it was possible to obtain an inversion of temperature in the storage tank and so observe and study its behaviour in the presence of locally disturbed stratification.

ECONOMIC ASPECT

Technical and economical analysis of the water containers involves different factors: the thickness of the walls, the type of material, the configuration, the insulation which obviously influence the cost. A system of bed rock with the same capacity of accumulation costs about half that of a system of accumulation with the water in steel tanks. It is also necessary to consider that because of the high cost of the distribution of air system with respect to that of water the whole system will have a total cost comparable or even higher than that of a water system of the same heat capacity.

The cost of the components of a water accumulator system and a bed rock system up to 254.000 kcal, are compared in table II, with a temperature difference of 10°C between the air which enters and exits. For a water accumulator system, Galloway[9]) specifically calculated the annual cost per kWh as equal to 9c/kWh, taking into account the invested capital and the running costs.(See Tab.III).

Cost of components a hot water storage system to store 254.000
Kcal (Δ T=10°C)

Component	Cost ($)
Water (23.000 1)	-
Stainless steel tank (27.000 1)	7200
Insulating material	500
	7700

Cost of components of rock bed storage system to store 254.000
Kcal (Δ T=10°C)

Component	Cost ($)
Granite stones (114.000Kg,medium diameter 50mm)	1250
Container	950
	2200

TABLE II

Capital and operating cost for hot water storage system

Capital	Cost ($,1978)
Collectors	660
Pump,valve,piping,controls	190
Baseboard heating system hot water	1200
Felon panel covers	100
Heat storage tank (1$/gal)	12500
	14650

Operating Cost	
Power (pump),peak load pricing	150
Maintenance	161
Taxes and insurance	439
Financing	1216
	1966

Total cost per year 9.0 c/kWh

TABLE III

REFERENCES

1) - C.Wyman,J.Castle,F.Kreith; Solar Energy vol.24,517,(1980).

2) - J.D.Walton; Proceedings of the First Seminar on Solar Energy Storage, THermal Storage, 4-8 September 1978, Trieste (Italy).

3) - M.Riaz, Solar Energy Vol.21, 123,(1978)

4) - H.C.White,S.A.Korpela; Solar Energy vol.23, 141,(1979).

5) D.J.Close, R.V.Dunkle; Solar Energy vol.19,233(1977).

6) Stato dell'arte dei sistemi di accumulo dell'energia termica per impianti ad energia solare. C.N.R., PFE Sottoprogetto Energia Solare

7) Z.Lavan, J.Thompson; Solar Energy vol.19, 519,(1977).

8) W.B.Veltkamp; Proceedings of the International TNO Symposium: "Thermal Storage of Solar Energy", Amsterdam 5-6 November 1980, C.Den Ouden.

9) T.R.Galloway; "Paraffin wax heat storage for solar heated homes", Contract No. W-7405-Eng-48.

LOW TEMPERATURE LATENT HEAT THERMAL ENERGY STORAGE

A. Abhat

Institut für Kernenergetik und Energiesysteme (IKE)
University of Stuttgart, Stuttgart, Federal Republic
of Germany

ABSTRACT

The present paper reviews heat-of-fusion storage materials
and heat exchangers for low temperature latent heat storage in the
temperature range 0-120 °C. The melting and freezing behaviour of
commercial paraffins, fatty acids, inorganic salt hydrates and
eutectica investigated using differential scanning calorimetry
and thermal analysis techniques is discussed. Following a brief
presentation of "active" and "passive" heat exchangers, the heat
transfer considerations in a latent heat store are described in
detail through the examples of a finned heat pipe and a finned-
annulus heat exchanger. Finally, some data pertaining to the
current economics of latent heat stores are provided.

INTRODUCTION

Efficient and economical heat storage is the key to the
effective and widespread utilization of solar energy for low temperature thermal applications. Amongst the various heat storage
techniques of interest, latent heat storage is particularly
attractive due to its ability to provide a high energy storage
density and its characteristics to store heat at a constant temperature corresponding to the phase transition temperature of the
heat storage substance.

The term "Latent Heat Storage", as we generally understand it
today, applies to the storage of heat as the latent heat of fusion
in suitable substances that undergo melting and freezing at a
desired temperature level. Consequently it is also often

called the 'Heat-of-Fusion' storage. Typical heat-of-fusion storage substances well-known to all of us are ice, paraffins or Glauber Salt. The term 'latent heat storage' may also be applied to include the heat stored in substances, such as Diaminopentaerythritol, wherein heat is stored as the heat of crystallization, as the substance is transformed from one solid phase to another. The stored heat is recovered in a likewise manner as the original solid phase is obtained back. Excluded in the present definition of 'latent heat storage' is, however, the heat stored in materials that undergo a liquid-to-vapor phase transition, e.g. water-to-steam. Although the latter phase transitions are associated with a latent heat of phase transition that is almost an order-of-magnitude higher than that for solid-to-liquid or solid-to-solid phase change, the practical problems of storing a gaseous phase and the necessity of pressurized containers for this purpose rule out their potential utility.

The present lecture shall be restricted to the discussion of heat-of-fusion storage, a technique which is also of the greatest current practical value. A temperature range of 0 - 120 °C to cover a variety of low temperature applications, such as domestic hot water production, direct or heat-pump assisted space heating, green house heating, solar cooling, etc., shall be considered. It should, however, be emphasized here, that although heat storage in solid-solid phase transitions is much less understood today, it does hold out future promise. For the sake of completeness, therefore, a brief discussion of this technique has been included in Appendix I.

BASIC REQUIREMENTS OF A LATENT HEAT THERMAL ENERGY STORE [1]

Any latent heat thermal energy storage system must possess at least the following three components:
1) a heat storage substance [1] that undergoes a solid-to-liquid phase transition in the operating temperature range and wherein the bulk of the heat added is stored as the latent heat of fusion.
2) a containment for the storage substance, and
3) a heat exchanging surface for transferring heat from the heat source to the heat storage substance and from the latter to the heat sink, e.g. from the solar collector to the heat storage substance to the load loop.

The development of a latent heat thermal energy storage (LTES) system hence involves the understanding of two essentially diverse

---

[1] the following abbreviations shall henceforth be used:
   LTES = latent heat thermal energy storage
   PCM = phase change (heat-of-fusion) storage material

subjects: heat storage materials and heat exchangers. The flow-chart in Figure 1 provides an overview of the different stages that may be involved in the development of a LTES system and of the specialized problems that need to be tackled. The problems concerning heat-of-fusion storage materials and heat exchangers shall be discussed in detail in this paper.

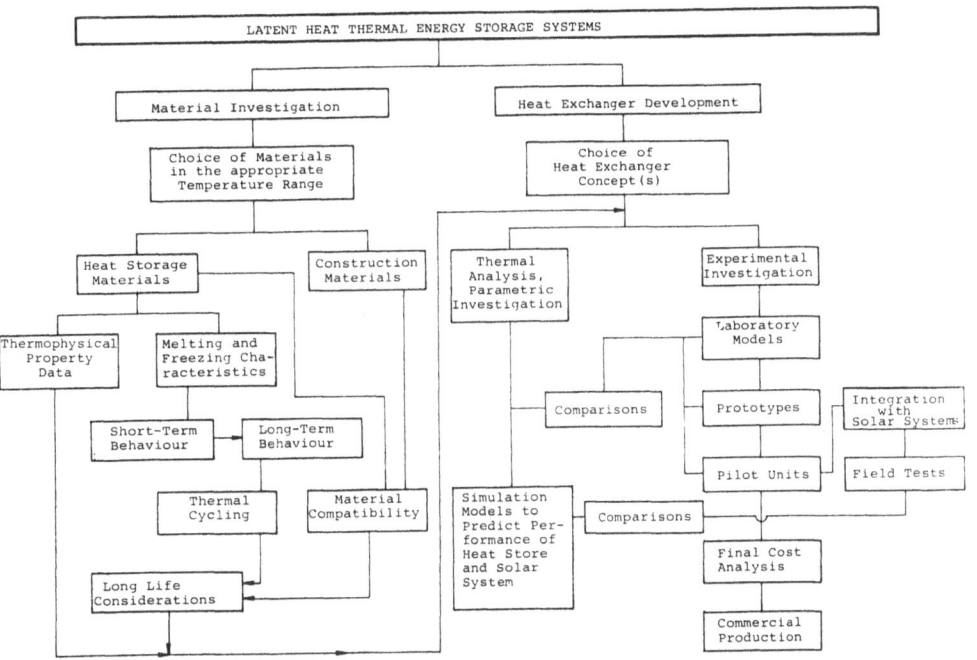

<u>Fig. 1:</u> Flowchart showing Different Stages involved in the Development of a Latent Heat Storage System

HEAT-OF-FUSION STORAGE MATERIALS

Desired Material Properties

A large number of organic and inorganic substances are known to melt with a high heat of fusion in any required temperature range, e.g. 0 - 120 °C. However, for their employment as heat storage materials in LTES systems, these phase change materials must exhibit certain desirable thermodynamic, kinetic and chemical properties. Moreover, economic considerations of cost and large-scale availability of the materials must be considered. The various criteria that govern the selection of phase change heat storage materials are summed in Table 1 on the following page.

A) *Thermodynamic criteria*

The phase change material should possess :

- a melting point in the desired operating temperature range
- high latent heat of fusion per unit mass, so that a lesser amount of material stores a given amount of energy
- high density, so that a smaller container volume holds the material
- high specific heat to provide for additional significant sensible heat storage effects
- high thermal conductivity, so that the temperature gradients required for charging and discharging the storage material are small
- congruent melting : the material should melt completely so that the liquid and solid phases are identical in composition. Otherwise, the difference in densities between solid and liquid will cause segregation resulting in changes in the chemical composition of the material
- small volume changes during phase transition, so that a simple containment and heat exchanger geometry can be used.

B) *Kinetic criteria*

The phase change material should exhibit :

- little or no supercooling during freezing. The melt should crystallize at its thermodynamic freezing point. This is achieved through a high rate of nucleation and growth rate of the crystals. At times, the supercooling may be suppressed by introducing nucleating agents or a *cold finger* in the storage material.

C) *Chemical criteria*

The phase change material should show :

- chemical stability
- no chemical decomposition, so that a high LTES system life is assured
- non-corrosiveness to construction materials
- the material should be non-poisonous, non-flammable and non-explosive.

D) *Economic criteria*

The phase change material should be :

- available in large quantities
- inexpensive.

Candidate Heat Storage Materials

It is quite apparent that no single material can fully satisfy the long list of criteria listed in Table 1. Trade-offs are hence made in the selection of candidate phase change heat storage materials in a desired operating temperature range. Within the operating

- 36 -

temperature range of 0 - 120 °C, candidate phase change heat storage materials are grouped into the families of organic and inorganic compounds and their eutectica, as seen in Figure 2. Sub-families of organic compounds include paraffin and non-paraffin organics.

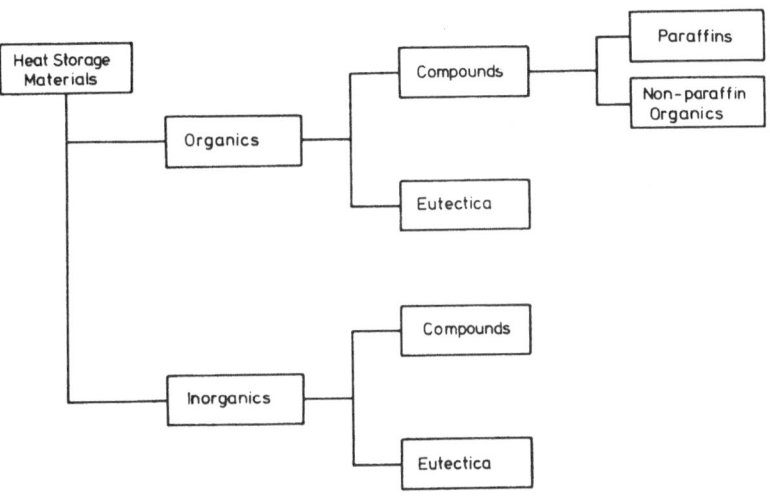

Fig. 2: Heat Storage Material Families

Figures 3(a) and (b) respectively show the latent heat of fusion per unit mass and per unit volume of some phase change heat storage materials of interest in the temperature range 0 - 120 °C /1, 24/. Of particular interest is Figure 3(b), which provides information on the compactness of a LTES system for a given amount of energy storage. Figure 3(b) shows that the organic compounds have a volumetric latent heat storage capacity in the range 125 - 200 kJ/dm³, whereas that of the salt hydrates is almost twice as much - between 250 - 400 kJ/dm³.

Notwithstanding the disadvantages concerning volume requirements, the organic substances serve as important heat storage materials due to the several desirable properties they possess in comparison with inorganic compounds. Some of these advantages includes their ability to melt congruently, their self-nucleating properties and their capability to freeze without supercooling, and their compatibility with conventional materials of construction.

We shall now examine some of the important heat storage material groups in some detail.

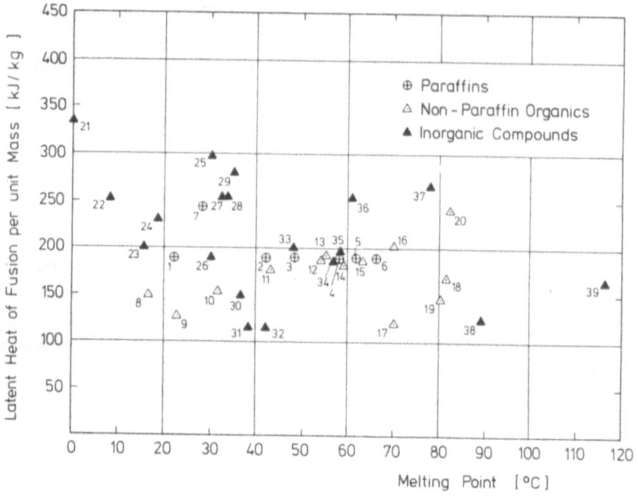

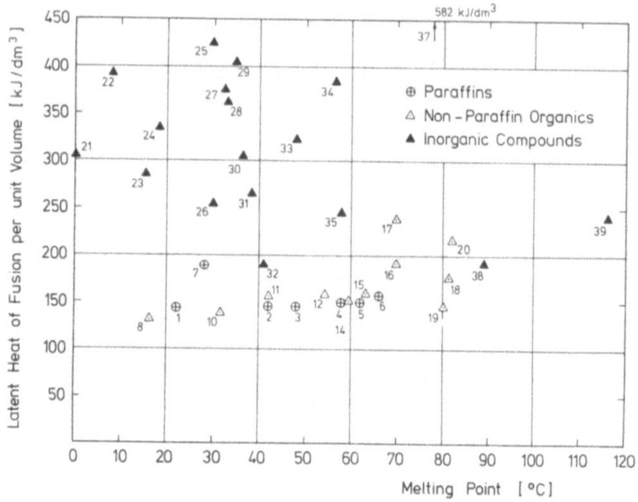

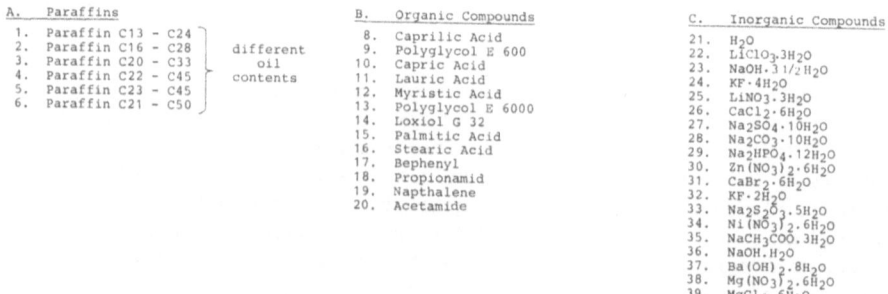

A.	Paraffins		B.	Organic Compounds	C.	Inorganic Compounds
1.	Paraffin C13 - C24		8.	Caprilic Acid	21.	H_2O
2.	Paraffin C16 - C28	different	9.	Polyglycol E 600	22.	$LiClO_3 \cdot 3H_2O$
3.	Paraffin C20 - C33	oil	10.	Capric Acid	23.	$NaOH \cdot 3\,1/2\,H_2O$
4.	Paraffin C22 - C45	contents	11.	Lauric Acid	24.	$KF \cdot 4H_2O$
5.	Paraffin C23 - C45		12.	Myristic Acid	25.	$LiNO_3 \cdot 3H_2O$
6.	Paraffin C21 - C50		13.	Polyglycol E 6000	26.	$CaCl_2 \cdot 6H_2O$
			14.	Loxiol G 32	27.	$Na_2SO_4 \cdot 10H_2O$
			15.	Palmitic Acid	28.	$Na_2CO_3 \cdot 10H_2O$
			16.	Stearic Acid	29.	$Na_2HPO_4 \cdot 12H_2O$
			17.	Bephenyl	30.	$Zn(NO_3)_2 \cdot 6H_2O$
			18.	Propionamid	31.	$CaBr_2 \cdot 6H_2O$
			19.	Napthalene	32.	$KF \cdot 2H_2O$
			20.	Acetamide	33.	$Na_2S_2O_3 \cdot 5H_2O$
					34.	$Ni(NO_3)_2 \cdot 6H_2O$
					35.	$NaCH_3COO \cdot 3H_2O$
					36.	$NaOH \cdot H_2O$
					37.	$Ba(OH)_2 \cdot 8H_2O$
					38.	$Mg(NO_3)_2 \cdot 6H_2O$
					39.	$MgCl_2 \cdot 6H_2O$

Fig. 3: Latent Heat of Fusion of Different Heat Storage
Media (a) per unit Mass (b) per unit Volume /1,24/

Paraffins. Paraffins, as we normally understand them, are substances having a waxy consistency at room temperature. Chemically speaking, paraffins are - like other mineral oil products - complicated mixtures of several organic compounds. Unlike other products like lubricating oils, however, paraffins contain one major component called alkanes, characterized by $C_nH_{2n}+2$. Thus pure paraffins contain only alkanes in them, for example the well-known paraffin octadecane, $C_{18}H_{38}$. The melting point of the alkanes increases with the increasing number of carbon atoms; alkanes containing 14 to 40 C-atoms possess melting points between 6 and 80 °C and are termed as paraffins. Depending on the chain length of the alkane in the paraffins, paraffins may be even-chained (n-paraffin) or odd-chained (iso-paraffin).

In their solid phase, paraffins are generally found in two allotropic modifications that differ in their physical properties and the crystal structure. The primary modification existing at higher temperature, i.e. slightly above the melting point of the substance, is soft and plastic, the individual crystals are needle-shaped. The secondary modification existing at lower temperatures, i.e. below the melting point, is hard and brittle, the crystals are disc-shaped. The transition from one crystal form to another is reversible /29/.

Paraffins qualify as heat-of-fusion storage materials due to their availability in a large temperature range and their reasonably high heat of fusion. Furthermore, they are known to freeze without supercooling /4/. Due to cost considerations, however, only technical grade paraffins may be used as PCMs in latent heat stores. Table 2 lists some technical grade materials, which are essentially paraffin mixtures and are not completely refined of oil. Manufacturers' data on their physical properties as well as cost data are also included in Table 2.

Non-paraffin organics. Although a number of non-paraffin organics have been included in Figs. 2(a) and (b), we shall restrict our discussion to the group of materials called 'fatty acids'. Other non-paraffin organics may, however, be assumed to have properties similar to those of the fatty acids.

Fatty Acids: Fatty acids are organic compounds characterized by $CH_3(CH_2)_{2n}COOH$ with heat of fusion values comparable to that of paraffins. Some fatty acids of interest to low temperature LTES applications are listed in Table 3.

Fatty acids are known to possess a reproducible melting and freezing behaviour and freeze with little or no supercooling /4,20/. They hence qualify as good PCMs. Their major drawback, however, is their cost which is 2 - 2.5 times greater than that of paraffins.

Table 2: Physical Properties and Cost Data of Some Paraffins (Manufacturers' Data)

Paraffin Type/No.	Distribution of C-Atoms	Oil Content	Freezing Point/Range	Heat-of-Fusion		Density at		Specific Heat at 100 °C	Thermal Conductivity (solid phase)	Cost (1979)	Ref.
		%	°C	kJ/kg	kJ/dm³	20 °C	70 °C	kJ/kg K	W/m K	DM/kg	
- [1]	C14	-	4.5	165	-	-		-	-	1.20 [2]	22
-	C15 - C16	-	8	153	-	-		-	-	0.50 [2]	22
5913 [3]	C13 - C24	20	22 - 24	189	144	0.900	0.760	2.1	0.21	0.50	4
Octadecane	C18	-	28	244	189	0.814	0.774	-	0.15	150.0	4
6106 [3]	C15 - C28	5	42 - 44	189	145	0.910	0.765	2.1	0.21	0.70	4
P116 (Sunoco)	-	-	47	210	165	-	0.786	-	-	0.44 [2]	22
5838 [3]	C20 - C33	0.5	48 - 50	189	145	0.912	0.769	2.1	0.21	1.00	4
6035 [3]	C22 - C45	4	58 - 60	189	150	0.920	0.795	2.1	0.21	0.60	4
6499 [3]	C21 - C50	3	66 - 68	189	157	0.930	0.830	2.1	0.21	0.80	4

[1] - implies data not available

[2] cost estimates are for 1974 (Ref. 24)

[3] manufacturers' of technical grade paraffins 5913, 6106, 5838, 6035 and 6499: 'ter Hell Paraffin, Hamburg, FRG

Table 3: Physical Properties and Cost Data of Some Fatty Acids (Literature Values and Manufacturers' Data)

Material	Melting Point/Range	Heat-of-Fusion		Density		Specific Heat	Thermal Conductivity		Cost (1979)	Ref.
	°C	kJ/kg	kJ/dm³	kg/dm³		kJ/kg K	W/m K		DM/kg	
Caprilic acid	16.3	149	128	1.033 0.862	(10 °C) (80 °C)	- [1]	0.148	(20 °C)	-	4
Capric acid	31.5	153	136	0.886	(40 °C)	-	0.149	(40 °C)	-	4
Lauric acid	42 - 44	178	155	0.870	(50 °C)	1.6	0.147	(50 °C)	2.50	4
Myristic acid	54	187	158	0.844	(80 °C)	1.6	-		2.50	4
Palmitic acid	63	187	159	0.847	(80 °C)	-	0.165	(70 °C)	2.30	4
Stearic acid	70	203	191	-		-	0.172	(70 °C)	2.00	4

[1] - implies data not available

Salt Hydrates. Salt hydrates, characterized by M.nH$_2$O, where M is an inorganic compound, form an important class of heat storage substances due to their high volumetric latent heat storage density (see. Fig. 3 (b)). In fact, their use as PCMs has been propagated as early as 1947 /25/. Table 4 provides a list of some salt hydrates melting in the temperature range 0 - 120 °C along with their thermophysical properties and cost data.

Table 4: Physical Properties and Cost Data of Some
Salt Hydrates (Literature Values and Manufacturers' Data)

Material	Melting Point	Heat-of-Fusion		Density	Specific Heat	Thermal Conductivity	Cost (1979)	Ref.
	°C	kJ/kg	kJ/dm³	kg/dm³	kJ/kg K	W/m K	DM/kg	
H$_2$O	0	333	306	0.917 (0 °C) 0.998 (20 °C)	2.09 (solid) 4.18 (liquid)	2.2 (solid) 0.6 (20 °C)	*	24
KF·4H$_2$O	18.5	231	336	1.455 (18 °C) 1.447 (20 °C)	1.84 (solid) 2.39 (liquid)	_ 1)	-	24
CaCl$_2$·6H$_2$O	29.7	171	256	1.710 (25 °C) 1.496 (liquid)	1.45 (solid)	-	0.36	4
Na$_2$SO$_4$·10H$_2$O	32.4	254	377	1.485 (solid)	1.93 (solid)	0.544	0.10	4
Zn(NO$_3$)$_2$·6H$_2$O	36.4	147	304	2.065 (14 °C)	1.34 (solid) 2.26 (liquid)	-	2.40	4
Na$_2$S$_2$O$_3$·5H$_2$O	48.0	201	322	1.600 (solid)	1.46 (solid) 2.39 (liquid)	-	0.30	4
Ba(OH)$_2$·8H$_2$O	78.0	267	581	2.180 (solid)	1.17 (solid)	-	1.75	4
MgCl$_2$·6H$_2$O	116.0	165	239	1.57 (20 °C) 1.442 (78 °C)	1.72 (solid) 2.82 (liquid)	-	0.20	24

1) - implies data not available   * negligible

The major problem in using salt hydrates as PCMs is that most of them melt incongruently i.e. they melt to a saturated aqueous phase and a solid phase which is generally a lower hydrate of the same salt. Due to density differences, the solid phase settles out and collects at the bottom of the container, a phenomenon called decomposition. Unless special measures are taken, this phenomenon is irreversible, i.e. during freezing, the solid phase does not combine with the saturated solution to form the original salt hydrate. Another important problem with salt hydrates is their poor nucleating properties resulting in supercooling of the liquid salt hydrate prior to freezing. Suitable measures must be adopted to eliminate supercooling or reducing it to a minimum. Typical methods suggested in the literature for this purpose are

(1) addition of nucleating agents that have a crystal structure similar to that of the parent substance /26/; (2) using a "cold finger" in the PCM; (3) promoting heterogenous nucleation by using (rough) metallic heat exchanging surfaces immersed in the salt hydrate /6/. We shall examine the influence of these techniques on the melting and freezing characteristics of salt hydrates in greater detail at a later stage.

The selection of a salt hydrate as a PCM can be eased by an understanding of the binary phase diagrams, M - H_2O, for which the reader is referred to References 5,30. To illustrate the use of the phase diagrams, two examples have been included here:
(a) $Zn(NO_3)_2$ - H_2O system, and
(b) Na_2SO_4 - H_2O system.

(a) $Zn(NO_3)_2$ - H_2O System. Figure 4 shows the phase diagram for $Zn(NO_3)_2$ - H_2O between the temperature limits -40 °C to +80 °C. If a liquid mixture of $Zn(NO_3)_2$ and H_2O is cooled below 80 °C, 5 different hydrated salts of $Zn(NO_3)_2$ as shown in Figure 4, crystallize out of solution. Of these, $Zn(NO_3)_2 \cdot 9H_2O$ has an incongruent melting point, whereas the remaining 4 hydrates have a congruent melting point. The 6-hydrate, with a melting point of 36.4 °C, additionally has a high latent heat of fusion and is hence well-suited as a heat storage material.

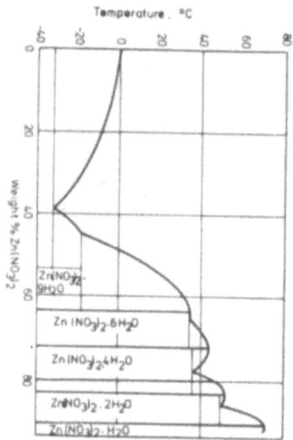

Fig. 4: Phase diagram for $Zn(NO_3)_2$-H_2O system /17/

(b) Na_2SO_4 - H_2O System. The second example we wish to discuss here is the well-known storage material sodium sulphate-10-hydrate or Glauber salt. This material has been extensively investigated so far for use in LTES systems. Fig. 5 shows a partial phase diagram of the sodium sulphate-water system /10/. $Na_2SO_4 \cdot 10H_2O$ decomposes peritectically on heating to 32.4 °C to yield anhydrous sodium sulphate and a saturated solution of Na_2SO_4 in water. The solid crystals containing 44 % anhydrous Na_2SO_4 and 56 % water by weight change to a mixture of 15 % anhydrous Na_2SO_4 and 85 %

saturated solution of Na_2SO_4 in water ($10H_2O$). The 15 % anhydrous Na_2SO_4 remains insoluble and settles down as a white bottom sediment. It is not possible to dissolve this material by the usual expedient of increasing the temperature, because the highest solubility occurs at the melting point with solubility decreasing at higher temperatures. Due to the large density difference between the saturated solution (1350 kg/m³) and the anhydrous solid Na_2SO_4 (2550 kg/m³), gravity induced separation results. Consequently melting is partly incongruent due to segregation /10/.

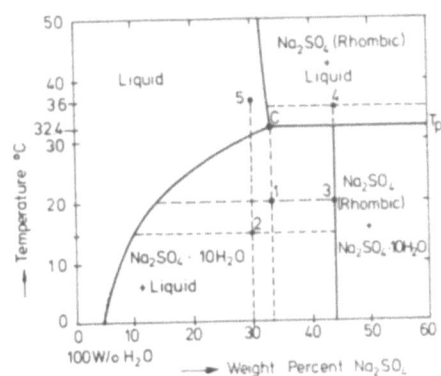

Fig. 5: Partial Phase Diagramm for Na_2SO_4 – H_2O
system /10/

Now, if the mixture of the saturated solution and anhydrous salt is cooled below 32.4 °C, sodium sulphate can be absorbed in the solution only as rapidly as water can diffuse through solid sodium sulphate decahydrate to the anhydrous sodium sulphate particles on which the decahydrate particles form. Since peritectic solidification reactions are characteristically much slower than congruent solidification or eutectic solidification, and because the rate limiting process here is solid state diffusion, even stirring cannot significantly affect the rate of absorption of Na_2SO_4, which therefore remains settled at the bottom of the container.

If the phase diagram for the Na_2SO_4 – H_2O system is now carefully followed, it can be seen that the solid state diffusion as a reaction step can be eliminated by using a mixture of decahydrate and water as the starting material (68.2 % $Na_2SO_4.10H_2O$ and 31.8 % H_2O by weight). At 15 °C (point 2) this mixture consists of 58.1 % $Na_2SO_4.10H_2O$ and 41.9 % solution of composition 10.6 % Na_2SO_4 and 89.4 % H_2O by weight. When the temperature of the mixture is raised, the solubility of decahydrate increases with the increase in temperature /10/.

    With the composition chosen, all the decahydrate dissolves when the system is heated above 32 °C and no segregation hence takes place. The overall composition of the solutuion then is 30 % Na_2SO_4 and 70 % H_2O by weight. Now if the solution is cooled from 36 °C (point 5 in Figure 5), decahydrate crystals should begin to separate, as the solution reaches the liquidus line (at 30 °C). With further reduction in temperature, the system enters into a two-phase region, $Na_2SO_4.10H_2O$ and liquid. The amount of each constituent can be easily calculated at any temperature by using the lever rule. The stored thermal energy in the system is released when the $Na_2SO_4.10H_2O$ crystals separate from the solution /10/.

    <u>Eutectica of Organic and Inorganic Compounds</u>. Eutectica of organic or inorganic compounds qualify as latent heat storage materials as they possess a fixed melting/freezing point. Some candidate materials having acceptable values of heat-of-fusion are listed in Table 5. It should be mentioned here that the search for eutectics is rather recent, and only limited data on thermophysical properties and cost are presently available.

<u>Table 5:</u> Eutectica of Organic and Inorganic Compounds

Material (weight % of compound in brackets)	Melting Point °C	Heat-of-Fusion kJ/kg	kJ/dm³	Cost (1974) DM/kg	Reference
$CaCl_2$ (48 %) NaCl (4.3 %) KCL (0.4 %) H_2O (47.3 %)	26.8	– [1]		–	22
$Ca(NO_3)_2 \cdot 4H_2O$ (47 %) $Mg(NO_3)_2 \cdot 6H_2O$ (33 %)	30	136	228	0.32	22
Urea (37.5 %) Acetamide (63.5 %)	53	–		–	4
$Mg(NO_3)_2 \cdot 6H_2O$ (53 %) $Al(NO_3)_2 \cdot 9H_2O$ (47 %)	61	148	249	–	12

[1] - implies data not available

    <u>Other compounds</u>. A material group that deserves mention for use in LTES systems is the clathrate hydrate. True clathrate hydrates are continuous solid water structures containing closed cavities within which are guest molecules that do not interact strongly with water. These guest molecules act to stabilize the "ice" structure. There are also other structures in which

the guest molecules participate in the water lattice directly, these substances are known as semi-clathrates and include hydrates of amines and tetra alkylammonium salts /22/.

Examples of clathrate hydrates are /22/: $SO_2 \cdot 6H_2O$ $(mp^{1)}$, 7 °C; $\Delta H_F^{1)}$ = 247 kJ/kg), $C_2H_4O \cdot (6.9H_2O)$ (mp, 11.1 °C).
Examples of semi-clathrate hydrate are /22/: $(CH_3)_3N \cdot 10^1/_4H_4O$ (mp, 5.9 °C, ΔH_F = 239 kJ/kg), $Bu_4NOH \cdot 32H_2O$ (mp, 30.2 °C).
Like salt hydrates, clathrate hydrates may melt congruently or incongruently. They also tend to supercool and hence suitable nucleating agents must be found to promote crystallization.

Melting and Freezing Characteristics

By melting and freezing characteristics of phase change heat storage materials, we understand the behaviour exhibited by these materials during heating and cooling, e.g. melting and freezing ranges, congruency of melting, nucleation characteristics, supercooling of the melt and stability to thermal cycling.
A comprehensive knowledge of the melting and freezing behaviour of heat-of-fusion storage substances and particularly its reproducibility as a consequence of repeated melting and freezing of the substances is hence essential for the assurance of the long-term performance of a latent heat store.

Measurement Techniques. Essentially two measurement techniques are employed for the determination of the melting and freezing characteristics: (1) Differential Scanning Calorimetery (DSC), and (2) Thermal Analysis (TA). Table 6 provides details of the two measurement techniques and a listing of data that may be evaluated.
The two techniques distinguish themselves in terms of the type of measurements made, the quantity of the sample used in the tests and the speed with which results can be obtained. For example, the DSC provides quick and reliable results in the form of Energy-Time diagrams (Thermograms) using very small quantities of the sample (ca. 1-10 mg). Evaluation of the thermograms yields rather precise values of the phase transition temperatures during melting and freezing of the sample, the heats of fusion and solidification and the specific heat variation as a function of temperature. The DSC is, however, a severe test for substances that supercool, since the supercooling tendencies are maximized due to the small quantity of the samples and the poor nucleation conditions in the DSC testing pans. Consequently, for materials like salt hydrates, the DSC fails to provide meaningful information on the degree of

---

1) mp = melting point
    ΔH_F = heat-of-fusion

TABLE 6: Details of the Measurement Techniques /8/

MEASUREMENT TECHNIQUE	APPARATUS DESIGN	QUANTITY OF HEAT STORAGE SUBSTANCE USED IN THE TESTS (ORDER OF MAGNITUDE)	MEASUREMENTS MADE	DATA EVALUATED
DIFFERENTIAL SCANNING CALORIMETRY (DSC)	PERKIN-ELMER CALORIMETER, MODEL DSC-2	1-10 MG	TEMPORAL VARIATION OF THE THERMAL ENERGY EXCHANGED WITH THE SAMPLE TO ENABLE IT TO UNDERGO HEATING OR COOLING AT A CONSTANT PRE-DETERMINED RATE. THE OUTPUT OF THE MEASUREMENTS IS THE ENERGY-TIME DIAGRAM, ALSO CALLED THERMOGRAM	1) FORM OF THE ENDOTHERMIC OF EXOTHERMIC PEAKS ON THE THERMOGRAM 2) MELTING POINT/RANGE 3) FREEZING POINT/RANGE 4) DEGREE OF SUPERCOOLING 5) HEATS OF FUSION AND SOLIDIFICATION 6) SPECIFIC HEAT AS A FUNCTION OF TEMPERATURE
THERMAL ANALYSIS (TA)	TEST TUBES OR GLASS AMPULES	10 G	TEMPORAL VARIATION OF TEMPERATURE AT SELECTED LOCATIONS WITHIN THE HEAT STORAGE SUBSTANCE RESULTING FROM ENERGY INPUT/OUTPUT TO THE SAMPLE. THE OUTPUT OF THE MEASUREMENTS IS THE TEMPERATURE-TIME DIAGRAM	1) SHAPE OF THE TEMPERATURE-TIME DIAGRAM, MELTING AND FREEZING PLATEAUS, HOMOGENEITY OF THE ORIGINAL SUBSTANCE FOLLOWING MELTING, ETC. 2) MELTING POINT/RANGE 3) FREEZING POINT/RANGE 4) DEGREE OF SUPERCOOLING 5) CONGRUENT/INCONGRUENT MELTING 6) DECOMPOSITION OF THE ORIGINAL SUBSTANCE, FORMATION OF NEW PHASES, SEGRAGATION EFFECTS
	FLAT GLASS CONTAINERS	100 G		
	LABORATORY MODELS OF PASSIVELY OPERATING LATENT HEAT STORAGE SYSTEM DESIGNS: 1) MODULAR FINNED HEAT PIPE HEAT EX-CHANGER /2/ 2) FINNED-ANNULUS HEAT EXCHANGER ELEMENT /7/	1-10 KG		

supercooling and on the freezing point of the substance.

The Thermal Analysis technique, on the other hand, involves
the determination of Temperature-Time (T-t)diagrams. The T-t diagrams recorded during the melting and freezing of the sample are
also termed as the Heating and Cooling curves respectively. The TA-
technique uses about 10 g to a few kilograms of the sample, depending on the apparatus used, and is hence slower. With proper care,
the rate of heating and cooling can be well controlled and relatively accurate results can be gained. The tests in glass containers,
however, suffer from the disadvantage that metallic surfaces that
generally form the heat exchanging surface within a latent heat
store or other means that are employed to improve heat transfer in
a latent heat store are absent. Nucleating conditions different to
those in actual LTES systems hence exist. Experiments done in test
models are the most accurate as conditions similar to those in large-
scale LTES systems are simulated within them. These tests are also
the slowest in comparison to the other techniques due to the large
quantities of phase change materials they require (1-10 kg).

Thermal cycling of the sample to obtain data on the reproducibility of the melting and freezing behaviour of the substance
may be undertaken using both measurement techniques. Any variations
observed in the form of the endothermic or exothermic peaks on the
DSC-thermograms or alterations in the values of the thermophysical
properties of the thermally cycled heat storage materials serve as
a measure of the change in their melting and freezing characteristics. In the TA-measurements, on the other hand, changes in the
form of the temperature-time diagrams or in the heat transfer rate
in or out of the sample provide information on the influence of
thermal cycling.

We shall now discuss some results for the melting and freezing
characteristics of PCMs belonging to three material groups:
paraffins, fatty acids and salt hydrates.

Paraffins: (a) TA-Measurements. Fig. 6 shows typical heating
and cooling curves obtained during TA-measurements with a technical
grade n-paraffin, paraffin C16-C28 (a paraffin mixture). As seen in
Fig. 6 (b), paraffins exhibit two freezing ranges: a narrow freezing range existing for a short period of time and a larger freezing range that occurs for a longer period of time. The two freezing ranges are respectively considered to signify a liquid-to-
solid transition and a solid-to-solid transition. A part of the
total latent heat of fusion is stored in the substance during each
of these transitions.

Not all commercially available paraffins (all of these are
essentially paraffin mixtures) display a freezing behaviour as
shown in Figure 6. For the sake of comparison, the cooling curves
for three different paraffins are presented in Figure 7. The paraffins in this diagram vary extensively in their freezing interval
which is bounded by the points marked Q and P on the diagram which

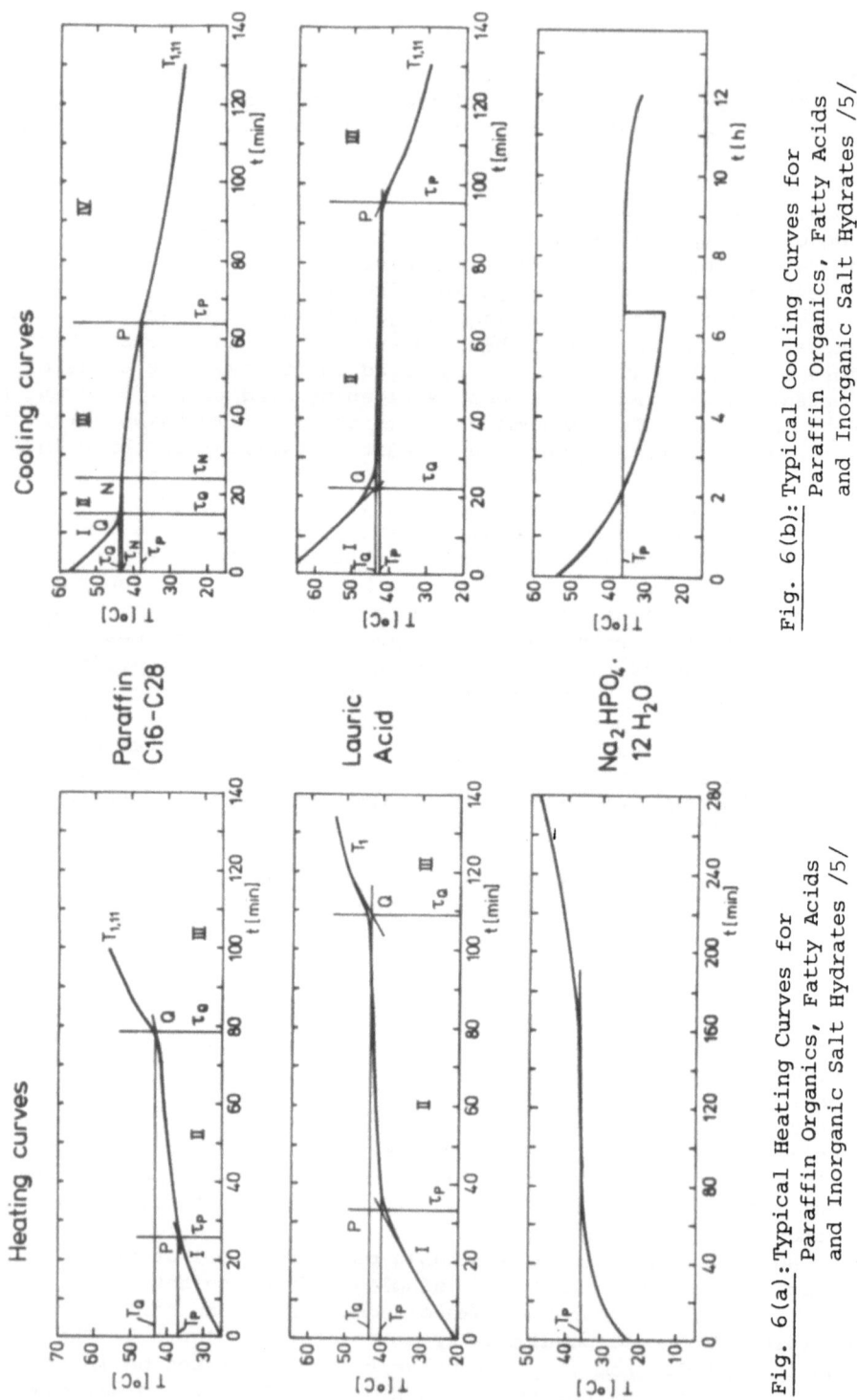

Fig. 6(b): Typical Cooling Curves for
Paraffin Organics, Fatty Acids
and Inorganic Salt Hydrates /5/

Fig. 6(a): Typical Heating Curves for
Paraffin Organics, Fatty Acids
and Inorganic Salt Hydrates /5/

- 48 -

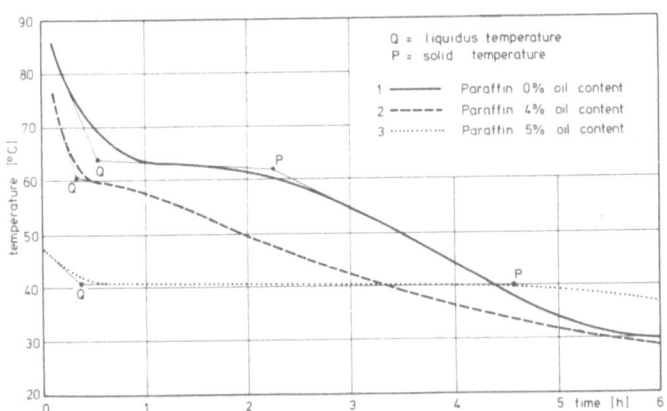

Fig. 7: Cooling Curves for Paraffins with
Different Oil Contents /19/

signify the beginning and the end of the freezing process. Sample
No. 2 with 4 % oil content, for example, has too large a freezing
interval which renders it completely unfit for use in a LTES
system. In fact, the point P marking the end of freezing can hardly
be determined for this paraffin.

An important observation with paraffins is the large difference
between the experimentally measured freezing range and the manufacturer's data. The values for Paraffin C16-C28 are shown in Table 7
below.

Table 7: Melting and Freezing Range of Paraffin C16-C28 /5/

Manufacturer's recommended freezing range	42 – 44 °C
Measured freezing range	38 – 43 °C
Measured melting range	36 – 43 °C

This result is of particular importance to the design and
operating of a LTES system, which calls for an exact knowledge of
the phase transition temperature range of the heat storage material.
Further quantification of these results can be carried out using

the DSC-technique.

(b) DSC-measurements. Typical thermograms for n- and iso-paraffins are presented through the example of Paraffin 5838 and Paraffin 6403 in Figs. 8 and 9 respectively. The n-paraffin exhibits a sharp peak around 46 °C and secondary peaks at temperatures between 12 °C and 46 °C, whereas the iso-paraffin exhibits one peak which is, however, spread over a large temperature range of about 43 K. Evaluation of the peaks shows that both paraffins undergo solid-liquid and solid-solid phase transitions. The DSC technique additionally provides a quantitative measurement of the energy content associated with each of the phase transitions.

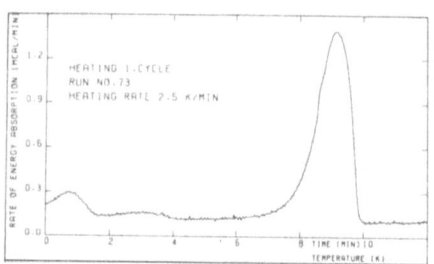 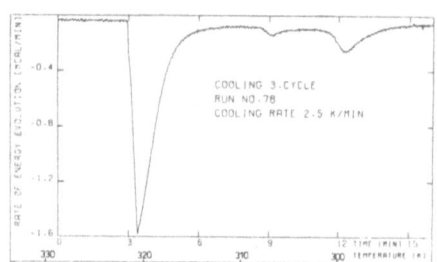

Fig. 8: DSC-Thermograms for n-Paraffin 5838:(a) Heating, 1st Cycle (b) Cooling, 3rd Cycle /7/

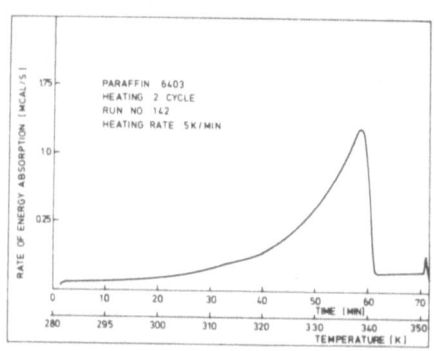 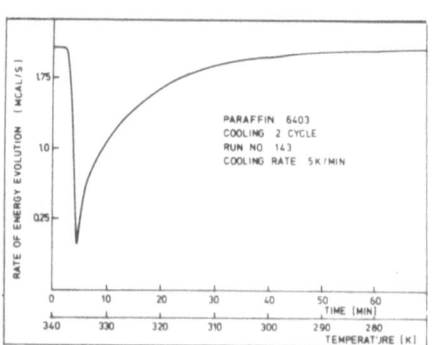

Fig. 9: DSC-Thermograms for iso-Paraffin 6403:(a) Heating, 2nd Cycle, (b) Cooling, 2nd Cycle /7/

Some results from the DSC-measurements of the phase transition temperatures and enthalpies of commercial paraffins are presented in Table 8. For the sake of comparison, results from measurements with 99 % pure octadecane, a research grade paraffin, are also included in the Table (from /7/).

Table 8: Measured Values of Phase Transition Temperatures and Associated Phase Transition Enthalpies for Paraffins /7/

PARAFFIN		MANUFACTURER'S DATA			MEASURED VALUES						
TYPE	CHAIN STRUC-TURE	OIL-CON-TENT	FREEZING RANGE	HEAT OF FUSION	PHASE TRAN-SITION TEMPE-TURE RANGE	SOLID-LIQUID PHASE TRANSITION		SOLID-SOLID PHASE TRANSITION		PHASE TRANSITION ENTHALPY	
					PT	PT_1	ΔH_1 [1]	PT_2	ΔH_2 [1]	TOTAL $\Delta H = \Delta H_1 + \Delta H_2$	PROPORTIONS $\frac{\Delta H_1}{\Delta H}$ $\frac{\Delta H_2}{\Delta H}$
		/%/	/°C/	/kJ/kg/	/°C/	/°C/	/kJ/kg/	/°C/	/kJ/kg/	/kJ/kg/	/kJ/kg/ %
6106	N	5	42 - 44	189	19 - 44	40.7 - 44	129.8	19 - 40.7	49.2	179	72.5 27.5
5838	N	0.5	48 - 50	189	12.7 - 48.3	46.2 - 48.3	134.4	12.7 - 46.2	63.0	197.4	68.0 32.0
6035	ISO	4	58 - 60	189	-8 - 64.4	39 - 64.5	85.7	-8 - 39	83.2	168.9	50.7 49.3
6403	ISO	0.5	62 - 64	189	23 - 65.6	51.7 - 65.6	129.8	23 - 51.7	59.2	189	68.7 31.3
6499	ISO	3	66 - 68	189	-6 - 71.6	2)	2)	2)	2)	145	2) 2)
OCTADECANE 3)			28 - 29	246	27 - 28.5	27 - 28.5	230	-	-	230	100 -

1) ΔH_1 AND ΔH_2 REPRESENT THE ENTHALPIES ASSOCIATED WITH THE SOLID-LIQUID AND SOLID-SOLID PHASE TRANSITIONS RESPECTIVELY

2) PARAFFIN 6499 EXHIBITED A PEAK SPREAD OVER A LARGE TEMPERATURE RANGE OF ABOUT 78 K, SO THAT THE TWO PHASE TRANSITIONS COULD NOT BE DISTINGUISHED FROM EACH OTHER

3) N-OCTADECANE (99 % PURE) WAS EMPLOYED AS A REFERENCE PARAFFIN DURING THE INVESTIGATIONS

For all commercial paraffins tested, the deviation between the manufacturer's data and the measured values is noteworthy. All paraffins exhibited two phase transitions. Whereas the n-paraffins experienced the primary phase transition (solid-liquid) in a narrow temperature range of about 2 K, the corresponding temperature range for iso-paraffins was large ($\geq$14 K). The research grade n-octadecane, on the other hand, exhibited only one solid-liquid phase transition in a narrow temperature range of 1.5 K. Evaluation of the thermograms furthermore showed that while the total phase transition enthalpy of the commercial paraffins agreed closely with the manufacturer's data, the enthalpy associated with the solid-solid phase transition was rather significant (ca. 30 to 50 % of the total).

The above results are of particular importance for the choice of paraffins for low temperature applications, wherein the temperature excursions of the store are generally limited to 10-15 K about the melting point. Only n-paraffins may hence be selected. Although the oil content of the paraffins does not play any significant role, the amount of heat stored would be dependent on the phase transition enthalpy of the primary solid-liquid transition, which should be measured for the paraffin in question.

Fatty Acids: (a) TA-measurements. The melting and freezing characteristics of fatty acids are presented through the example of lauric acid in Fig. 6 (page   ) which contains the heating and colling curves for this material. The freezing plateaus are long and flat and no supercooling is evident, though in some cases, a small degree of supercooling (~ 0.5 K) has been measured. This behaviour is representative of all fatty acids and also of polyethylene glycols with melting points between 15 °C and 70 °C /20/.

(b) DSC-Measurements. Fatty acids possess good melting and freezing characteristics, as may be seen in Fig. 10 from the sharp peak traced on the DSC-thermograms during the heating and cooling of lauric acid.

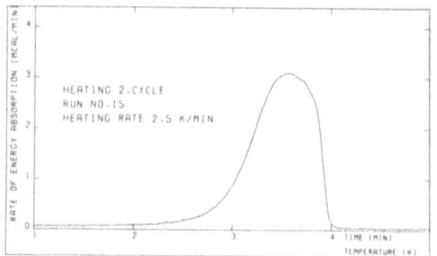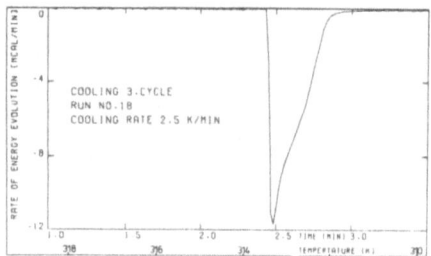

Fig. 10: DSC-Thermograms for Lauric Acid:(a) Heating, 2nd Cycle, (b) Cooling, 2nd Cycle /7/

Results from the DSC-measurements with the two fatty acids tested are summarized in Table 9. The difference between the melting and freezing points given in the Table is a measure of the supercooling of the substances.As mentioned earlier, yet smaller amounts of supercooling (ca. 0.5 K) were observed in TA-tests undertaken using somewhat larger quantities of the substance. The small amount of supercooling during the freezing of fatty acids does not hamper their potential use as heat storage substances.

Although fatty acids are good heat storage substances, they are somewhat too expensive for large-scale heat storage applications. They are, however, far cheaper than research grade paraffins and are recommended for the function tests of new passively-operating heat-of-fusion storage system designs. Recent discussions with manufacturers indicate that the cost of some fatty acids can be brought down by about 50 percent as a result of large scale production. For LTES systems to be operated within a narrow temperature range, the cost of heat storage material per kilojoule of stored energy may then compare fairly well for fatty acids and commercial paraffins.

Table 9: Results from DSC-Measurements with
Fatty Acids /8/

HEAT STORAGE SUBSTANCE	NUMBER OF THERMAL CYCLES PERFORMED	MEASURED VALUES			
		MELTING POINT /°C/	FREEZING POINT /°C/	HEAT OF FUSION /KJ/KG/	HEAT OF SOLIDIFICATION /KJ/KG/
LAURIC ACID	3	43.5 ± 0.05	39.9 ± 0.05	169.3 ± 2.0	168.8 ± 2.5
PALIMITIC ACID	5	61.2 ± 0.07	59.9 ± 0.13	196.1 ± 2.0	197.0 ± 3.0

Salt Hydrates (a) TA-Measurements. Typical heating and
cooling curves for salt hydrates are shown through the example
of $Na_2HPO_4.12H_2O$ in Fig. 6. Melting of the material takes place
at a constant temperature, whereas freezing is associated with
appreciable supercooling. Freezing curves for several other salts
obtained from tests in glass capsules /27/ are presented in
Fig. 11. For all these materials the melt does not freeze at its
thermodynamic freezing point, but is supercooled by several degrees below the freezing point. The supercooled liquid hence
exists in a highly metastable state. Formation or introduction of
a single crystal nucleus into the melt causes a spontaneous crystallization of the whole melt.
       In the case of salt hydrates, the geometry of the test appara-
tus has been found to have a significant influence on the melting and
freezing behaviour of the material /8/. Table 10 compares results from
TA-measurements with the salt hydrates carried out in three different experimental apparatuses. Poor nucleation conditions in the
glass test tube apparatus (Apparatus A) - which is also the most
commonly suggested apparatus in the literature - cause substantial
supercooling of the salt hydrates. In fact, the degree of supercooling observed during the cooling of $Na_2S_2O_3.5H_2O$ was so large
(>40 K) that no freezing of the substance occurred within the
temperature range used in the tests. As a consequence, the tests
were terminated after merely 2 thermal cycles.
       The amount of supercooling could be reduced by using glass
containers (Apparatus B) containing a shallow bed of the salt hydrate. The larger glass surface aided in somewhat improving the
nucleation conditions in the melt and the shallow bed assisted
in eliminating segregation effects in the molten material.
       Both apparatuses A and B are, however, far remote from prac-
tical latent heat store designs. Experiments to investigate the
melting and freezing characteristics of salt hydrates were hence
undertaken in laboratory models (Apparatus C1, C2) of two latent
heat stores that have been recommended for large scale applications /2, 7/. The heat storage concepts selected employ large
metallic heat exchanger surfaces (fins) for heat transfer into
the poorly conducting heat-of-fusion storage substance, and are
suitable for passively operating latent heat stores. Results

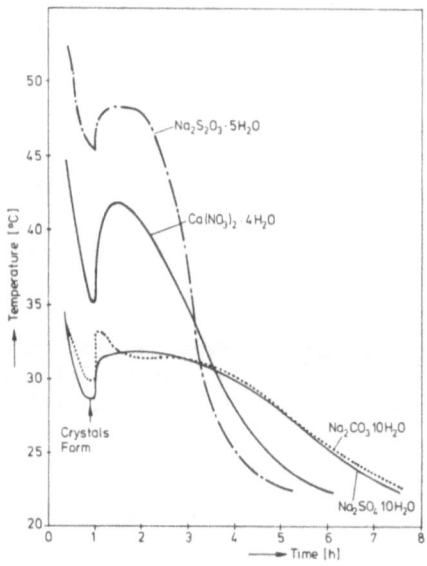

Fig. 11: Comparison of Cooling Curves for Various
Salt Hydrates /27/

Table 10: TA-Measurements with Salt Hydrates /8/

SALT HYDRATE	LITERATURE VALUE OF FREEZING POINT /°C/	TEST APPARATUS USED +	NO. OF THERMAL CYCLES PERFORMED	MEASURED VALUES (AVERAGES) DEGREE OF SUPERCOOLING /K/
$CaCl_2 \cdot 6H_2O$	29.7	A	10	25
		B	50	5
		C1	100	1
$Zn(NO_3)_2 \cdot 6H_2O$	36.4	A	10	8
		C2	20	4
$Na_2S_2O_3 \cdot 5H_2O$	48.0	A	2	>40
		B	40	11
		C2	20	5

+ TEST APPARATUS DESIGNS: A – GLASS TEST TUBES

B – GLASS CONTAINERS

C1 – LABORATORY MODEL OF THE FINNED HEAT PIPE HEAT EXCHANGER DESIGN

C2 – LABORATORY MODEL OF THE FINNED ANNULUS HEAT EXCHANGER ELEMENT

contained in Table 10 now indicate a very favourable freezing behaviour of the salt hydrates, characterized by their small degree
of supercooling. This improvement, in comparison to the values measured using Apparatus A and B, results from the presence of the
aluminum surfaces contacting the salt hydrate, which strongly promote heterogenous necleation in the molten salt.

(b) DSC-Measurements. The calorimetric measurements with salt
hydrates are illustrated here through the example of the incongruently melting calcium chloride 6-hydrate. Experimentation with this
substance was undertaken in two operation modes: Mode 1 - tests in
hermetically sealed pans, and Mode 2 - tests in non-hermetically
sealed or 'open' pans, wherein a hole was punched in the hermetically sealed pans.

Figs. 12 and 13 respectively present typical thermograms obtained during the heating and cooling of $CaCl_2 \cdot 6H_2O$ in hermetically

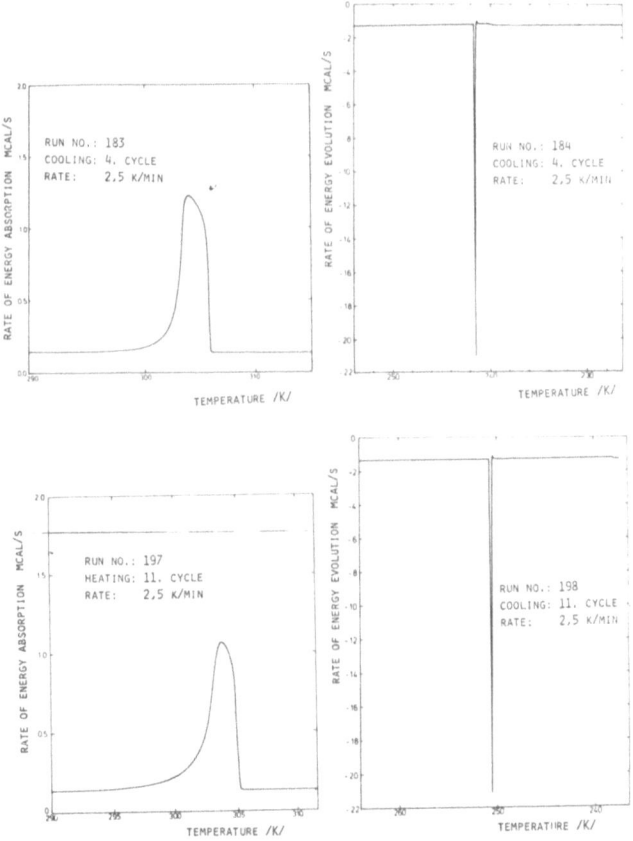

Fig. 12: DSC-Thermograms for $CaCl_2 \cdot 6H_2O$ from Measurements
in Hermetically Sealed Pans (Operation Mode 1) /8/

sealed and open pans /6, 8/. Whereas tests in the hermetically
sealed pans (mode 1) delivered thermograms with a single endothermic or exothermic peak which was reproducible in form even
after thermal cycling upto 11 cycles, operation in open pans
(mode 2) yielded thermograms with a single endothermic peak during
the first melting process but with two or more peaks during subsequent cooling and heating processes. The salt hydrate $CaCl_2.6H_2O$
thus undergoes decomposition during tests in the open pans. By the
fifth thermal cycle, a clear secondary peak is observed at 45 °C,
signifying the partial decomposition of $CaCl_2.6H_2O$ into its lower
form $CaCl_2.4H_2O$.

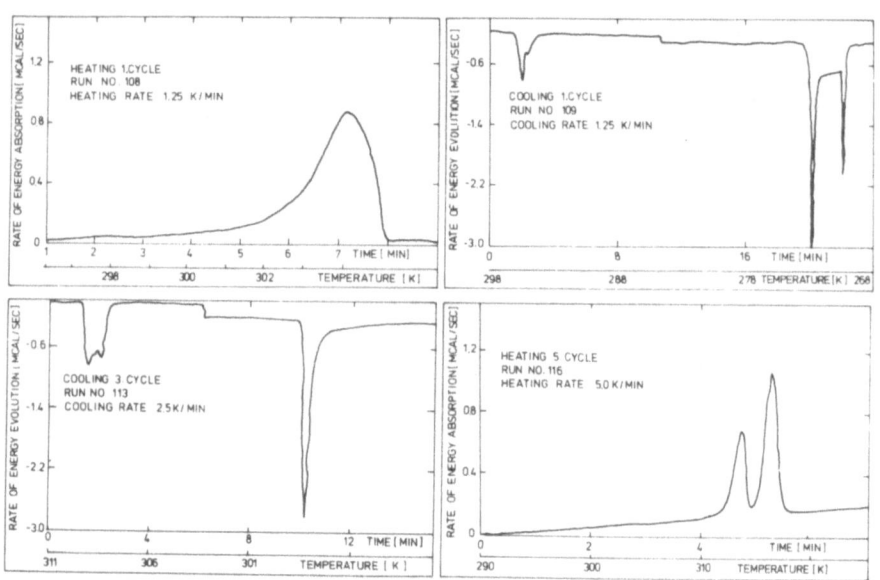

Fig. 13: DSC-Thermograms for $CaCl_2.6H_2O$ from Measurements
in Non-Hermetically Sealed or 'Open' Pans
(Operation Mode 2) /8/

Results pertaining to the melting point and the heat of fusion
of four salt hydrates tested are summarized in Table 11. The values
of the thermophysical properties of the substances as obtained from
measurements with various samples tested in hermetically

Table 11: DSC Measurements with Salt Hydrates /8/

SALT HYDRATE	LITERATURE VALUE OF		OPERATION MODE+	MEASURED VALUES						
	MELTING POINT	HEAT-OF-FUSION		NO. OF SAMPLES	TOTAL NO. OF THERMAL CYCLES	RANGE OF MEASURED VALUES		AVERAGE VALUES		
						MELTING POINT (PRIMARY TRANSITION)	HEAT-OF-FUSION	MELTING POINT	HEAT-OF-FUSION	
	/°C/	/kJ/kg/				/°C/	/kJ/kg/	/°C/	/kJ/kg/	
$CaCl_2 \cdot 6H_2O$	29.7	171	1 2	3 3	18 16	28.5 - 30.2 22.5 - 47.0	140 - 180 30 - 185	29.0 *)	160 *)	
$Zn(NO_3)_2 \cdot 6H_2O$	36.4	147	1	5	29	36.0 - 37.0	135 - 175	36.5	155	
$Na_2S_2O_3 \cdot 5H_2O$	48.0	201	1 2	1 1	4 2	48.5 - 49.0 +)	190 - 200 +)	48.7 +)	195 +)	
$Ni(NO_3)_2 \cdot 6H_2O$	56.7	188	1	3	13	53.5 - 57.0	145 - 160	55.2	152	

NOTES: + OPERATION MODE: 1-TESTS IN HERMETICALLY SEALED PANS
2-TESTS IN OPEN PANS

*) SCATTER TOO LARGE TO OBTAIN MEANINGFUL AVERAGES

+) NO EVALUATION POSSIBLE

+ AVERAGE OF 8 THERMAL CYCLES

sealed pans (mode 1) are seen to fall in a small range and the average values agree fairly well with literature values. On the other hand, results from measurements in open pans show a large scatter as a consequence of the decomposition of the salt hydrates.

In large scale systems, the use of salt hydrates is hence recommended only in hermetically sealed or encapsulated heat-of-fusion storage system designs. It is furthermore recommended to use the salt hydrates in an atmosphere free of air /8/.

Modified Salt Hydrates. The discussion in the preceding section has been restricted to the investigation of the behavioural characteristics of "pure" salt hydrates i.e. in which no foreign substances are intentionally added to prevent segregation and/or to eliminate supercooling of the substance prior to crystallization. There have been several attempts in the literature to "modify" salt hydrates to obtain a reproducible melting and freezing behaviour. Although only limited success has been reported with the methods employed, we shall discuss them briefly here for the sake of completeness.

a) Suspension Media. Addition of suspension media or thickening agents to the salt hydrate to prevent separation of the solid and liquid phases has been recommended /12, 28/. The use of a thickener also assists in suspending the nucleating agents within the heat storage medium bulk, which otherwise tend to collect at the container bottom due to density differences.

Thickening agents, however, displace a part of the salt hydrate in the heat store, so that the volumetric heat storage capacity of the heat store is reduced. Furthermore, they work towards a lowering of the melting point of the heat storage substance. Table 12 lists some suspension media recommended for different salt hydrates /12/.

Table 12: Some Suspension Media for Use with Salt Hydrates /12/

HEAT OF FUSION MATERIALS	SUSPENSION MEDIA
$CaCl_2.6H_2O$	Hydroxy Ethyl Cellulose
$CaCl_2.8H_2O$	Polyvinyl Alcohol
$Ca(NO_3)_2.4H_2O$	Polyacrylic Acid
$NaCO_3.19H_2O$	Polyethylene Oxide
$Na_2HPO_4.12H_2O$	Starch
$Na_2S_2O_3.5H_2O$	Wood Pulp
$Na_2SO_4.10H_2O$	Clay (Bentonite)

(b) Nucleating Agents. A nucleating agent is a material having a crystal structure similar in lattice spacing to that of the heat storage substance. They serve as nuclei for the PCM crystals to grow on them during freezing of the PCM and are also termed as "seed-crystals". For good results, measures should be taken whereby the nucleating agents are homogenously dispersed within the bulk PCM. Table 13 provides a list of some of the commonly recommended nucleating agents for use with salt hydrates.

(c) Extra Water Principle. Extra water may be added to a salt hydrate to allow dissolution of the anhydrous salt in the water at the melting point of the salt hydrate, so that the heat storage medium becomes a saturated salt solution at the melting point. During cooling, the solubility of the salt in water for temperatures below the melting point decreases with decreasing temperatures resulting in crystallization of the salt hydrate. Thus at temperatures below the melting point, the storage medium consists of a salt hydrate solid phase and solution. Upon repeated heating and cooling, the salt hydrate and its phases thus completely go into solution or crystallization out of solution takes place. Soft stirring of the medium is, however, necessary to overcome the density differences between the salt solution and the solid phases /15/.

Table 13: Some Nucleating Agents for Use with
Salt Hydrates

Heat Storage Substance (PCM)	Nucleating Agent	Reference
$LiClO_3 \cdot 3H_2O$	$KClO_4$, Na_2SiF_6, K_2SiF_6, $BaSiF_6$	24
$KF \cdot 4H_2O$	pumice stone	24
$CaCl_2 \cdot 6H_2O$	$BaCO_3$, $SrCO_3$, BaF_2, SrF_2	12, 24
$Na_2SO_4 \cdot 10H_2O$	$Na_2B_4O_7$	26
$Zn(NO_3)_2 \cdot 5H_2O$	ZnO, $Zn(OH)_2$	20
$KF \cdot 2H_2O$	Al_2O_3	24

Using a mixture of 68.2 weight % $Na_2SO_4 \cdot 10H_2O$ and 31.8 weight % H_2O, Biswas /10/ reports good results in comparison to pure Glauber salt. Nucleation of the decahydrate occured readily, even without the addition of borax. Furbo /15/ reports some preliminary results with several different salt hydrates and finds them attractive in use in comparison to sensible heat storage in water.

The major disadvantage of the extra water principle is, however, the loss in volumetric heat storage capacity in relation to that of the pure salt hydrate. For example, with Biswas' composition of $Na_2SO_4 \cdot 10H_2O$ and H_2O, a mass 50 percent larger and a volume 71 percent larger than for an ideally efficient system based on pure $Na_2SO_4 \cdot 10H_2O$ would be required to store the same amount of heat /10/.

(d) Other Techniques. Another technique to avoid phase separation in incongruent melting salt hydrates is that suggested by Carlsson, et.al. /11/, who chemically modify an incongruent heat-of-fusion system to make it behave as a congruent system. Through the addition of 2 weight percent $SrCl_2 \cdot 6H_2O$, the melting point of the resultant salt hydrate has been altered so that the melting point maximum for $CaCl_2 \cdot 6H_2O$ coincides with the point where equilibrium between $CaCl_2 \cdot 6H_2O$, $CaCl_2 \cdot 4H_2O$ and the solution exists. Upon melting of the modified salt hydrate, the peritectic point is thus by-passed and the formation of the lower salt hydrate ($CaCl_2 \cdot 4H_2O$ in this case) is avoided.

Thermal Cycling

One of the most severe tests that phase change heat storage materials must undergo is thermal cycling involving repeated melting and freezing of the heat storage materials. For example, for a 20 year life of a one-day heat store, the phase change material experiences one melting-freezing cycle daily or a total of 7,365 cycles during the system life.

The influence of thermal cycling on the phase change material characteristics must be measured experimentally. A large gap still exists today in this area. Limited thermal cycling (120 cycles) carried out with paraffins and lauric acid exhibited no degradation of materials /4/. Test results for 1000 heating-cooling cycles with $Na_2SO_4.10H_2O$ are also available /28/. Two material samples of $Na_2SO_4.10H_2O$ were used in these tests - one sample comprising the phase change material and 3 % Borax by weight as nucleating agent, and the second sample comprising the same constituents as above plus 8 % thickener that formed a thixotropic gel. The results of these thermal cycling tests are presented in Figure 14. The material with thickener showed no degradation in properties following cycling. However, the material without the thickener contained approximately 30 to 35 % liquid at the end of the cooling cycle. A sediment layer formed which remained throughout the testing period. The non-thickened material is also seen in Figure 14 to reach a higher temperature during its heating cycle, due to the unavailability of the stratified lower layer.

Limited thermal cycling (upto 90 cycles) of $Na_2HPO_4.12H_2O$ showed that the material melts congruently when it is not contaminated with the heptahydrate, i.e. when the formation of the heptahydrate has been prevented by appropriate nucleation /28/.

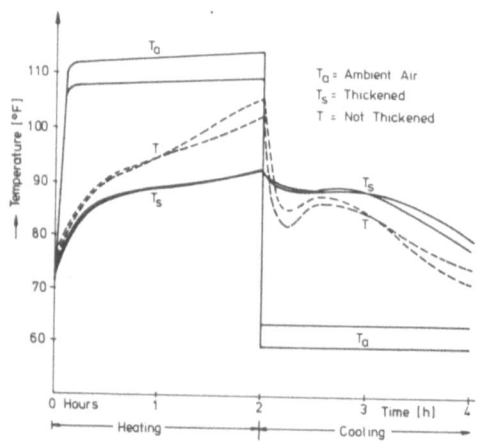

Fig. 14: Thermal Cycling Tests with $Na_2SO_4.10H_2O$ /28/

Compatibility with Materials of Construction

Knowledge regarding the compatibility of phase change heat storage materials with conventional materials of construction is of particular importance to the assurance of the life of a LTES system. Only limited compatibility data is presently available and the results reported here are those gained from tests conducted at our institute /18, 19/.

Figure 15 shows the combinations of the phase change materials and metals, which were subjected to experiments. Metal samples having dimensions of 30 x 25 x 2 mm were cleaned and wet-polished with 1000 grain abrasive paper. The samples were immersed in the liquid phase of the storage material contained in air-tight bottles, one sample to a bottle. The bottles were placed in a thermostatically controlled water bath whose temperature was maintained constant at 20 K above the melting point of the corresponding storage material.

Heat Storage Material \ Metal	Al 99.5	AlMg 3	Cu 99.9	Stainless Steel 1.4301	Mild Steel 1.0330
Sodiumthiosulphate-5-hydrate (48°C)*		x	x	x	x
Sodiumhydrogenphosphate-12-hydrate (35°C)	x	x	x	x	x
Calciumchloride-6-hydrate (30°C)	x	x	x	x	x
Loxiol G 32 (58°C)		x	x	x	
Lauric acid (44°C)		x	x	x	

* melting point in brackets

Fig. 15: Heat Storage Materials and Materials of Construction selected for the Corrosion Investigations /18/.

The metal samples were removed from the bottles after predetermined time intervals and cleaned. Gravimetric analysis prior to and following the corrosion tests provided the mass loss m(g). Using DIN 50905, the reduction in sample thickness Δs (μm) and the corrosion rate v (g m^{-2} d^{-1}) or (mm/a) were then computed. Measurements were carried out using various samples of the same metal, which were in contact with the same storage material for different periods of time. Information on the temporal variation of the corrosion behaviour could thus be obtained. In accordance with Dechema tables /13/, a metal experiencing a thickness reduction of $\leq$0.1 mm/a was then termed corrosion resistant, and of $\leq$1.0 mm/a fairly corrosion resistant. For cases where the corrosion rate attained a linear value, an estimate of the lifetime was possible.

In addition to the gravimetric analysis, optical and scanning electron microscopy techniques were employed to investigate the sample surface and cross-sections. The products of corrosion were furthermore chemically analysed.

The results of these investigations are summed up in Table 14. The organic materials were found to be compatible with the metals tested. With the salts, however, one needs to be careful as preferential compatibility was observed. Stainless steel is the only metal that was found compatible with all phase change materials tested. Copper exhibited rapid corrosion when immersed in sodiumthiosulphate-5-hydrate and a black layer of CuS was seen to form just 10 days after contact. The mass loss of the sample after 300 days of contact was found to be 8.17 g and the thickness reduction 610 μm. A photograph of the corroded sample taken after 50 days contact is presented in Figure 16.

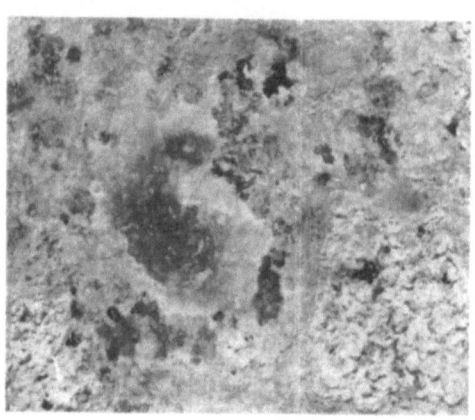

Fig. 16: Surface of Cu99.9 Sample after 50 d in $Na_2S_2O_3 \cdot 5H_2O$ /18/

Aluminum and aluminum alloy $AlMg_3$ were found incompatible with sodiumhydrogenphosphate-12-hydrate and were covered with a white layer of aluminum hydroxide $Al(OH)_3$ after a short period of contact time. In general, $AlMg_3$ was found to be more sensitive due to its magnesium content. A scanning electron microscopy investigation showed a surface attack after 20 days contact. After 50 days of contact, however, pitting corrosion with trans- and inter-crystalline cracks were seen on the aluminum surface, while $AlMg_3$ showed fewer cracks and tended more towards shallow pit formation. These sample surfaces are seen in the scanning electron microscope photographs of Figures 17 and 18 for the two samples after 80 d and 105 d in contact with $NaHPO_4 \cdot 12H_2O$ respectively.

<u>Table 14:</u> Results of Corrosion Investigations /18,19/

Heat Storage Material / Metal	Al 99.5	Al Mg 3	Cu 99.9	Stainless Steel 1.4301	Mild Steel 1.0330
$Na_2S_2O_3 \cdot 5H_2O$	+	+	–	+	+
$Na_2HPO_4 \cdot 12H_2O$	–	–	+	+	+
$CaCl_2 \cdot 6H_2O$	–	–	+	+	+
Loxiol G 32		+	+	+	
Lauric acid		+	+	+	

<u>Explanation of Symbols:</u>

+	resistant	–	corrosion rate < 0.1 mm/a
⊕	fairly resistant	–	corrosion rate < 1.0 mm/a
⊖	not particulary resistant	–	corrosion rate < 25–30 mm/a
–	unusable		

$CaCl_2 \cdot 6H_2O$ was also found corrosive to aluminum and its alloy $AlMg_3$. Local corrosion in the form of both pitting corrosion and shallow pit formation was also observed in this case.

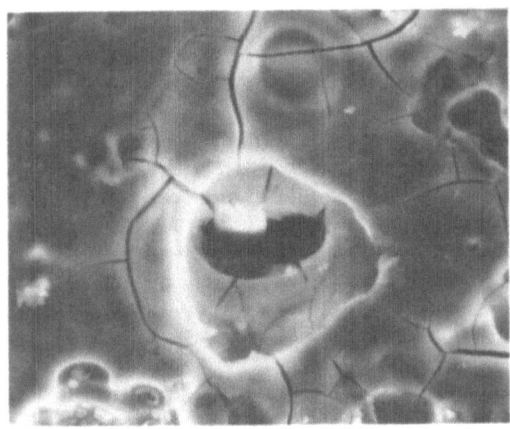

<u>Fig. 17:</u> SEM Photograph of Al99.5 Sample Surface, showing Pitting Corrosion after 80 d in $Na_2HPO_4 \cdot 12H_2O$ /18/

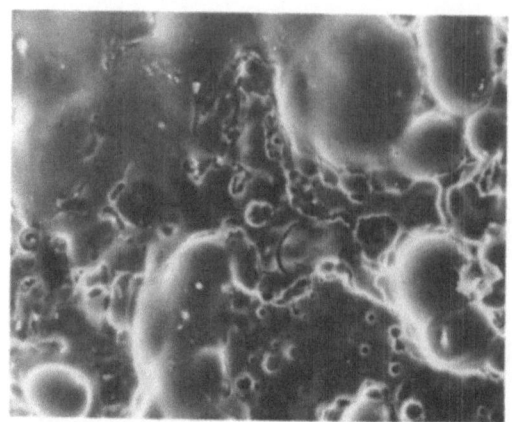

Fig. 18: SEM Photograph of AlMg$_3$ Sample Surface showing
Shallow Pit Formation after 105 d in Na$_2$HPO$_4$.12H$_2$O
/18/

HEAT TRANSFER CONSIDERATIONS

We have seen earlier that a LTES system must possess a heat
exchanger for transferring heat from the heat source to the heat
storage substance and from the latter to the heat sink. The type
of the heat exchanging surface itself plays an important role in
the design of LTES systems, as it strongly influences the temperature gradients for charging and discharging of the storage.

Heat exchanger requirements

The LTES heat exchanger must fulfill the following requirements:
1) It should provide for a high effective heat transport rate to
   allow rapid charging and discharging of the storage. This is a
   very pressing requirement for latent heat stores as the thermal conductivity of most phase change heat storage materials
   is extremely low – most of these materials possess insulating
   properties. A high effective heat rate can be obtained either
   by embedding a metallic filler of high thermal conductivity
   within the heat storage medium or by introducing natural or
   forced convection effects in the storage medium e.g. forced
   convection through stirring of the medium.
2) It should permit only small temperature gradients for charging
   and discharging of the storage. This effect may be achieved
   by providing a substantially large heat transfer surface and
   small heat transfer paths in the storage medium.
3) The heat exchanger should guarantee a high thermal diffusivity
   and a high heat diffusivity.

The type of heat exchanger geometries that fulfill these requirements are discussed in the next section.

Type of heat exchanger geometries

LTES heat exchangers fall in 2 categories:
1) "passive", in which the PCM is at rest and/or the heat exchanger has no moving parts and
2) "active", in which the PCM is set in motion and/or the heat exchanger has moving parts.

Passive heat exchangers typically comprise of tubes of small diameter (30-50 mm) or flat pans (20-30 mm deep) within which the phase change material is filled. The tubes can be bundled together as in shell-and-tube type heat exchangers with the heat transfer fluid flowing in the gaps formed between the tubes. Such a configuration is shown in Figure 19a. Figure 19b shows an arrangement wherein the material is sealed into pan-shaped containers. 294 such pans measuring approximately 520 x 520 x 25 mm and containing 3,000 kg of $Na_2S_2O_3.5H_2O$ were stacked together to form a latent heat thermal store for the solar heating system at the University of Delaware /27/. A 10 mm gap was provided between adjacent pans for the circulation of the heat transfer fluid (air).
Active heat exchangers are those in which means are generally provided to stir the phase change heat storage material. This is done to improve the heat transfer rate within the storage material, as well as to prevent segregation of phases - a phenomenon typical of inorganic salt hydrates. Two types of active heat exchangers are shown schematically in Figures 19c and 19d. The heat exchanger in Figure 19c is a direct contact heat exchanger, so called as the storage material and the heat transfer fluid are brought in direct contact with each other /14, 21/. Hot oil heated e.g. using a solar collector is sprayed through fine pores in the solid phase of the phase change material. The large number of oil droplets provide an enhanced heat transfer surface and rapidly transfer their heat to the surrounding salt thereby melting it. When the salt has partially melted, the oil spray creates a turbulence effect in the melt that not only improves the heat transfer rate in the medium, but also prevents segregation of the phases. To remove heat from the storage material, cold oil is sprayed through the liquid phase of the storage material. Thereby heat is removed from the material and salt crystals are formed in the liquid bulk which fall on the container bottom and collect there.
Figure 19d is a novel conception of an active heat exchanger - the so called 'rolling cylinder' system /16/. The concept comprises of a cylinder filled with sodium-sulphate decahydrate (Glauber salt) that serves as the heat storage material. The cylinder is rotated about its longitudinal axis at about 3 rpm. The rotation is considered to provide just enough stirring action to keep the tem-

A. Passive Heat Exchangers

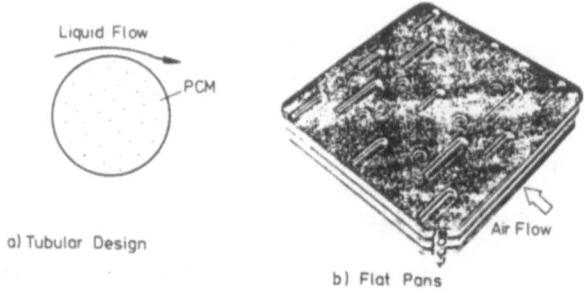

Liquid Flow

PCM

a) Tubular Design

Air Flow

b) Flat Pans

(PCM* within Tube or Pan)

B. Active Heat Exchangers

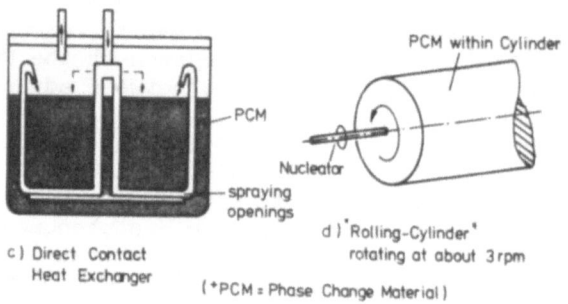

PCM within Cylinder

PCM

Nucleator

spraying openings

c) Direct Contact Heat Exchanger

d) 'Rolling-Cylinder' rotating at about 3 rpm

(*PCM = Phase Change Material)

Fig. 19: Type of Heat Exchanger Geometries considered for LTES Systems

perature of the Glauber salt uniform and very close to the wall temperature. The researchers involved with the development of this concept believe that under these conditions, the material crystallizes on nuclei in the liquid and not on the cylinder walls. To promote crystal formation, a thin tubular device, called a nucleator is inserted through one end of the rotating vessel. The nucleator contains seed crystals which act as nucleating agents to initiate the crystallization process when the temperature of the liquid salt drops below its freezing point.

Several other types of passive and active heat exchangers for LTES systems on which R&D work is being undertaken at different institutions are shown in Appendix II. In the present lecture, we shall discuss passive heat exchangers in some depth through the example of two concepts that are under development at the Institut für Kernenergetik und Energiesysteme (IKE), University of Stuttgart.

Passive Heat Exchangers

Heat Exchanger Concepts
A. The Modular Finned Heat Pipe Heat Exchanger /1, 2/ Figure 10 shows
a schematic of a single heat exchanger module. The module essentially
comprises of a container (1) of square cross-section provided with a
heat pipe (2) along its longitudinal axis. The container, as well as
the heat pipe, are divided into three regions - region A, B and C
in Figure 1 - by separation walls (5). One of the regions, B, is
filled with a heat-of-fusion type storage substance and is called
the Storage Chamber. The remaining two regions, A and C, are respectively in contact with the fluid, e.g. water or air, flowing through
the solar collector and heating load loops, and are termed here as
the Heat Source and Heat Sink regions respectively.

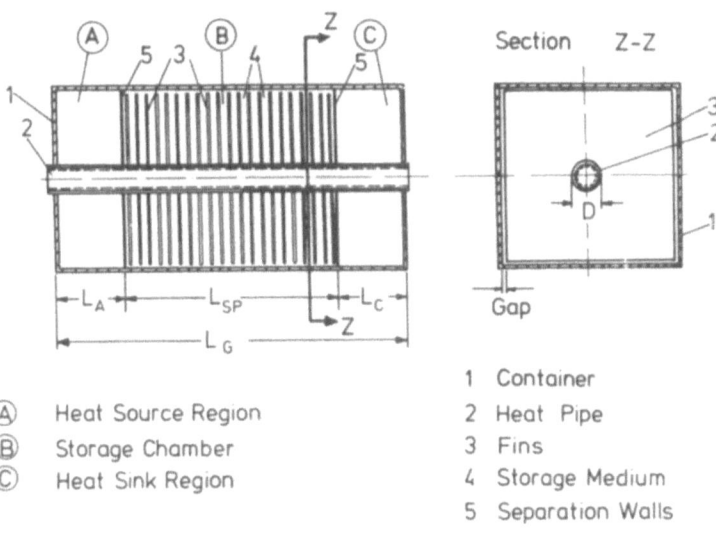

A   Heat Source Region
B   Storage Chamber
C   Heat Sink Region

1   Container
2   Heat Pipe
3   Fins
4   Storage Medium
5   Separation Walls

Fig. 20: Schematic showing the Modular Finned Heat Pipe
Heat Exchanger Concept /1, 2/.

The heat pipe length within the Storage Chamber is provided
with closely and equally spaced square fins of large cross-section
(3), made from a material of high thermal conductivity, e.g. aluminum. The Storage medium (4) fills the free volume between the
fins. Thus embedded within the storage medium, the fins compensate for its generally poor thermal conductivity and simultaneously ensure small heat flow paths within the LTES system.
The use of the heat pipe offers several advantages and renders flexibility in operation and application. Some of these are:

a) Additional heat exchangers are eliminated. The heat pipe
    length within the Heat Source and Heat Sink regions may
    be suitably finned depending on whether air or liquid is
    employed as the heat transfer medium in these regions.
b) The heat pipe transports heat under very low temperature gra-
    dients so that an almost isothermal heat source in contact
    with the fins and the thermal storage medium within the Sto-
    rage Chamber is attained.
c) The heat flux transformation capability of the heat pipe can
    be utilized to give low heat flux densities within the Storage
    Chamber for large heat flow rates in the Heat Source/Heat Sink
    section.
d) The heat pipe can operate undirectionally as a thermal diode.
    For certain applications where a heat source or heat sink
region external to the storage unit are considered advantageous,
the finned heat pipe may be replaced by a finned tube, through
which the heat transfer fluid, e.g. solar collector liquid, flows.
In this case, however, the fluid flowing within the tube experiences a temperature gradient in the direction of flow, so that the
isothermality of the heat-carrying tube is lost.

Arrangement of the modules. Different arrangements of the
modules are possible depending on the space available. Two possible arrangements are shown in Figure 21. Figure 21a shows a so-called
Box-arrangement, whereby 4 heat pipes are attached to plate fins
of square cross-section. Figure 21b shows a Stack arrangement,
which can be particularly useful when only limited floor space is
available. If modules are stacked against a wall, the insulation
effect of the wall can additionally be used to the advantage of
reducing heat losses from the modules to the surroundings.

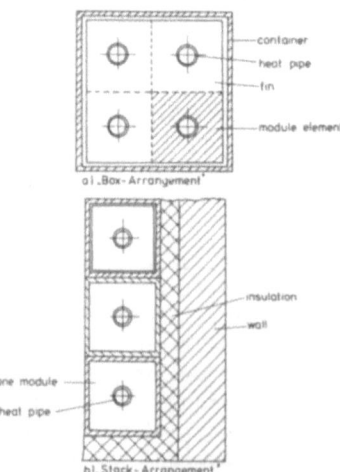

Fig. 21: Arrangement of the Modules /1/

B.   The Finned Annulus Heat Exchanger /6, 7/.

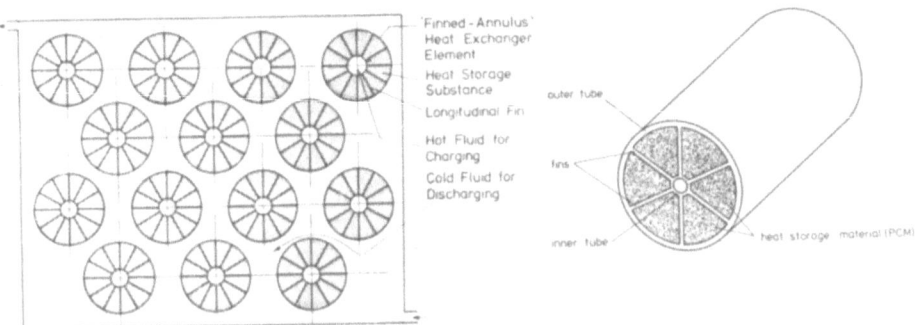

Fig. 22: (a) Schematic of the Finned-Annulus Heat Exchanger Concept for a Heat-of-Fusion Store
(b) Isometric View of One Heat Exchanger
Element /6, 7/.

Fig. 22a shows a schematic of the finned-annulus heat exchanger
concept. The heat exchanger consists of a number of 'finned-annulus'
heat exchanger elements bundled together in a conventional shell-
and-tube pattern. Each element (Fig. 22b) comprises of an inner tube
and an outer tube maintained in thermal contact with each other
through longitudinal fins made from a good heat conducting material,
e.g. aluminum. The region in the annulus and between the fins is
filled with a heat-of-fusion storage material. Hot liquid from a
heat source, e.g. solar collector, flows within the inner tube of
an element, whereas the cooler liquid from a heat sink, e.g. the
load of a heating system, flows around the outer tube.

The finned-annulus geometry offers a number of advantages for
use in a heat-of-fusion store. The fins provide an extended heat
transfer surface, they ensure mechanical stability, and they restrict
the storage medium to small volumes. Depending on the thermal conductivity of the fins and of the inner/outer tubes, the tube surface
may itself serve as a fin to the longitudinal fin. The heat exchanger thus permits not only charging and discharging, but also
simultaneous charging and discharging. Furthermore, the finned-
annulus heat exchanger concept permits the operation ot the heat
storage system as a hybrid latent heat-sensible heat store, whereby the sensible heat is stored in the liquid surrounding the
finned-annulus elements in the containment.

Performance Investigation. The performance investigation of
a latent heat store includes:
(1) a study of the melting and freezing behaviour of the PCM
(2) assessment of thermal performance in terms of:
    o temperature distribution within the storage medium,
       fins and heat transfer walls at any time, t:
    o the position of the solid-liquid interface at any time, t:
    o the heat transfer rate during charging/discharging for
       given initial and boundary conditions
    o the storage charging/discharging time, t_{max}, and
    o the corresponding maximum temperature gradient in the
       storage chamber, ΔT_{max}, required (at t_{max}) for heat flow
       into the storage medium as a function of heat flux density.
(3) determination of the heat storage capacity for a given operating
    temperature range and as a function of the number of charge/
    discharge cycles experienced by the heat store.
    A thermal analysis of the selected heat exchanger concept and
an experimental investigation of appropriate laboratory models and/
or prototype units of the same for the assessment of the thermal
performance of the latent heat store are recommended.

Melting and Freezing Behaviour of the Heat Storage Material
The melting and freezing behaviour of the PCM in heat exchangers
employing finned surfaces is shown here for the finned-annulus
heat exchanger (Concept B). For this purpose, a Test Model, shown
in Fig. 23, was used. Both ends of the cylindrical storage chamber of the model were closed with plexiglas flanges to allow for
visual observations and for photographing the motion of the solid-
liquid interface. A number of PCMs, such as eicosane (a paraffin),
lauric and myristic acids (fatty acids) and $CaCl_2.6H_2O$ and
$Na_2S_2O_3.5H_2O$ (salt hydrates) were employed as heat storage substances in the tests.

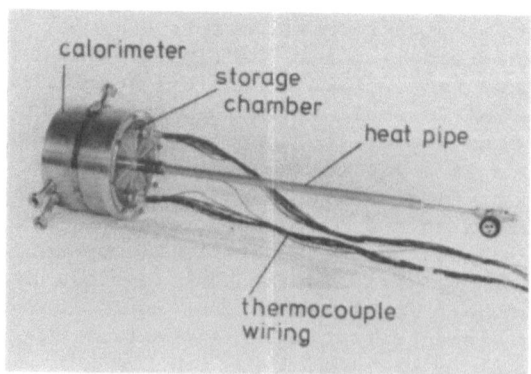

Fig. 23: Test Model of a Finned-Annulus Heat Exchanger
        Element

Figs. 24 and 25 respectively present a sequence of pictures taken during charging and discharging of the test model. The heat storrage medium here is eicosane, a paraffin melting at 37 °C. Charging of the heat store is performed by flowing hot liquid at 50 °C at the evaporator end of the heat pipe, while discharging takes place by flowing cold liquid at 15 °C around the outer tube of the element. The pictures indicate that thermal convection effects eventually set in within the segments of the top hemisphere during charging, inspite of the fact that the fins assist in suppressing convection. During discharging, on the other hand, the heat transfer mechanism is that of pure heat conduction and the frozen front moves symmetrically in all segments. A quantitative assessment of the rate of motion of the phase change interface could be obtained from an evaluation of the photographs and the variation of the solid/liquid fraction with time could thus be determined /9/.

Thermal Analysis. A thermal analysis of the selected heat exchanger geometry is necessary to determine the influence of the variation in geometric dimensions on the thermal performance of the latent heat store. Due to the complicated geometry of the heat exchanger and the non-linearities introduced in the mathematical equations by the moving solid/liquid phase interface, analytical solutions are seldom possible and a numerical analysis is normally undertaken.

Fig. 26 shows a flow chart of the thermal analysis performed for the two heat exchanger concepts. A heat conduction model is employed. We shall restict our discussion here to the analysis of the finned heat pipe heat exchanger (Concept A) /5/.

Parameters Considered: The influence of various geometrical parameters such as fin height, fin thickness, fin spacing and void fraction on the thermal performance was investigated. Heat flux densities were assumed to vary between 1 W/cm² and 6 W/cm² (based on the fin cross-sectional area). The largest module size, and hence the module storage capacity, for acceptable maximum temperature gradients of 5 K and 10 K within the storage chamber was computed.

Two thermal storage substances of particular interest to LTES systems for solar heating applications were selected for the analysis:
(i) Paraffin-white, a paraffin (melting range 48 °C – 55 °C) and
(ii) Sodiumthiosulphate-pentahydrate, an inorganic salt hydrate (melting range 48 °C – 49 °C).

Besides the large difference in their melting ranges these two substances differ markedly in their volumetric storage capacities and thermal conductivities. For the same volume, the heat-of-fusion of the salt is approximately 2.5 times that of the paraffin, whereas its thermal conductivity is roughly 3 times as large.

Fig. 24: Sequence of Photographs Depicting Charging of
the Test Model with Eicosane

Fig. 25: Sequence of Photographs Depicting Discharging
of the Test Model with Eicosane.

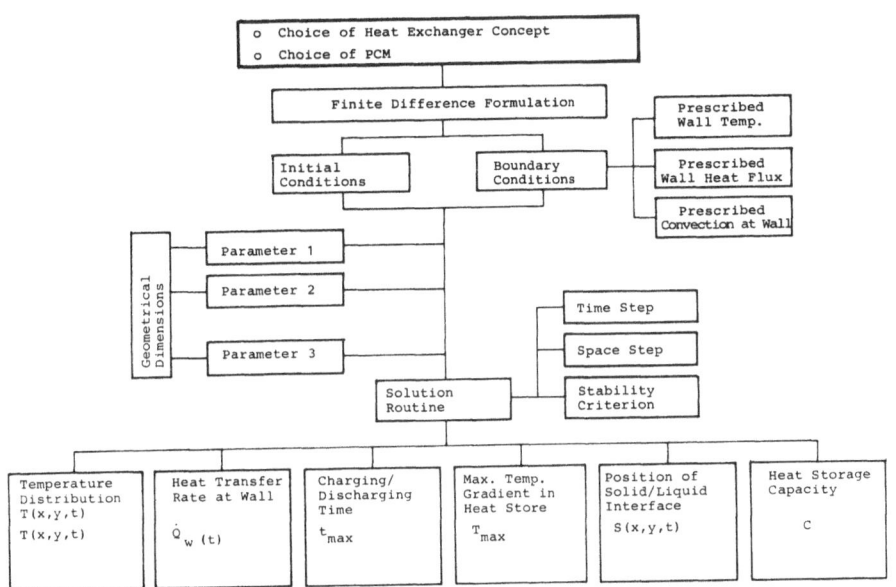

```
┌─────────────────────────────────────────┐
│ o  Choice of Heat Exchanger Concept      │
│ o  Choice of PCM                         │
└─────────────────────────────────────────┘
```

Fig. 26: Flowchart of the Thermal Analysis

Results from the Analysis /2, 5/

(1) Thermal Performance Indicators. Of particular interest to LTES
design are the heat transfer rate, the storage charging/dischar-
ging time, the maximum temperature gradients required for charging
or discharging the heat store, and the heat storage capacity.
Assuming that the PCM does not degrade during melting and freezing,
the charging/discharging time becomes a function of the total
system storage capacity and the heat transfer rate. The maximum
temperature gradient, ΔT_{max}, on the other hand, depends on the
heat transfer rate and heat exchanger geometry - in the present
case on the fin dimensions, viz. fin height (DFIN), fin thickness
(SR), fin spacing (DR) and the void fraction (SF).

In actual practice, the LTES system heat exchanger for low
temperature solar heating applications is constrained to operate
within small temperature swings. Consequently the maximum temperature gradient ΔT_{max} is fixed in advance and we proceed to seek
the relationship between the fin geometry and the heat flux input
rate.

For the computations described here, two values of ΔT_{max}:
ΔT_{max} = 5 K and ΔT_{max} = 10 K were selected. For a heat flux input
$\dot{q}''_{FIN}$ of 4 W/cm² and assuming a void fraction SF of 85 %, the fin
dimensions (i.e. fin thickness SR and fin height DFIN) were com-

puted with the aid of the thermal analysis. The results are summed in Table 15 for both paraffin-white and sodium thiosulphate. Also included in the Table are the computed values of the module storage capacity per meter length of the module and the relative storage capacity for the different cases considered. For the different cases included in Table 15, the variation in module capacity is observed to be almost six-fold, from a minimum value of 0.31 kWh for paraffin-white (ΔT_{max} = 5 K, SR = 0.9 mm) to a maximum value of 1.91 kWh for thiosulphate (ΔT_{max} = 10 K, SR = 0.3 mm). The example thus demonstrates the importance of the permissible temperature gradients and fin dimensions on the heat storage capability of a single module.

Table 15: Module Capacities for Selected Temperature
Gradients and Fin Dimensions /2/

Thermal Storage Substance	Allowable ΔT_{max} /K/	SR /mm/	DFIN /mm/	C^+ /kWh/	Relative Storage Capacity[++]
Paraffin-white	5	0.3	105	0.41	1.32
		0.9	92	0.31	1.0
	10	0.3	146	0.79	2.55
		0.9	131	0.64	2.06
Sodium-thiosulphate-pentahydrate	5	0.3	107	1.03	3.32
		0.9	96	0.93	3.0
	10	0.3	141	1.91	6.15
		0.9	135	1.75	5.64

[+] C is the module storage capacity per meter length based on the heat-of-fusion of the storage substance only

[++] The relative storage capacity is defined as the ratio of the storage capacity of any given module to that of a reference module. A paraffin-filled module, for which ΔT_{max} = 5 K and SR = 0.9 mm, is taken as the reference module here. The relative storage capacity of such a module is hence = 1.0.

(2) Influence of the melting range: The temperature gradients discussed above have been computed using the upper limit of the melting range of the storage substance as the reference temperature. The melting range itself has hence not been explicitly accounted for. In actual practice, the presence of a melting range has an adverse effect on the thermal performance, as its magnitude dictates the temperature swings that a LTES system experiences during heat addition or removal. The total temperature gradient required for storage charging would then be,

$$\Delta T_{total} = \Delta T_{max} + \Delta T_{melting\ range}$$

Figure 27 shows a comparison of ΔT_{max} and ΔT_{total} for a representative module (DFIN = 140 mm, SF = 85 %, SR = 0.5 mm) filled with paraffin and thiosulphate. Due to the large melting range (7 K) of paraffin-white, rather high total temperature gradients are noticed for complete charging of a module employing paraffin as the storage medium. In other words, such a module would be suitable for low temperature applications only if the magnitude of the heat flux density is low, in the range of 1-2 W/cm² .

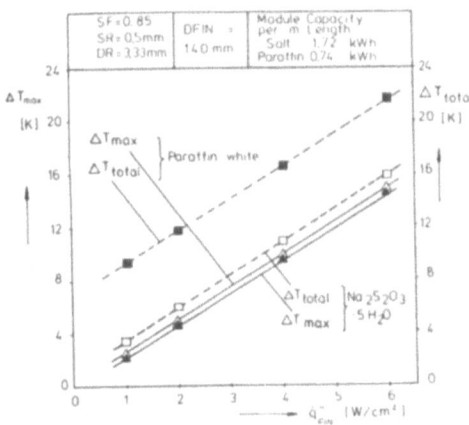

Fig. 27: Influence of Melting Range on the Maximum Temperature Gradient.

Experimental Investigation. The results presented in this paper are those obtained from experimentation with the first prototype of the finned heat pipe heat exchanger (Concept A) /3/. For results with the finned annulus heat exchanger the reader is referred to References 7, 9 .

Fig. 28 shows a photograph of the experimental set-up. The prototype heat store is a 3 m long unit, the storage chamber being 2 m long and the heat source/sink regions 0.5 m long each. The storage chamber has a cross-section of 220 x 250 mm and is filled with ca. 60 kg of lauric acid, which acts as the heat storage substance. The prototype is installed at a small incline of 1° with respect to the horizontal, so as to utilize the thermal diode action of the heat pipe. The various parameters varied during the experiments, and the range of variation, are listed in Table 16. Prior to actual testing the heat losses from the prototype heat store as a function of its outer skin temperature were carefully measured. For most tests, the average heat loss rate during the period of the test was found to vary between 75-150 kJ/h, and the average heat loss between 1000-2000 kJ.

Fig. 28: Photograph of the Experimental Set-up for Testing the Prototype Unit of a Latent Heat Store using the Finned Heat Pipe Heat Exchanger /3/.

Table 16: Parameters Considered during the Experiments and their Range of Variation

Parameter	Range of Variation/Values
Inlet temperature of circulating heat transfer fluid during charging	50 °C, 60 °C, 70 °C
Inlet temperature of circulating heat transfer fluid during discharging	30 °C, 22.5 °C, 15 °C
Mass flow rate of circulating heat transfer fluid during charging and discharging	45 l/h, 90 l/h

Results from the Experiments /3/

<u>Charge/Discharge Cycle</u> Fig. 29 presents a typical charging/discharging cycle obtained with lauric-acid. The freezing plateau is sharp, though the material exhibits a very small amount of supercooling ΔT_U - on the order of 0.5 K - during freezing. The difference in charging and discharging times on the diagram is due to the fact that the temperature potential ΔT^* [+] allowed during the heating and cooling processes differ considerably.

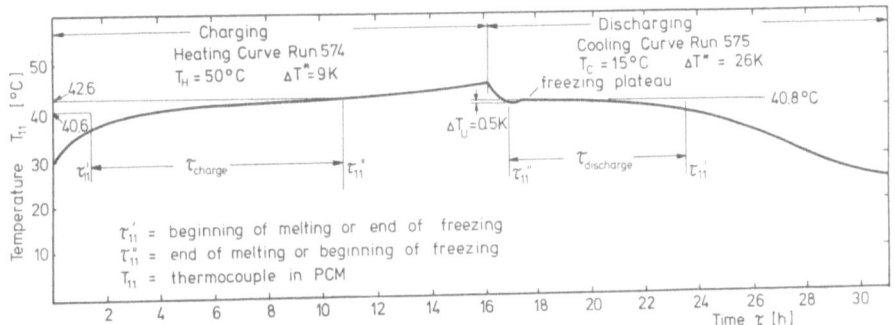

Fig. 29: A Typical Charging/Discharging Cycle. The Phase Change Material is Lauric Acid

<u>Axial Temperature Profile Along Heat Pipe</u> Figs. 20 (a) and (b) show the axial temperature profile along the heat pipe wall, as it exists during charging and discharging of the heat store. Within the storage chamber, the heat pipe is seen to behave rather isothermally within 2 K, so that a uniform phase change of the PCM within the storage chamber can be expected.
<u>Charge/Discharge Time</u> The charge time τ_{charge}, and discharge time $\tau_{discharge}$, for the heat store, as read out from diagrams similar to Fig. 29, are shown in Fig. 31. The charge/discharge times are plotted here as a function of the maximum temperature potential available for heat flow, ΔT^*. The charging time is somewhat smaller, which is due to the higher charging rate for the same temperature potential ΔT^*, for reasons to be seen later.

[+] The temperature potential ΔT^* is defined as the difference between the circulating fluid inlet temperature and the phase transition temperature of the PCM. It represents the driving potential during a charging or discharging process.

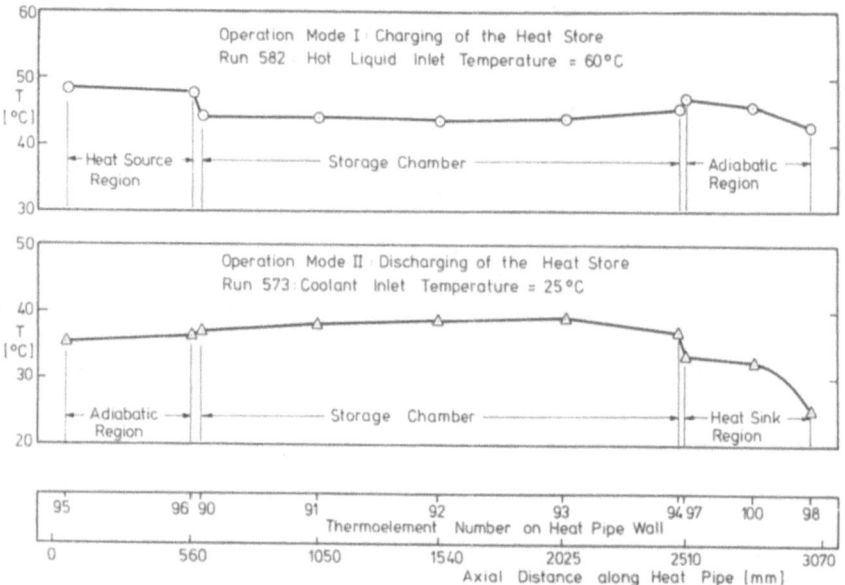

Fig. 30: Axial Temperature Profile along the Heat Pipe
during (a) Charging (b) Discharging

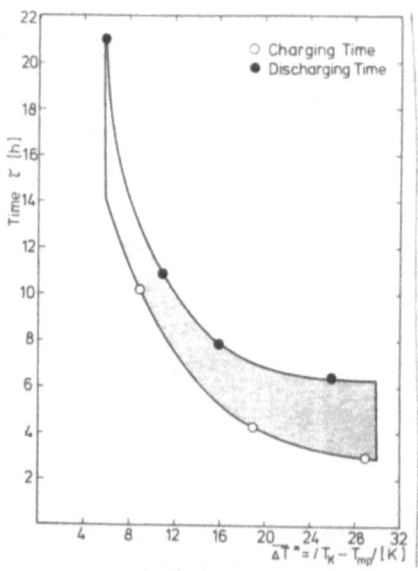

Fig. 31: Time for Charging and Discharging of the Prototype Store as a Function of the Driving Temperature Potential ΔT*

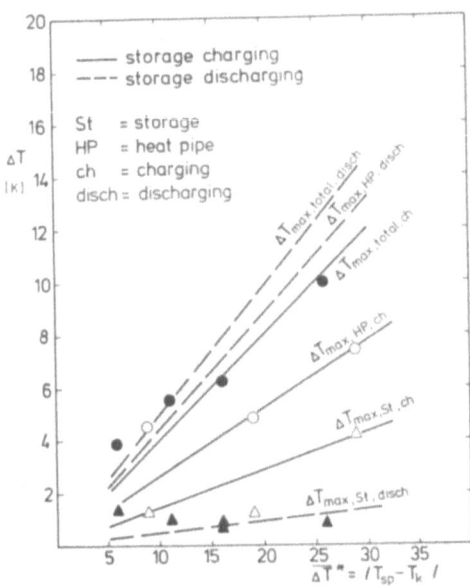

<u>Fig. 32:</u> Maximum Temperature Gradients in the Storage
Chamber and Heat Pipe for Charging and Discharging of the Prototype Heat Store as a
Function of the Driving Temperature Potential
for Heat Flow.

One of the important characteristics underlining the performance of a LTES system is the maximum temperature gradient required
for charging and discharging the heat storage medium that it contains. Fig. 32 shows these temperature gradients in the storage
chamber and along the heat pipe wall. The maximum temperature gradient in the storage chamber $\Delta T_{max,ST}$ is defined as the difference
between the heat pipe wall temperature and the PCM temperature
measured at the remotest point in the chamber. For either of the
operating modes of storage charging and discharging $\Delta T_{max,ST}$ is
rather small reaching a maximum of 4 K for extreme operating conditions. However, along the heat pipe wall in the heat source and
heat sink regions, the temperature gradients are somewhat higher.
This is due to the fact that the heat exchangers in the heat source
and sink regions are not optimally designed. Improvement in the external heat transfer coefficients in these region would bring about
a reduction in the corresponding temperature gradients.

Charging/Discharging Rate
———————————————————————————

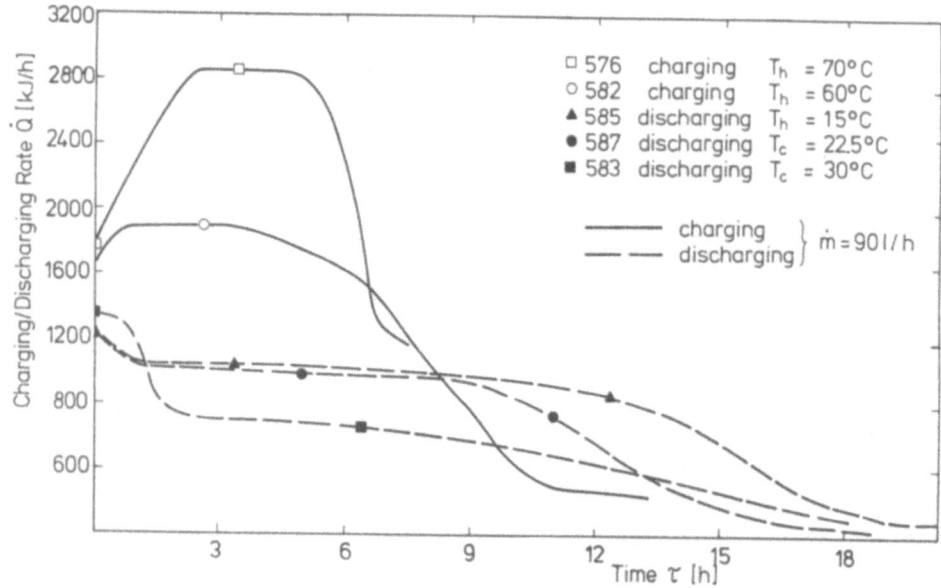

Fig. 33: Charging and Discharging Rate as a Function
of Time. The Parameter is the circulating
Fluid Inlet Temperature

Another major characteristic of a thermal store is the charging and discharging rates it permits and their variation with time.
Fig. 33 presents the charging and discharging rate as a function
of time. The parameter here is the circulating liquid inlet temperature, denoted in the diagram as T_h for hot liquid and T_c for the
coolant.

Fig. 33 shows that the charging and discharging rates are
strongly dependent on the fluid inlet temperature. During operation
the fluid inlet temperature governs the heat pipe saturation temperature, which,in turn, affects the transport capability of the heat
pipe. Through the choice of water as the working fluid, the heat
pipe operation is very sensitive in the temperature range considered
in the tests, so that an almost 4 - fold variation in the heat transfer rate during charging and discharging is observed as the fluid
inlet temperature is varied between 15° and 75 °C. For a fluid inlet temperature of 70 °C, charging rates upto 2840 kJ/h (approximately 800 W) could be measured. Moreover in the region of latent
heat addition and removal, the heat transfer rate is rather constant
indicating a uniform change of phase along the length of the storage chamber.

Simultanous Charging and Discharging of the Heat Store and
Heat Storage Capacity Further to investigating the charging and
discharging mechanisms, the prototype heat store was tested for
its capability to be by-passed in case of simultaneous supply and
removal of energy. This operation mode was simulated by adding
energy in the heat source region with a fluid at a constant inlet
temperature of 60 °C. Simultaneously, a coolant at 30 °C was made
to circulate in the heat sink region for 7.8 h following commence-
ment of the test. The PCM was initially solid at 20 °C. Fig. 34
shows an energy balance of the heat store during simultaneous
charging and discharging. A considerable amount of the heat added
is directly removed, so that the storage is to a good extent by-
passed.

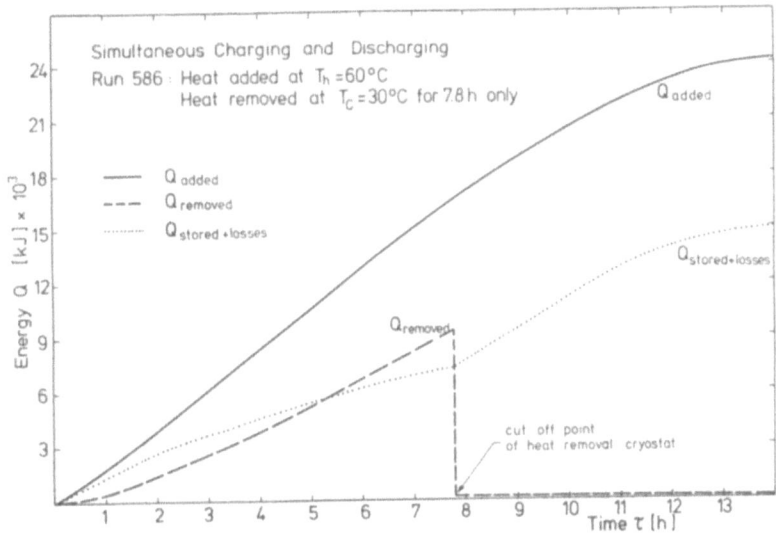

Fig. 34: Energy Balance During Simultaneous Charging
and Discharging.

The energy balance in Fig. 34 also provides an indication of
the total amount of heat stored in the prototype unit. Allowing
approximately 2000 kJ for losses, the energy stored is in the
range of 13,000 kJ, of which the latent heat content is 10,500 kJ.
Reproducible values of the heat content of the prototype store
within ± 15 percent were obtained from the various test runs.

Integration of the LTES in a solar space heating and hot water
production system

Figure 35 shows a system schematic of a possible solar heating system using the finned heat pipe LTES concept discussed
above. The heat pipe is divided in 4 regions: region A is in contact
with the solar collector fluid, region B is the storage chamber,
region C is connected to the hot water loop and region D to the
space heating loop. The respective inlet and outlet temperature
are marked on the diagram.

Heat entering region A is transported axially along the
heat pipe to the thermal store B, or to the load loops via the
heat exchangers in the regions C and D. If a simultaneous supply
of solar energy and load demand exists, the storage is more-or-
less bypassed, as the heat transfer in regions C and D is far better
than in region B. On the other hand, when little or no load exists
in regions C and D the thermal store in region B is charged.

When no solar energy is available, the collector loop in region A is shut off. The storage then supplies its heat to the regions C and D with the help of the heat pipe and, in turn, gets
discharged. In case more energy is needed than the storage can deliver, the auxiliary heaters are automatically turned on.

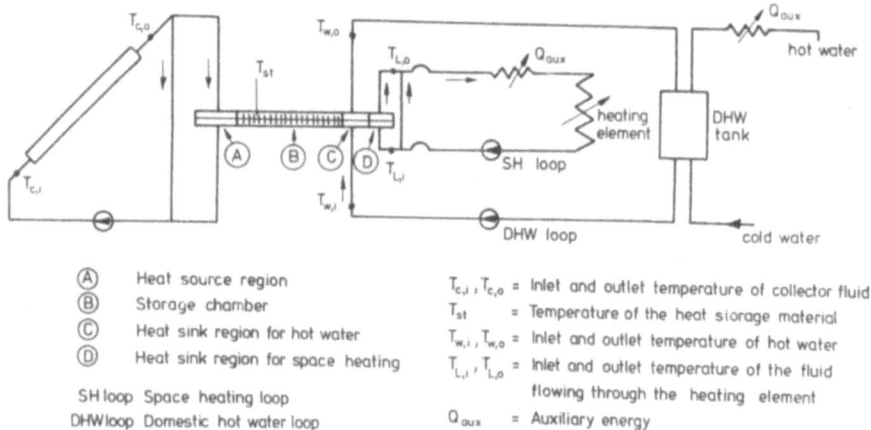

Ⓐ	Heat source region
Ⓑ	Storage chamber
Ⓒ	Heat sink region for hot water
Ⓓ	Heat sink region for space heating

SH loop Space heating loop
DHW loop Domestic hot water loop

$T_{c,i}$, $T_{c,o}$ = Inlet and outlet temperature of collector fluid
T_{st} = Temperature of the heat storage material
$T_{w,i}$, $T_{w,o}$ = Inlet and outlet temperature of hot water
$T_{L,i}$, $T_{L,o}$ = Inlet and outlet temperature of the fluid
flowing through the heating element
Q_{aux} = Auxiliary energy

Fig. 35: Integration of the Finned Heat Pipe Exchanger
in a Solar Heating System /4/.

COST OF LATENT HEAT STORAGE

Cost calculations for low temperature latent heat stores are
available only for a few units, all of these manufactured in

the USA. Table 17 shows the costs quoted by the manufacturers. Per MJ of stored energy, the costs are seen to vary over a large range, between ca. $ 7.50 and $ 13.50.

Table 17: Costs of Low Temperature Latent Heat Stores /23/

Manufacturer	Type of heat ex-changer	PCM	Unit capacity /MJ/	Unit cost /US$/	Specific cost /US $/MJ/
Valmont Industries Valley, NE, USA	Passive (Trays)	$Na_2SO_4.10H_2O$	1.6	12.50	7.81
PSI, Fenton, MI, USA	Passive (Energy Storage Rods)	Thermol 81, containing $CaCl_2.6H_2O$	2.34	29.80	12.34
Calmac Corp., Engelwood, NJ, USA	Passive (Heat Bank)	a) $Na_2S_2O_3.5H_2O$ b) 116 °C Salt c) 7 °C Inorg. Eutectic	400.9	3000.00	7.48
TES, San Diego, CA, USA	Passive	$Na_2S_2O_3.5H_2O$	263.8	3500.00	13.27
OEM Products, Plant City, FL, USA	Active (Direct Contact)	$Na_2SO_4.5H_2O$	258.7	2800.00	7.81

SUMMARY

Latent heat storage in the temperature range 0-120 °C is of interest for a variety of low temperature applications, such as space heating, domestic hot water production, heat-pump assisted space heating, green house heating, solar cooling, etc. The development of dependable heat storage systems for these and other applications requires an understanding of two essentially diverse subjects: heat storage materials and heat exchangers. Both these aspects have been discussed in considerable detail in the paper.

Inexpensive commercial paraffins, fatty acids, inorganic salt hydrates and eutectica of organic and inorganic compounds are the important material families, to which candidate heat-of-fusion storage materials with phase transition temperatures within the temperature range 0-120 °C belong. While the inorganic salt hydrates are generally preferred due to their higher volumetric heat-of-fusion in comparison to the organics, they usually suffer from the disadvantages of supercooling and decomposition.

The knowledge of the melting and freezing characteristics

of the heat storage materials, their ability to undergo thermal
cycling and their compatibility with construction materials is
essential for determining the life of a latent heat store. Using
two different measurement techniques - differential scanning calorimetry and thermal analysis - the melting and freezing behaviour
of the heat storage materials is determined. Results from measurements undertaken with representative organic and inorganic substances are discussed in the paper. Commercial paraffins are characterized by two phase transitions - a solid-liquid and a solid-
solid phase transition - which may be spread over a large temperature range depending on the paraffin concerned. N-paraffins are
preferred in comparison to their iso-counterparts, as the desired
solid-to-liquid phase transition is generally restricted to a
narrow temperature range. Fatty acids are organic materials with
excellent melting and freezing characteristics and may have a good
future scope, if their cost can be brought down. Inorganic salt
hydrates, on the other hand, must be carefully screened for congruent, "semi-congruent", or incongruent melting substances with
the aid of phase diagrams. Incongruent melting salt hydrates may
be "modified" to overcome decomposition by adding suspension media,
or extra water or other substances to shift the peritectic point.
The use of salt hydrates in hermetically sealed containers is recommended. Moreover, the employment of metallic surfaces to promote heterogeneous nucleation in a salt hydrate is seen to reduce
the supercooling of most salt hydrates to a considerable extent.
Some results from thermal cycling and corrosion tests to assist
in the choice of the more appropriate materials are also included
in the paper.

Due to the extremely poor thermal conductivity of the heat
storage materials, heat exchanger design plays a major role in
ensuring the desired operation of a latent heat store. Broadly
speaking, heat exchangers are classified as "active" or "passive",
depending on whether the heat-of-fusion material is in some kind of
motion or completely at rest. The different heat exchanger types
presently under development are the direct contact heat exchanger
("active"), macroencapsulation of the storage materials in trays,
tubes or spheres ("passive"), or bulk storage in finned-tube/ex-
tended surface heat exchangers ("passive"). Whatever be the type
of heat exchanger it must, for the low temperature applications,
ensure a high charging/discharging rate for a small driving temperature potential for heat flow.

The various heat transfer considerations in a latent heat
store are illustrated in the paper through the example of two
novel finned-tube heat exchanger designs: (1) a modular finned heat
pipe heat exchanger and (2) a finned-annulus heat exchanger. A
thermal analysis based on two-dimensional conduction heat transfer in a composite medium, of which one of the medium undergoes
melting or freezing is discussed. The influence of the variation
of the geometric dimensions on the thermal performance of the latent heat store is described. The motion of the phase change inter-

face during charging and discharging of the finned-annulus heat store , as obtained using photographic techniques, is presented. Furthermore, results from experimentation showing a typical charge/discharge cycle, as well as charge/discharge time, charge/discharge rates and the temperature gradients in a finned heat pipe latent heat store are provided.

The cost of a latent heat store is one of the major deciding factors in its employment for a particular application. Cost data has, however, started to pour in only recently. Based on costs quoted by some manufacturers in the USA, the cost of low temperature latent heat stores is seen to vary between $7.50 to $13.50 per megajoule of stored energy.

REFERENCES

/1/ Abhat, A., S. Aboul-Enein und G. Neuer, Latentwärmespeicher
zur Verwendung in Solar-Energiesystemen für Wohngebäude.
VDI-Berichte Nr. 288, 1977, pp. 97-104

/2/ Abhat, A.: Performance Studies of a Finned Heat Pipe Latent
Heat Thermal Energy Storage System in SUN, Mankind's Future
Source of Energy, (ed: de Winter, F. and Cox, M.), Pergamon
Press, 1978, pp. 541-546

/3/ Abhat, A.: Experimentation with a Prototype Latent Heat Thermal Energy Storage System, International Solar Energy Congress,
Atlanta, Georgia, USA, May, 1979

/4/ Abhat, A., D. Heine, M. Heinisch, N. Malatidis, G. Neuer:
Entwicklung modularer Wärmeübertrager mit integriertem La-
tentwärmespeicher, Final Report, BMFT Project No. ET 4060 A,
IKE, Stuttgart, December 1979

/5/ Abhat, A.: Short Term Thermal Energy Storage, Revue Phys.
Appl., Vol. 15, 1980, pp. 477-501

/6/ Abhat, A., S. Aboul-Enein, N. Malatidis and G. Neuer: Latentwärmespeicher für solare Heizungssysteme, in VDI-Status-
bericht "Sonnenenergie", 1980, pp. 375-394

/7/ Abhat, A., S. Aboul-Enein and N. Malatidis: Heat of Fusion
Storage Systems for Solar Heating Applications in Thermal
Storage of Solar Energy (ed. C. den Ouden), Martinus Nijhoff
Publishers, The Hague, 1981, pp. 157-171

/8/ Abhat, A. and N. Malatidis: Determination of Properties of
Heat-of-Fusion Storage Materials For Low Temperature Applications, 1st IEA Conference on New Energy Conservation Technologies, Berlin, 6-10 April, 1981

/9/ Aboul-Enein, S. and A. Abhat: Experimental Investigation and
Analysis of a Finned Annulus Heat Exchanger for Heat-of-
Fusion Storage Applications, Proceedings, Second World Con-
gress of Chemical Engineering, Montreal, Canada, Vol. 2,
October 1981, pp. 146-150

/10/ Biswas, D.R.: Thermal Energy Storage using Sodium Sulphate
Decahydrate and Water, J. Solar Energy, Vol. 19, 1977,
pp. 99-100

/11/ Carlsson, B., H. Stymne and G. Wettermark: An Incongruent
Heat-of-Fusion System - $CaCl_2.6H_2O$ - made Congruent through
Modification of the Chemical Composition of the System,
J. Solar Energy, Vol. 23, 1979, pp. 343-350

/12/ Chahroudi, D.: Suspension Media for Heat Storage Materials,
Proceedings, Workshop on Solar Energy Subsystems for the
Heating and Cooling of Buildings, Charlottensville, Virginia

/13/ Dechema Werkstoff Tabellen, Chemische Beständigkeit,
Dechema Frankfurt

/14/ Edie, D.D. and S.S. Melsheimer: An Immiscible Fluid Heat-of-
Fusion Energy Storage System, in Sharing the Sun (ed. K. Boer),
American Section of International Solar Energy Society, Vol. 8,
1976, pp. 262-272

/15/ Furbo, S.: Heat Storage with an Incongruently Melting Salt
     Hydrate as Storage Medium Based on the Extra Water Principle,
     in Thermal Storage of Solar Energy (ed. C. den Ouden),
     Martinus Nijhoff Publishers, The Hague, 1981, pp. 135-146
/16/ General Electric's - Rolling Cylinder - Heat Storage Device,
     Mechanical Engineering, March 1978, pp. 55
/17/ Gmelin's Handbuch der anorganischen Chemie, Verlag Chemie
     GmbH, Berlin
/18/ Heine, D.: Korrosionsuntersuchungen an Materialien für den
     Einsatz in Latentwärmespeichern, VDI-Bericht Nr. 288, 1977
     pp. 105-110
/19/ Heine, D. and Abhat, A.: Investigation of Physical and Che-
     mical Properties of Phase Change Materials for Space Heating/
     Cooling Applications, in SUN: Mankind's Future Source of
     Energy (ed. de Winter, F. and Cox, M.), Pergamon Press, 1978,
     pp. 500-506
/20/ Lane, G.A. and Glew, D.N.: Heat-of-Fusion Systems for Solar
     Energy Storage, Proceedings of the Workshop on Solar Energy
     Storage Subsystems, Charlottensville, Virginia, 1975, pp. 43-55
/21/ Lindner, F.: Physikalische, chemische und technologische Grund-
     lagen der Latentwärmespeicherung, in Grundlagen der Solar-
     technik, Proceedings of the Meeting of German Solar Energy
     Society, Stuttgart-Fellbach, 1976, pp. 205-235
/22/ Lorsch, H.G., Kaufmann, K.W., and Denton, J.C.: Thermal
     Energy Storage for Solar Heating and Off-Peak Airconditioning,
     Energy Conversion, Vol. 15, 1975, pp. 1-8
/23/ Michaels, A.I.: An Overview of the USA Program for the Develop-
     ment of Thermal Energy Storage for Solar Energy Applications,
     in Thermal Storage of Solar Energy, (ed. C. den Ouden), Mar-
     tinus Nijhoff Publishers, The Hague, 1981, pp. 79-90
/24/ Schröder, J.: R and D of Systems for Thermal Energy Storage
     in the Temperature Range from -25 °C to 150 °C, Seminar,
     New Ways to Save Energy, Commission of the European Communi-
     ties, Brussels, Belgium, October 1979
/25/ Telkes, M.: Solar House Heating - A Problem of Heat Storage,
     J. Heating and Ventilating, Vol. 44, 1947. pp. 68-75
/26/ Telkes, M.: Nucleation of Supersaturated Inorganic Salt Solu-
     tions, Industrial and Engineering Chemistry, Vol. 44, No. 6,
     1952, pp. 1308-10
/27/ Telkes, M.: Solar Energy Storage, ASHRAE Journal, 1974
/28/ Telkes, M.: Thermal Storage of Solar Heating and Cooling,
     Proceedings of the Workshop on Solar Energy Storage Subsystems,
     Charlottensville, Virginia, 1975, pp. 17-23
/29/ Teubel, J., W. Schneider und R. Schmiegel: Erdölparaffine,
     VFB Deutscher Verlag für Grundstoff, Leipzig, 1965
/30/ Zief, M. and Wilcox, W.R.: Fractional Solidification, Marcel
     Dekker, Inc., New York, Vol. 1, 1967

APPENDIX I: HEAT STORAGE IN SUBSTANCES UNDERGOING SOLID-SOLID PHASE TRANSITIONS

Storage of heat as the heat of phase transition in substances that undergo solid-solid phase transitions at a desired temperature is a relatively new area of research. Michaels /A1/ reports investigations in this direction by D. Benson at SERI, Colorado and lists three best candidates found to date:

1. **Pentaerythritol:** $T_t = 188^{\circ}C$, $\Delta H_t = 323$ kJ/kg, Cost = $1.28/kg, equivalent to $3.96/MJ.

2. **Pentaglycerine:** $T_t = 81^{\circ}C$, $\Delta H_t = 216$ kJ/kg, No cost given.

3. **Neopentyl Glycol:** $T_t = 43^{\circ}C$, $\Delta H_t = 130$ kJ/kg, cost = $0.80/kg, equivalent to $6.22/MJ.

Fittipaldi /A2/ identifies organometallic compounds termed "layer perovskites", which undergo fully reversible solid-solid phase transitions with a wide choice of transition temperatures in the range 0-120 °C, and which are associated with reasonably high phase transition enthalpies. These materials are characterized by the general formula $(n - C_n H_{2n+1} NH_3)_2 MX_4$, where M is a bivalent metal and X a halogen. Typical compounds studied have M = Mn, Cr, Hg, Fe, Co, Zn, X = Cl and n varying between 8 and 18. These substances are termed "layer perovskites" as for M = Mn, Cn, Hg and Fe, their crystal structure consists of layers similar to those present in the mineral Perovskite, CaT_iO_3 /A2/. Some relevant thermodynamic data for the layer perovskites and their comparison with normal paraffins is given in Table 1 below.

Table 1: Some relevant thermodynamic data for the layer perovskites and their comparison with normal paraffins /A2/

LAYER PEROVSKITES	NORMAL PARAFFINS
Solid-solid phase transition	Solid-liquid phase transition
Transition temperature range: 0 - 120 °C	Transition temperature range: 10 - 70 °C
Transition enthalpy: 10-35 cal g^{-1}	Transition enthalpy: 30-50 cal g^{-1}
Change of volume at the transition: 5-10%	Change of volume at the transition: 15%
Density (at 25°C): 1.1 - 1.5 g cm^{-3}	Density (molten): 0.8 - 0.9 g cm^{-3}
Specific heat at constant pressure: 0.4 cal g^{-1} K^{-1}	specific heat at constant pressure: 0.4 cal g^{-1} K^{-1}
Thermal conductibility (at 25°C): 2×10^{-2} W cm^{-1} K^{-1}	Thermal conductibility (at 25 °C): 0.5-2.0×10^{-3} W cm^{-1} K^{-1}

References

/A1/ Michaels, A.I.: An Overview of the USA Program for the Development of Thermal Energy Storage for Solar Energy Applications in Thermal Storage of Solar Energy (ed. C. den Ouden), Martinus Nijhoff Publishers, The Hague, 1981, pp. 79-90

/A2/ Fittipaldi, I. et al.: Solid-Solid Phase Transitions, ibid., pp. 309-324

APPENDIX II. SOME "PASSIVE" AND "ACTIVE" HEAT EXCHANGER CONCEPTS FOR LOW TEMPERATURE LATENT HEAT STORAGE SYSTEMS

"PASSIVE" HEAT EXCHANGERS

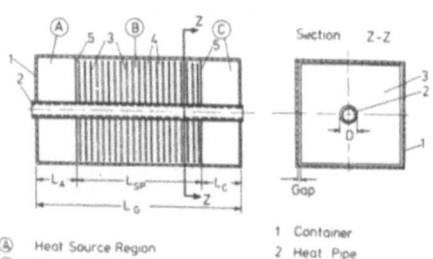

Section Z-Z

Ⓐ Heat Source Region
Ⓑ Storage Chamber
Ⓒ Heat Sink Region

1 Container
2 Heat Pipe
3 Fins
4 Storage Medium
5 Separation Walls

A-1

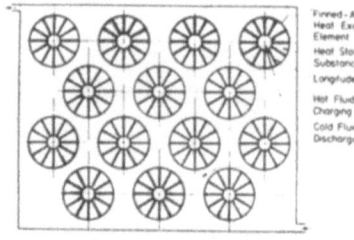

Finned-Annulus Heat Exchanger Element
Heat Storage Substance
Longitudinal Fin
Hot Fluid for Charging
Cold Fluid for Discharging

A-2

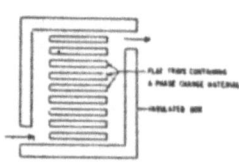

A-3

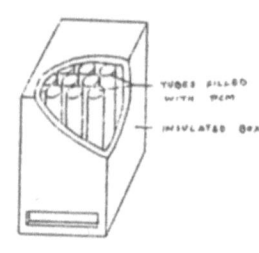

A-4

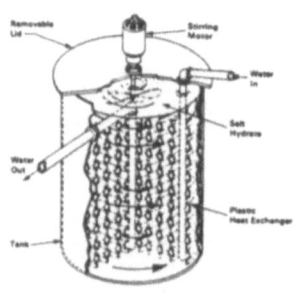

A-5

A-6

"ACTIVE" HEAT EXCHANGERS

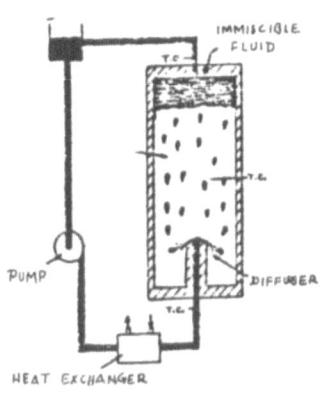

A-7

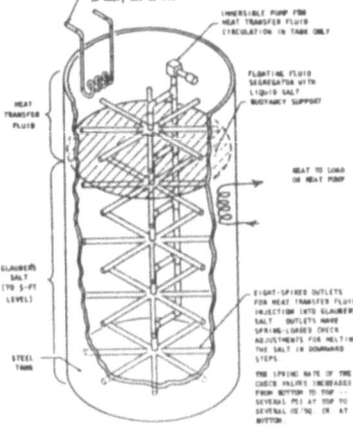

A-8

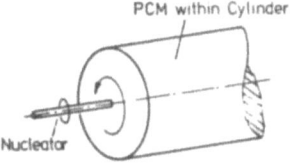

PCM within Cylinder

A-9

Fig. No.	Heat Exchanger Concept	Institution/Organization responsible for development
A-1	Modular Finned Heat Pipe Exchanger (Several PCMS)	IKE, Stuttgart, FRG
A-2	Finned-Annulus Heat Exchanger (Several PCMS)	IKE, Stuttgart, FRG
A-3	Plastic Trays Containing Glauber Salt	Valmont, NE, USA
A-4	Plastic Tubes Containing Thermal 81 ($CaI1_2 \cdot 6H_2O$)	PSI, MI, USA
A-5	"Heat-Bank" Latent Heat Store (Several PCMS)	CALMAC, NT, USA
A-6	Metal Cylinders Containing Sodium Acetate Trihydrate	TNO, Delft, NL
A-7	Immiscible Fluid - Heat of Fusion Storage System	Clemson, NC, USA
A-8	Direct Contact Heat Exchanger - "Heat Battery" Containing Glauber Salt	OEM, FL, USA
A-9	Rolling-Cylinder Containing Glauber Salt	General Electric, USA

CHEMICAL HEAT PIPES
(REVERSIBLE CHEMICAL REACTIONS AT MEDIUM/HIGH TEMPERATURE)

D. van Velzen

Commission of the European Communities
Joint Research Centre - Ispra Establishment
I-21020 Ispra (Va), Italy

ABSTRACT

Reversible chemical reactions can be used for the storage of thermal energy, as well as for upgrading of this energy by means of chemical heat pumps and heat transformers. A survey of the technological possibilities, the work in course and an outlook for the future is given. The use of catalytic homogeneous reactions at medium/high temperature generally involve low round-trip thermal efficiencies. The development of chemical heat pumps and heat transformers may offer a solution to this problem. The technological development of such devices is in a very early stage and there exists a considerable potential for interesting innovations.

1. INTRODUCTION

There is an increasing tendency to promote the introduction of alternative energy sources onto the energy market. The use of nuclear and solar energy as primary heat sources is one of the most advanced candidate possibilities to reach this goal.

Both aforementioned energy sources can produce thermal energy in the medium/high temperature range ($250-800^{\circ}C$) and they both require the development of suitable energy storage systems, as the production as well as the demand are time-dependent.

In the case of nuclear energy, the production rate will preferably be constant, and the demand is time-dependent, whereas in the case of solar energy, production as well as demand fluctuate over short and long periods of time. It follows that the cycle times and storage characteristics for the storage of nuclear and solar energy are not entirely equal and that there is a need for a flexible energy storage system working in the medium/high temperature range, which can serve in both cases.

In principle, thermal energy can be stored by:

- the use of the sensible heat of solids and liquids,
- the use of the latent heat of phase changes from the solid to the liquid state,
- using the bond energy of reversible chemical reactions.

The last method, chemical heat storage, which seems to be particularly suitable for the medium/high temperature range, due to the potential high storage density ($0.5-1$ MWh/m^3) and to the anticipated high reaction rates, will be the subject of this paper.

The basic idea is simple; when excess heat is available, an endothermic chemical reaction is carried out, thereby absorbing heat into the system. The reaction products are then separated, stored and eventually transported. When heat is needed, at another point of time or place, the reaction products are recombined and the heat of reaction again becomes available. This kind of application is called the "storage mode". Besides this rather straightforward type of operation, there are some additional, more sophisticated possibilities. Reversible chemical reactions can also be used to upgrade heat from a lower temperature level to a higher temperature. This operation is usually called "heat transformation".

Finally, the same chemical reactions can also be applied in the so-called "chemical heat pump" mode. Here a certain amount of high-quality heat is used in combination with an (otherwise useless) low temperature heat source to produce an increased amount of energy at a useful medium temperature level.

The aforementioned three modes of application of reversible chemical reactions for heat storage will be discussed below in detail, in separate chapters.

There is no doubt that heat stores, heat pumps and heat transformers using reversible chemical processes can be a promising means for the conservation and rational use of energy. The potential application is widespread, especially regarding the possibility of upgrading industrial waste heat which seems to be very attractive.

The technical development of chemical heat storage is now in a very early stage and we believe that considerable development possibilities are present in this field.

2. THE STORAGE MODE

The reversible chemical reactions suitable for heat storage at higher temperatures can be subdivided into catalytic reactions and thermal dissociation reactions. The catalytic reactions are usually gas/gas reactions and the use of a catalyst is required to obtain acceptably high reaction velocities for both the exothermic and the endothermic reaction. The system whereby the reaction products are subsequently transported in pipelines from the heat source to the consumer, is usually called a chemical heat pipeline or a chemical heat pipe.

The other class of reversible reactions is the thermal dissociation. Here the dissociation takes place by addition of heat to the original compound, which can be a solid or a liquid. In most practical cases, a gas is released and the depleted liquid or solid remains in the reactor. The reverse (exothermic) reaction occurs spontaneously if the equilibrium is changed by a temperature decrease or a pressure increase. Consequently, in this case the dissociation products have to be separated and individually stored to avoid an uncontrolled reverse reaction. A well-known example of this class is the reaction system

$$Ca(OH)_2 \longrightarrow CaO + H_2O \tag{1}.$$

2.1 Chemical Heat Pipes

The idea to use reversible chemical reactions for heat storage and transportation, i. e. as chemical heat pipes, has been launched in relationship with nuclear energy, especially the

HTGR (1) and also for use with solar collectors (2).

The primary advantage of such reaction systems is that long-term, ambient-temperature storage of the chemicals is possible without the thermal losses inherent in systems based on high-temperature latent or sensible heat storage. However, the conversion of heat to chemical energy and the reconversion back to heat involve other types of thermal losses. They are:

a) the maximum theoretical efficiency of any reversible process (availability losses),
b) unavoidable losses of energy due to the irreversible nature of any real process (e.g. separation losses),
c) losses resulting from the fact that economics and not efficiency determines the final design of the process.

The losses are considerable. Recent estimates of the round-trip efficiencies for a number of systems show figures between 30 and 50% (3).
It is of interest to investigate, with some practical examples, how these losses originate.

The most intensively studied reaction systems for chemical heat pipes are the following:

1) $CH_4 + H_2O \rightleftharpoons CO + 3 H_2$ (205 kJ/mol) $\qquad$ (2)

2) $SO_3 \rightleftharpoons SO_2 + 1/2 O_2$ $\qquad$ (98 kJ/mol) $\qquad$ (3)

3) $2 NH_3 \rightleftharpoons N_2 + 3 H_2$ $\qquad$ (46 kJ/mol) $\qquad$ (4)

The first one is the so-called Eva/Adam system developed by KFA Jülich, Germany (1, 4). This is without any doubt the most developed of all systems. A pilot plant circuit, called EVA I/ ADAM I has been in operation since February 1979 and there is a report on more than 800 hours of successful combined operation (4). In this plant the endothermic steam reforming reaction is carried out at 820°C and 31 bar. The reverse methanation occurs over a wide range of temperatures, between 300 and 650°C at a pressure of 27 bar.
The bulk volumetric flow rate of the plant is 700 m³/h at the outlet of the steam reformer. The total conversion rate is about 85 m³/h STP of methane reacted, involving approximately 185 kW

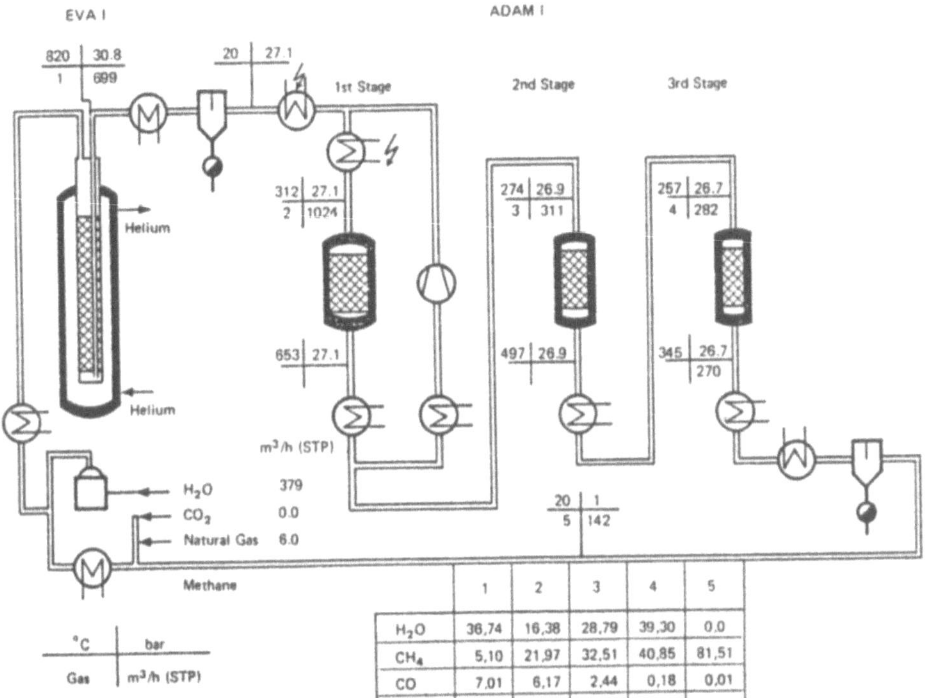

EVA I   ADAM I

820	30.8
1	699

20 | 27.1

1st Stage   2nd Stage   3rd Stage

312	27.1
2	1024

Helium

274	26.9
3	311

257	26.7
4	282

653	27.1

Helium

497	26.9

345	26.7
	270

m³/h (STP)

H₂O	379
CO₂	0.0
Natural Gas	6.0

20	1
5	142

Methane

°C	bar
Gas	m³/h (STP)

	1	2	3	4	5
H₂O	36,74	16,38	28,79	39,30	0.0
CH₄	5,10	21,97	32,51	40,85	81,51
CO	7,01	6,17	2,44	0,18	0,01
CO₂	5,90	7,05	5,40	3,40	2,85
H₂	42,89	43,78	25,54	10,41	4,49
N₂	2,37	4,65	5,33	5,86	11,14

Figure 1. Test results with combined energy transport system
       EVA I - ADAM I (Ref. 4)

of heat of reaction. The test results and a sketch of the plant are given in Fig. 1 (4).

The main result of this first successful pilot plant scale experiment is that the stability of the system over a prolonged time has been proved. The authors do not quote any figure for the thermal efficiency of the storage operation.

However, their material balances allow us to make estimations about the order of magnitude of the thermal losses. From the data given in Fig. 1 it appears that the methane conversion in the EVA I reactor is about 70% and that a large excess of water vapour is applied ($H_2O/CH_4 = 3.13$).

Moreover, there is a certain amount of CO_2 formed by the shift reaction. The stoichiometry of the storage reaction is therefore not exactly the one given in Eq. (2).

Neglecting the small amounts of CO_2 and H_2 present in the recycle methane, we obtain the following approximate material balance (in gmol/gmol CH_4 converted):

	In EVA I	Out EVA I	In ADAM I	Out ADAM I
H_2O	4.43	3.00	-	1.43
CH_4	1.42	0.42	0.42	1.42
CO	-	0.57	0.57	
CO_2	-	0.43	0.43	
H_2	-	3.42	3.42	

Both exit streams are stored and transported at 20°C and the water vapour is condensed. It will be impracticable to recover its latent heat of condensation.

The water required at the inlet of the EVA plant (4.43 gmoles/gmol CH_4) must however be evaporated, which will require at least 200 kJ/mol CH_4. This energy loss, together with the inevitable thermal losses causes the first law thermal efficiency of the process certainly to be lower than 50%. The first law definition of the efficiency does not account for the difference in heat quality between the charging and discharging reaction. Therefore, frequently also the "second law" efficiency is quoted, where the efficiency is calculated in terms of exergy (11), which is thought to suitably represent the "available work" balance. The second law thermal efficiency for the EVA/ADAM pilot plant conditions will probably be about 40%. In an independent study of the same reaction system, Stewart (5) quotes a second-law efficiency of 29% based on a detailed process design study.

The SO_3 system (Eq. (3)) has also been the subject of various investigations (6, 7, 8). A very detailed study about the feasibility of this system has been carried out by Lurgi on contract with the JRC. Here a concept of a process has been developed as a basis for the preliminary lay-out of a demonstration plant (8). A schematic drawing of the plant is given in Fig. 2.

The plant was designed for a thermal power input of approximately 2 MW to the energy storage. A maximum phase duration of 12 hours has been assumed.

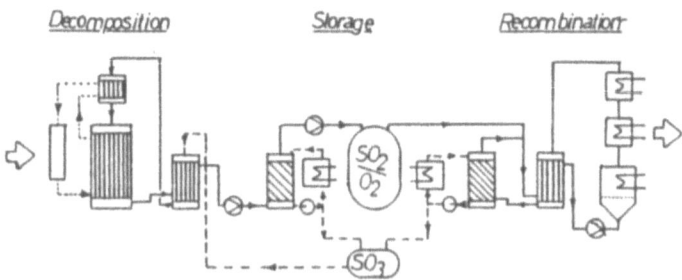

Figure 2. Lay-out of a demonstration plant for the SO_3/SO_2-
system (ref. 8)

During the charging period SO_3 is catalytically decomposed at
850°C and 6 bar. The SO_3 conversion is assumed to be about
50%. The excess SO_3 is condensed, the product gases stored in
a gas storage tank and, when needed, transported to the reverse
reactor.

During the discharging phase the back reaction is carried
out in a fluidized bed reactor at about 550°C and 3.7 bar. The
SO_2 conversion is here approximately 90%. Produced SO_3 containing about 1.5 wt% SO_2 is condensed, stored and used as
feedstock for the decomposition reactor.

The estimated first law thermal efficiency of the storage
operation is 43%, whereas the second law efficiency is about
35%. For the greater part the losses are caused by the condensation and evaporation of sulfur trioxide. Indeed, the correct
reaction equation for the charging reaction should read as follows:

$$2(SO_3)_1 \longrightarrow 2\ SO_3\ (g) \longrightarrow (SO_3)_g + SO_2 + 1/2\ O_2 \qquad (5).$$

The latent heat of evaporation for SO_3 is about 43 kJ/mol, i.e.
86 kJ are required for the evaporation in Eq. (5).

If there were no other losses than this latent heat of evaporation, the first law thermal efficiency would be $98/(98 + 86) =$ 53%. It follows that the evaporation/condensation of SO_3 accounts for more than 80% of the total losses.

In a similar, independent study, based on a preliminary process design for a solar plant, Smith (9) quotes a second law round-trip efficiency of 33%.

The third system, based on ammonia dissociation has received considerable study in Australia (10). Here the decomposition reaction will take place at 750°C and the back reaction at 500°C. Ammonia will be stored as a liquid, incurring the same type of thermal losses as occur in the SO_3-system. The reported overall second law thermal efficiency of the operation is about 27%.

It must therefore be concluded that the efficiencies of heat storage by chemical heat pipes are disappointingly low. Additionally, it must be anticipated that the required capital costs for the chemical equipment will be considerable, also due to the fact that the involved chemical reactors must be designed in such a way that they perform satisfactorily with varying throughputs. It must therefore be feared that the outlook for chemical heat pipelines for heat storage at medium/high temperature is not very promising.

2.2 Thermal Dissociation Reactions

In the case of a dissociation reaction, a gas or vapour is released from a solid or a liquid which remains in the primary store. There are three possibilities for gas storage:

i) Storage of gas in gaseous form, either at atmospheric pressure or compressed. Here thermal losses are not very important, but the energy storage density is small. For instance, in the case of atmospheric pressure storage, the storage density is only 0.5-1 kWh/m^3.

ii) Storage in condensed form. This way of operation can be applied when water, ammonia, methanol are the dissociated products. The disadvantage of this system is that to run the reverse reaction the condensed vapour has again to be eva-

porated. The latent heat of evaporation must then be considered as a thermal loss, similar to the aforementioned
examples of chemical heat pipes.
Therefore, the obtainable thermal efficiencies will be modest
(less than 50%).

iii) Storage of gas in resorption stages. The dissociated gas is
absorbed in another liquid or solid as a secondary store.
This method is especially suitable for chemical heat pumps
and heat transformers. Its principle is explained in Fig. 3.

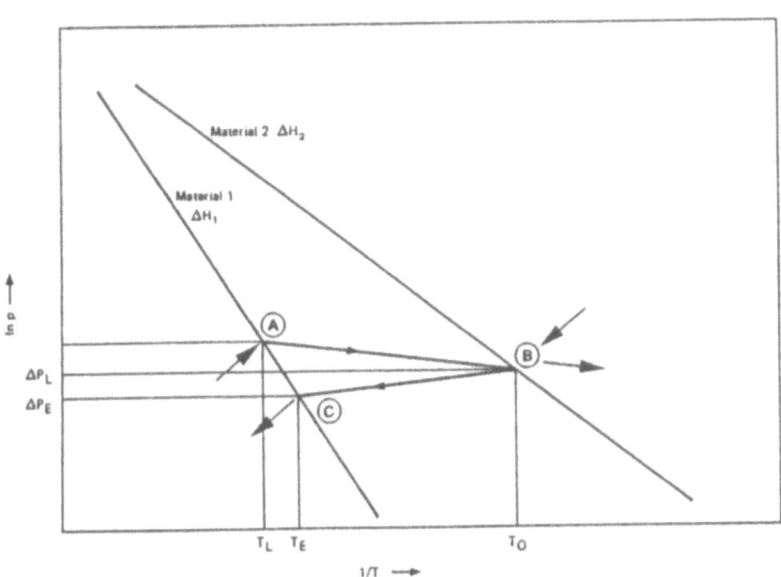

Figure 3. Chemical heat storage, gas resorbed in a secondary
store

The involved chemical components form pure phases, consequently the vapour pressure in the two vessels depends only on
the temperature and not on the extent to which the reaction has
proceeded. If we treat the vapour as an ideal gas, the following
formula will be valid:

$$\ln P = -\Delta H^o/RT + \Delta S^o/R \qquad (6).$$

The linear dependence of ln P as a function of $1/T$ is shown in Fig. 3. Here the dissociation curves are given for the primary storage material (No. 1, formation enthalpy ΔH_1) and the secondary storage material (No. 2, formation enthalpy ΔH_2). During charging (heat input: ΔH_1) at temperature T_L (point A) gas is released, flows from container No. 1 to No. 2. Here it is absorbed in material No. 2, where heat (ΔH_2) is released at temperature T_O (point B).
Obviously, $T_O < T_L$, and a certain driving pressure potential is required to obtain reasonable reaction velocities.

For the discharging operation, one has to apply heat (ΔH_2) to the secondary store to liberate the gas, which is fed back to reactor No. 1. Here the gas is absorbed in material No. 1, releasing heat (ΔH_2) at a temperature $T_E < T_L$ (Point C). The operations for the cases ii) and iii), i. e. storage in condensed and in resorption form, are very similar. Also here, considerable thermal losses due to the secondary binding and release of the working gas are inevitable.

A promising feature of the dissociation/resorption system may be, however, the possibility to use it as a heat transformation process or as a chemical heat pump. Here, the heat from a waste heat source is incorporated into the system, considerably increasing the thermal efficiency.

The work of Smith (9) has indicated that the economics of chemical energy storage, used in the storage mode, for solar thermal electric conversion is economically unattractive. His study comprised catalytic as well as thermal dissociation systems, most of them able to work in the medium/high temperature range.

For short-term energy storage, sensible heat storage systems will probably be preferable. In Smith's study, however, the application of heat transformers and chemical heat pumps at medium/high temperature has not been considered.

3. HEAT TRANSFORMATION

In Fig. 4 the principle of the heat transformation process is illustrated. The quoted example is working in the low tempera-

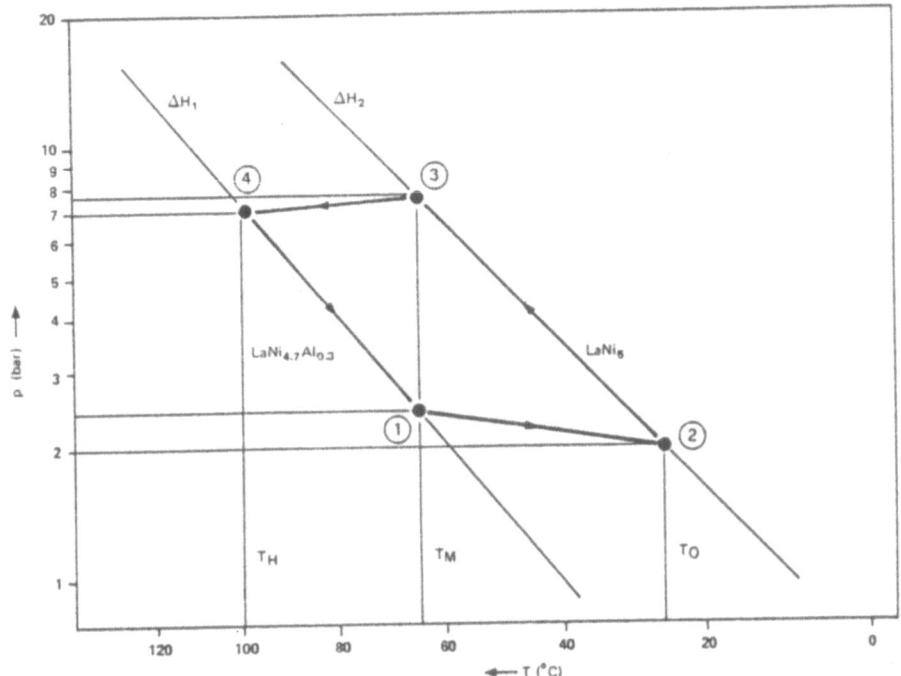

Figure 4. Single-stage metalhydride heat transformer (ref. 12)

ture range and it is claimed to be suitable for low temperature
process heat generation from waste heat (12). The same principles apply in the medium/high temperature range.

The operation is as follows. Heat from a low temperature
heat source is added to the primary stage at about 65°C. The
working gas is dissociated at a pressure of about 2.4 bar. It
flows to the secondary stage where it is absorbed in the secondary material, under the release of very low temperature heat
to the heat sink at T_O (= 26°C) (Point 2). So far, the charging
procedure has followed exactly the operation in the storage mode.
The difference now lies in the discharging phase. Instead of releasing the working gas at nearly the same pressure and temperature, the closed vessel is heated up to about 65°C. The pressure increases during heating to approximately 8 bar (Point 3).
When more heat is added and the pressure is maintained constant, the secondary compound dissociates, generating the working gas at 8 bar.

This allows the charging reaction to be carried out at a temperature of about 100°C (Point 4). This reaction is exothermic so that now ΔH_1 becomes available at a temperature higher than the one of the original heat source.

It is evident that the temperature levels for heat transformation depend on the properties of the chemical compounds. Heat transformation at much higher temperatures than quoted in the example of Fig. 4 is certainly feasible and applications in the medium/high temperature range can be very attractive.

Heat transformation can advantageously be applied to upgrade industrial waste heat, with emphasis on those cases where the upgraded heat can be recycled into the process, e.g. distillation and drying operations.

Neither compressors nor high quality primary energy are needed for the upgrading operation.

Finally, it must be again noted that a particular advantage of solid/gas dissociation reaction systems is that all chemical components form pure phases. Therefore, the pressure in the system depends only on the temperature and not on the extent to which the actual reaction has advanced. This feature allows the production of energy at a constant temperature without the use of complicated control devices.

4. CHEMICAL HEAT PUMPS

The chemical heat pump is at present receiving a lot of attention, especially for application in the low temperature range. As an example, in Sweden there are two full-scale pilot plants operating with the Na_2S/H_2O system (System Tepidus). The system is applied in the field of the use of solar energy in the household and in industrial buildings. The thermal energy becomes available at 60-65°C (13).

The chemical heat pump process is essentially the reverse of a heat transformation cycle (see Fig. 5). In the depicted example, the high temperature heat is available at 90°C and is used to dissociate the primary storage material at 5.5 bar. The working gas flows to the secondary reactor and is absorbed at

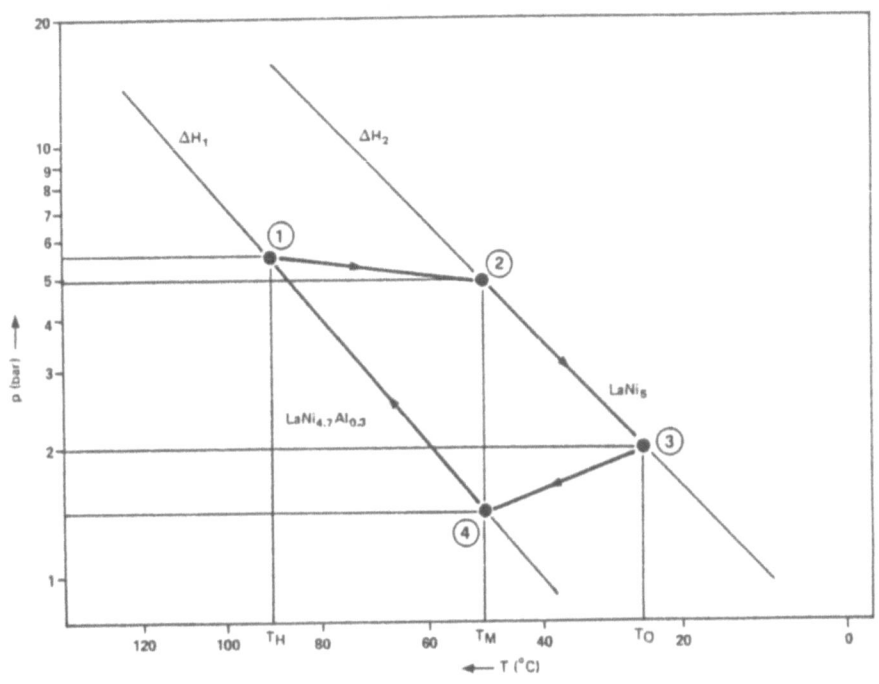

Figure 5. Single-stage metalhydride heat pump; reversed process
of ref. 12

50°C, whereby heat is released (Points 1 and 2). After having
charged the secondary storage material, it is cooled down to T_O.
Hereby the vapour pressure decreases from 5 to 2 bar. Under
these conditions the energy remains stored in the secondary
store. When heat is needed, the secondary material is dissociated
by allowing the working gas to flow to the primary vessel and by
using a low temperature waste heat input (ΔH_2, point 3). Thermal energy is released by the absorption of the working gas at
50°C and 1.5 bar (point 4). It follows that the chemical heat
pump is able to transform energy from a high temperature level
plus the energy from a waste heat source (at low temperature)
into energy at an intermediate level of temperature. Usually the
input of the low level energy is neglected in the energy balance
of such systems. Therefore, the thermal efficiency (first law) of
a chemical heat pump can be over 100%, i.e. the coefficient of
performance can be larger than unity.

There has been another approach for the design of chemical heat pumps and heat transformers. Fujii c. s. (14) proposed a closed triple-reaction cycle, with one endothermic and two exothermic reactions, represented schematically by:

i) $A + B \longrightarrow C + D$ $\Delta H_1 < 0$

ii) $C + E \longrightarrow B + F$ $\Delta H_2 > 0$

iii) $F + D \longrightarrow E + A$ $\Delta H_3 < 0$ (7).

Fig. 6 illustrates the supposed relationship among those reactions. The authors give some examples of potential systems, one using SnO, SnO. $2H_2O$ and SnS in a dissociation reaction system where water and H_2S were the working gases. It seems, however, that practical work on this system has been limited to thermogravimetrical and small scale laboratory tests.

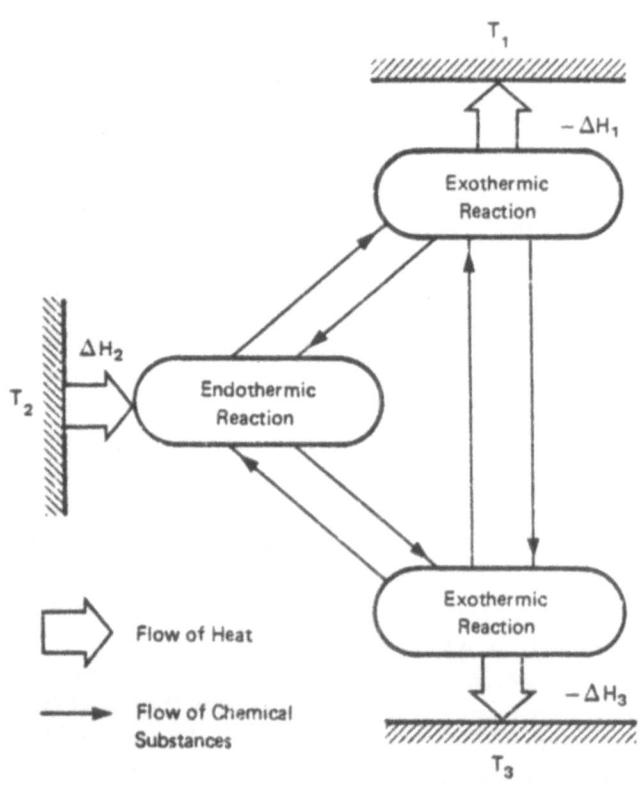

Figure 6. Triple-Reaction System (ref. 14)

The triple-reaction systems probably do not offer a valid alternative to the "conventional" heat pump cycles.

5. WORKING MATERIALS

The dissociation reaction systems suitable for application in chemical heat pumps can be divided into liquid-gas (wet) and solid-gas (dry) working material pairs. Wet systems are already widely applied in the low temperature region in the well known absorption refrigerators and absorption heat pumps. Common working materials are ammonia/water and water/lithium bromide. In these systems the maximum operation temperature is limited to about 180°C and 100°C, respectively.

Recently, the system H_2SO_4/H_2O is also receiving much attention (15, 16). An engineering model has been designed, fabricated and tested at the Rockwell Research Company (17). This plant consists of two glass acid tanks each loaded with about 65 kg of 93% H_2SO_4, a stainless steel water tank and two glass, shell and tube heat exchangers. The system has been in operation for 475 hours, working at a maximum temperature of 205°C. The attainable charge/discharge rate is 7 kW. The present proposals for this system are limited to working pressures equal and/or below atmospheric. This practically limits its application to the low/medium temperature range.

Dry systems are more promising for applications in the higher temperature region. Several candidate systems are under study.
The choice for a suitable working gas is ample and comprises:

- ammonia
- amines (especially methylamine)
- alcohols (methanol)
- water
- sulfur dioxide
- hydrogen.

Many feasible operating cycles can be conceived where the working temperatures are far above the operating temperatures of conventional compression and absorption devices. Especially metal-hydrogen reactions cover a very extensive range of possibilities. This wide temperature range is depicted in Fig. 7, where the reaction enthalpies are shown as a function of the equilibrium temperature at a hydrogen pressure of 1 bar. It follows that the potential working temperatures range from 0 to over 1000°C.

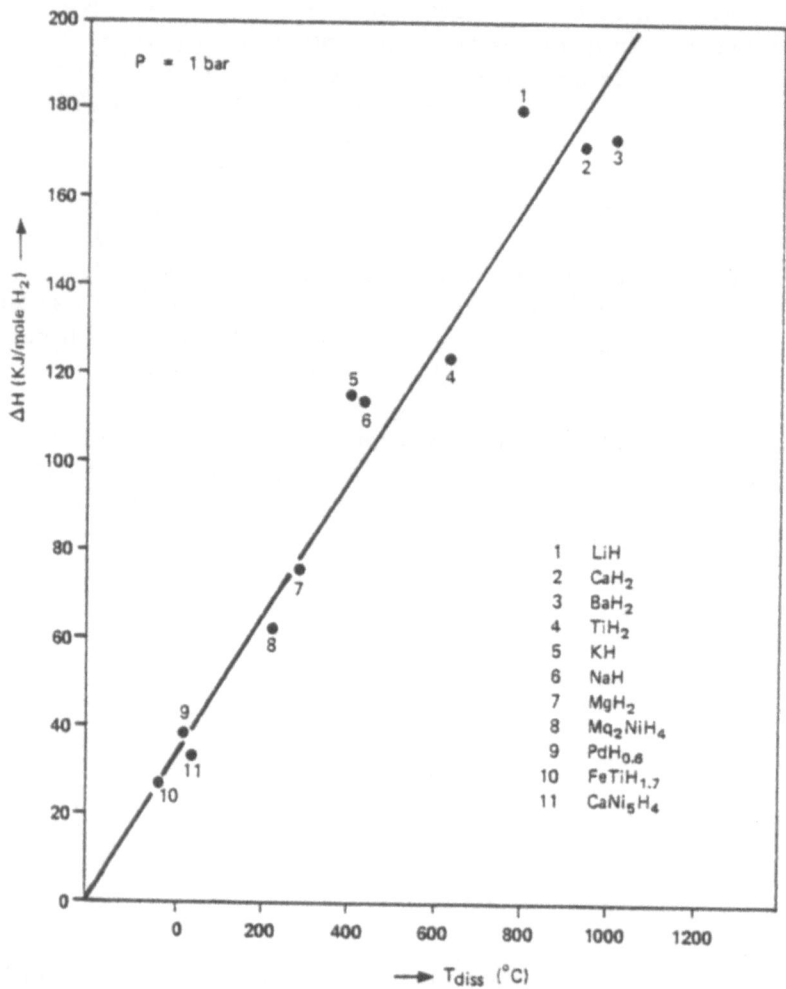

Figure 7. Thermal storage properties of metal-hydrides

A low temperature application for hydrides has been developed in the U.S., using $CaNi_5$ and $LaNi_5$ hydrides (18). Here the maximum working temperature will be around 150°C. A laboratory model of a medium temperature metalhydride cycle is in operation at the University of Stuttgart by Nonnenmacher c. s. (12).

The first test results of this system, using $CaNi_5$ and Mg_2Ni as low and high temperature storage materials, showed that heat upgrading from 263 to 348°C could be obtained. The theoretical maximum temperature under the test conditions was however 385°C. The reason for the discrepancy between the test results and theory lies in the fact that the hydrogen absorption in Mg_2Ni occurs at a much faster rate than the dissociation from $CaNi_5H_4$. This causes a high pressure drop in this operation step. After modification of the system, better results were obtained and the maximum temperatures were only 10 to 20°C below the theoretical values.

Also Buchner (19) has done much work on the development of low and high temperature metal-hydride systems, especially in connection with the developments of hydrogen-powered automobiles. Here mainly TiFe and Mg_2Ni-hydrides are applied.

Besides the work on metalhydrides there also are several activities and investigations of laboratory models of chemical heat stores, heat pumps and heat transformers using ammoniates (20, 21), alcoholates (22), hydrates (13, 23) and zeolites (24). Most of this work is in the low temperature range, but it must be noted that temperatures up to 300°C have been reached in the zeolite system (24). Here, a semi-technical laboratory model is operated, loaded with 17 kg zeolite granulate. The heat output amounts to 1 to 2 kW. The charging and discharging times were in the order of 2 to 6 hours.

6. PROBLEM AREAS

The survey of the work in course, given in the previous chapter, shows clearly that the development of chemical heat pumps and heat transformers for the medium/high temperature range is in a very early (laboratory-scale) stage. However, already these early results yield an indication of the anticipated

problem areas. For instance, we refer to the above mentioned
work of Nonnenmacher (12) where problems occurred due to
differences in reaction rates for the desorption and the absorption step. Indeed, heat and mass transfer problems are likely to
be of paramount importance in the development of dissociation
reaction systems. Methods to minimise the pressure drop and
the concentration gradient inside the solid material have to be
studied. Also the anticipated poor heat transfer properties of
solids are likely to create considerable problems.

Besides the heat and mass transfer there are other points
being of importance, such as:

- the occurrence of side reactions
- possible sagging or swelling of the solids
- corrosion problems
- costs.

The importance of possible side reactions is obvious. In the
$Ca(OH)_2$-system, it was found that cycling may continuously decrease the amount of water that can be reacted with the system.
After 140 cycles only 34% of the original capacity remained,
probably due to the presence of non-condensable gases (23).
Sagging and/or swelling of the solid causes the same type of
effect: the loss of absorption capacity and the deterioration of
the mass and heat transfer properties. Also corrosion and, last
but not least, cost problems are likely to be decisive factors for
the success of the application of reversible chemical reactions
for heat storage purposes.

7. CONCLUSIONS

Reversible chemical reactions for energy storage at medium/
high temperatures can be applied in the storage mode, or as
chemical heat pumps or heat transformers. The use of catalytic
reactions in the storage mode involves low round-trip efficiencies, the first law efficiencies lying around 40-50%, the second
law thermal yields being as low as 30-40%.

Also the use of solid-gas dissociation reaction systems in
the storage mode leads to low thermal efficiencies. Efforts to
improve these low efficiencies have to be concentrated on the
integration with other processes which can act as (waste) heat

sources or sinks. The development of chemical heat pumps and heat transformers may offer a solution to this problem. The development in this field has really only begun so that many new ideas can be expected. The potential of such cycles is enormous and comprises the storage of solar and nuclear energy as well as the recovery of industrial waste heat.

The technological development possibilities are really widespread, but it must be kept in mind that finally the future success of such systems will depend totally on economic and engineering factors. It is therefore necessary to consider the practical problems connected with the development of chemical heat pumps and transformers already at a very early stage.

REFERENCES

1. Hilberath, F. and Teggers, H., German Patent 129 8233 (15.1.1968)
2. Chubb, T.A., Sol. Energy, 17, 129-36 (1975)
3. Ianucci, J.J., Fish, J.D. and Bramlette, T.T., "Review and Assessment of Thermal Energy Storage Systems Based upon Reversible Chemical Reactions", SAND 79-8239, Sandia Laboratories, Livermore, CA (August, 1979)
4. Fedders, H. and Höhlein, B., "Operating a Pilot Plant Circuit for Energy Transport with Hydrogen-Rich Gas", 3rd World Hydrogen Energy Conference, Tokyo, Japan, June 1980. Proc. Vol. 2, p. 823-838
5. Stewart, J. et al., "An Assessment of the Use of Chemical Reaction Systems in Electric Utility Applications", Phase I Final Report, EPRI Contract RP 1086-1, Gilbert Associates, Reading, PA (December, 1978)
6. Hill, S.A., "Adoption of the Sulfur Oxide System for Chemical Storage of Solar Energy", Masters Thesis, University of California, Berkeley (1978)
7. Lynn, S. and Faso, A.S., "Evaluation of a Sulfur Oxide Chemical Heat Storage Process for a Steam Solar Electric Plant", Final Report, DOE contract W-7405-ENG-48, Lawrence Berkeley Laboratory, Berkeley, CA (1979)
8. Sander, U. and Rothe, U., "Feasibility Study and Preliminary Lay-out of a Minimum Size Demonstration Plant for Energy Storage and Transport by Means of Dissociation and Synthesis of SO_3". Final Report Study Contract No. 972-78-11

SISP D, Lurgi Frankfurt/JRC-Ispra (1980)

9. Smith, R. D., "Chemical Energy Storage for Solar Thermal Electric Conversion", 15th IECEC, Seattle, Wash., August 1980; Proc. Vol. 1, p. 243-247

10. Carden, P. O. and Williams, O. M., Int. J. of Energy Research, 2, 389-406 (1978)

11. Vakil, H. B. and Flock, J. W., "Closed Loop Chemical Systems for Energy Storage and Transmission", Final Report, ERDA contract EY-76-e-02-2676, General Electric Company, Schenectady, NY (February, 1978)

12. Nonnenmacher, A. and Groll, M., "Chemical Heat Storage and Heat Transformation Using Reversible Solid-Gas Reactions", International Conference on Energy Storage, Brighton (UK), April, 1981. Proc. p. 47-60

13. Bakken, K., "System Tepidus, High Capacity Thermochemical Storage/Heat Pump", Int. Conference on Energy Storage, Brighton (UK), April, 1981. Proc. p. 23-28

14. Fujii, S., Kameyama, H., Yoshida, K. and Kunii, D., J. Chem. Eng., Japan, 10, 224 (1977)

15. Clark, E. C. and Carlson, D. K., "Development Status and Utility of the Sulfuric Acid Chemical Heat Pump/Chemical Energy Storage System", 15th IECEC, Seattle, Wash., August, 1980, Proc. p. 926-931

16. McBride, J. C., "Chemical Heat Pump Cycles for Energy Storage and Transmission", Int. Conf. on Energy Storage, Brighton (UK), April, 1981. Proc. p. 29-46

17. Clark, E. C., "Final Report - Phase II Sulfuric Acid-Water Chemical Heat Pump and Storage System, DOE Contract No. EY-76-1185", Rocket Report No. RRC-79-R-627, September, 1978

18. Clinch, J. M., Gruen, D. M., Nelson, P. A., Blomquist, C. A., Horowitz, J. S., Lamich, G. J. and Sheft, I., "The Metal Hydride Chemical Heat Pump". Proceedings of the DOE Chemical Energy Storage and Hydrogen Energy Systems Contracts Review, Reston, Va., November, 1979, p. 84-88

19. Buchner, H., "The Hydrogen/Hydride Energy Concept", 2nd World Hydrogen Energy Conference, Zürich, Switzerland, August, 1978. Proc. p. 1749-1792

20. Raldow, W. M. and Wentworth, W. E., Solar Energy, 23, p. 75-79 (1979)

21. Taube, M. and Furrer, M., "Opportunities and Limitations for the Use of Ammoniated Salts as Carrier for Thermochemical Storage", Int. Seminar on Thermochemical Energy

Storage, Stockholm, January, 1980. Proc. p. 349-370

22. Offenhartz, P. O. D., Schwartz, D., Malsberger, R. E. and Rye, T. V., "Engineering Prototype Studies on the $CaCl_2 \cdot CH_3OH$ Chemical Heat Pump for Solar Air Conditioning, Heating and Storage", 15th IECEC, Seattle, Wash., August, 1980; Proc. p. 932-935

23. Bauerle, G. L., Pearlmann, H., Rosemary, J. K. and Springer, T. H., "Solar Energy Storage by Reversible Chemical Processes", Final Report, Sandia Contract FAO 92-7671, Rockwell International, Canoga Park, CA (January, 1979)

24. Alefeld, G., Bauer, H. C., Maier-Laxhuber, P. and Rothmeyer, M., "A Zeolite Heat Pump, Heat Transformer and Heat Accumulator", Intern. Conference on Energy Storage, Brighton (UK), April, 1981. Proc. p. 61-69.

HYDROGEN PRODUCTION AND STORAGE

Michel POTTIER

Direction des Etudes et Techniques Nouvelles
GAZ DE FRANCE
PARIS

Industrialized countries for their energy consumption have been supplied in the past by fossil fuels and progressively turn to nuclear energy.

Thermal energy is an intermediate between primary source and final useful carrier. Out of these electricity is the easiest to use but not suited to storage. In order to economically operate production and transmission systems, in front of custumers variable needs, energy storage is necessary. Energy storage can be achieved through chemical mean by hydrogen. This gaseous energy carrier could be produced from water through thermochemical cycles or electrolysis and is well suited to massive storage and transportation.

1 - INTRODUCTION

1.1. Energy

During a long time in the past energy has been afforded to industrialized countries through fossil fuels. Recently, some of these countries turned progressively to nuclear energy or ever to sun energy.

Energy is consumed either to deliver mechanical work or heating. The needs of customers are widely variable along the day for mechanical work or along the year for heating.

Production or transformation plants, transmission and distribution systems are necessary to afford customers with suited energy carriers from energy sources.

In front of customers variable needs and in order to economically operate production and transmission systems, energy storage is necessary.

1.2. Storage = Electricity, energy carrier very easy to use in any way, is not suited to storage.

Very often indeed electricity is industrially obtained through thermo dynamical engines by combustion of fossil fuels; thermal energy is an intermediate between primary source and final useful carrier. You know that thermal energy is not easy to store directly = heat losses cause progressive reduction of the amount and quality of energy available from storage in sensible heat.

Latent, or better, chemical storage shows no sensitivity to duration of storage for amount or quality of avaible stored energy. Chemical storage shows high energy content per unit mass and unit volume and is easy to use. Fossil fuels like coal, oil or natural gas are in the same time primary source, energy carrier and chemical storage products.

Faced with exhaustion or rising cost of oil, industrialized countries with poor national fossil energy resources should mostly turn to nuclear energy.

Nuclear power plants have to ensure the most efficient use of equipment in front of a highly variable demand. For this purpose chemical storage through hydrogen seems to be very well suited.

- production can be achieved from easily available non energetic raw product = water

- hydrogen shows a good ability for (transmission and) storage in large quantities.

2 - HYDROGEN PRODUCTION = hydrogen can be obtained from water by splitting

2.1. Thermodynamics of water splitting [1] : We state the main points here only briefly.

In order to split a water molecule into its elements, hydrogen and oxygen, it is necessary to give it at least the same amount of enthalpy (heat and free energy) as can be obtained when combining hydrogen and oxygen to form water. For liquid water at 298 K (25°C) the enthalpy of formation is H = 285.9 kJ/mole and the free energy G = 237.2 kJ/mole. This means that if a process of water splitting is run at 298 K an amount of energy equal to G must be delivered in the form of work, and the difference between H and G can be delivered in the form of heat. Work can be given to the water molecule as electricity (water electrolysis), or as radiation (photolysis, radiolysis). Work can also be provided by delivering high temperature heat to the water splitting system, and discharging part of this heat at lower temperature, e.g. at 298 K. This cannot be done directly with only the water molecule because it can not absorb heat (in addition to that required by its heat capacity) unless it is heated to very high temperatures (more than 2500 K; in this case thermolysis), but it can be done with heat at lower temperature with the aid of chemical compounds reacting in a closed system (thermochemical cycles). (See List 1)

As work is provided by heat flowing from a high temperature level to a lower one, a Carnot efficiency has to be taken into account for the evaluation of the amount of heat necessary for the delivery of the required amount of work.

The formation of water from its elements also implies a variation of entropy. From the relationship linking the free energy of a chemical reaction to its enthalpy and entropy change, G - H - T S, it follows that the amount of free energy necessary to split water decreases when the temperature increases and correspondingly the energy which can be delivered as heat increases, so there is an incentive to run water splitting processes isothermally at high temperatures because of the reduced work requirement. The conditions requiring only heat and no work would be reached at the temperature of the reversible reaction between hydrogen and oxygen, i.e. to first approximation the temperature of the hydrogen flame in oxygen.

	Hydrogen	Natural gas *
Density (relative to air)	0.0695	0.614
Specific gravity in kg/m³ at 0°C, 760mm. Hg	0.0893	0.790
Heat value in thermies/Nm³	3.05	9.64
Viscosity in micropoises at 20°C and 760 mm. hg	88.2	107.7
$Y = \dfrac{Cp}{Cv}$	1.41	1.30

Formula for estimating the compressibility coefficient:

$$Z = \frac{1}{1 + (a + bt)\dfrac{P}{1000}}$$

Where t = temperature (°C)

p = pressure (bars)

a	-0.6	2.56
b	0.001	-0.032

* Natural gas properties could change according to the content of nitrogen, ethane, etc.

TABLE 1

CHARACTERISTICS OF HYDROGEN AND NATURAL GAS

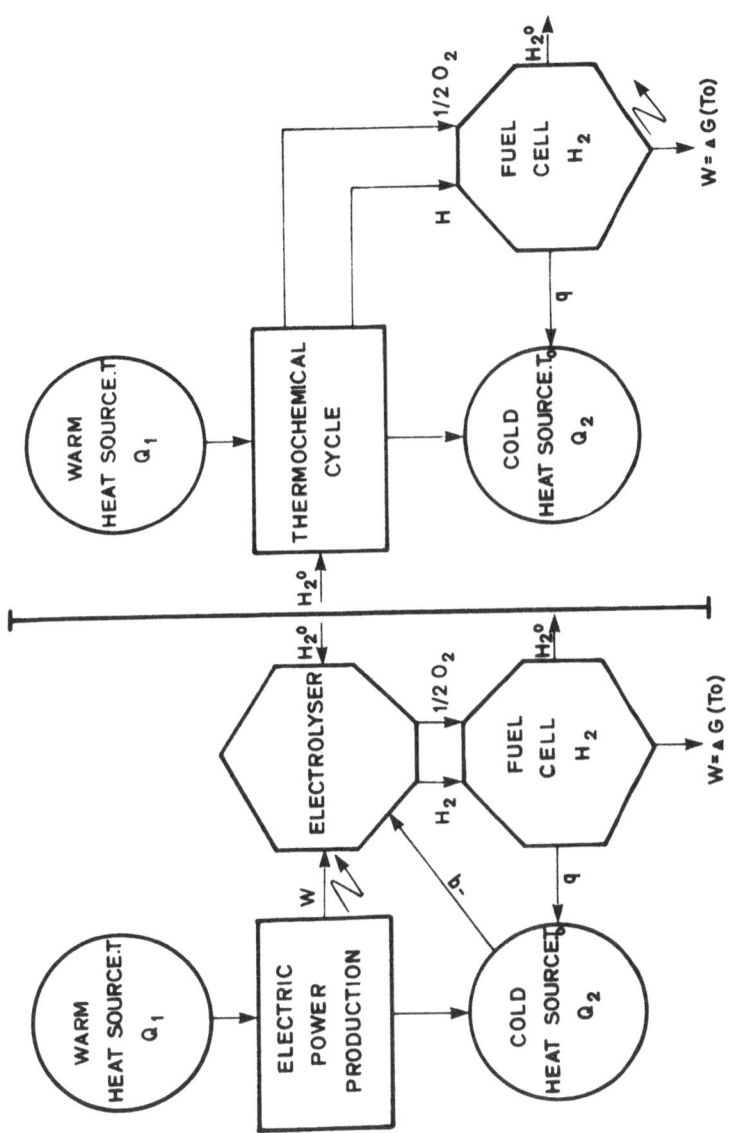

ELECTROLYSIS $\eta_{th} = \frac{[W]}{Q_1} = \frac{1 \text{To}}{T}$ THERMOCHEMISTRY

Fig.1 THEORITICAL EQUIVALENCE

Among the various possible processes for water splitting the two most studied now are the electrolysis and the thermochemical cycles. The necessary comparison between these processes can be made when the same boundary conditions are considered, i.e. when energy in the form of heat is available at a given temperature and has to be transformed into hydrogen. One way consists in using the heat for the production of electricity and then operating an electrolytic process; the other consists in using the heat for running the endothermic reaction of a thermo-chemical cycle (see fig.1). The theoretical efficiency is the same for both ways, because it depends only on the boundary conditions, (level of temperature of input and output heat) and the thermodynamic properties of water*. The pratical efficiency will depend on the irreversibilities introduced in the technical realization of each process, the investment cost will depend of technological encountred or technical state of the art for used devices.

2.2. Thermochemical water splitting - Principle [1], [2]

A thermochemical water splitting cycle is a kind of chemical engine, able to absorb heat at high temperature by endothermic chemical reactions and rejecting some heat at low temperature by exothermic reactions. It is governed by a Carnot law; the minimum amount of work that must be produced corresponds to the free energy of formation of water.

By this technique the water molecule is decomposed into its constituents through a chemical cycle which consumes only heat as energy input. It has already been said that for the direct decomposition of water, heat at extremely high temperature would be necessary. When the temperature is not high enough, the same result can be reached by absorbing heat in two or more steps (chemical reactions). The situation is similar to what occurs in green plants. The energy contained in one photon of the visible light is not strong enough to decompose the water molecule, but this goal is achieved by the use of two photons through the chlorophyll chain.

From the thermodynamics of thermochemical cycles it follows that the number of steps in a cycle is linked to the maximum temperature at which the heat is available : the steps can be fewer when the temperature is higher.

Within the limits set by the maximum outlet temperature of the HTGRs (about 1200K) it is recognized that at least 3 or 4 chemical reactions are necessary.

All the chemical reactions of a cycle operate with water afforded from outside the cycle and deliver oxygen and hydrogen. In order not to consume intermediate products very clean separation would have to be achieved between different steps of a cycle.

The separation steps usually require large amounts of energy, with consequences for the overall thermal efficiency, particularly when one of the major components undergoes a phase transformation.

The thermal matching with the primary heat source and all the heat recovery must be very well designed because the heat is the "raw material" of these plants and it must be used very carefully. Very good heat exchangers are expensive.

According all these aspects it does not seem that, for massive hydrogen production from nuclear energy, thermochemical cycles can compete for efficiency and cost with water electrolysis.

2.3. WATER ELECTROLYSIS [1], [3]

The electrolytic decomposition of water into H_2 and O_2 is an old process which was developed industrially in the first decades of this century, for the production of hydrogen for the chemical industry. This process became obsolete when cheaper hydrogen could be produced from fossil fuels. Hydrogen was nevertheless produced by water electrolysis for small processes requiring high purity gas, or when large amounts of cheap electrical energy were available, or for a combined heavy water production. Because of the small size of the market for electrolyzers, and consequently the very small research effort, very small improvements were gained in their performances during the last decades. Only after the sharp rise in the price of oil and of the other fossil fuels was there renewed interest in water electrolysis as a method for the production of a fuel from non-fossil energy sources.

The industrial installations existing today are all based on alkaline electrolysis; the operating conditions are mostly around 353 K (80°C), 1 bar pressure, cell voltage 1.8 - 2.2 V, and current density 2 kA/m^2. Specific production ratio is :

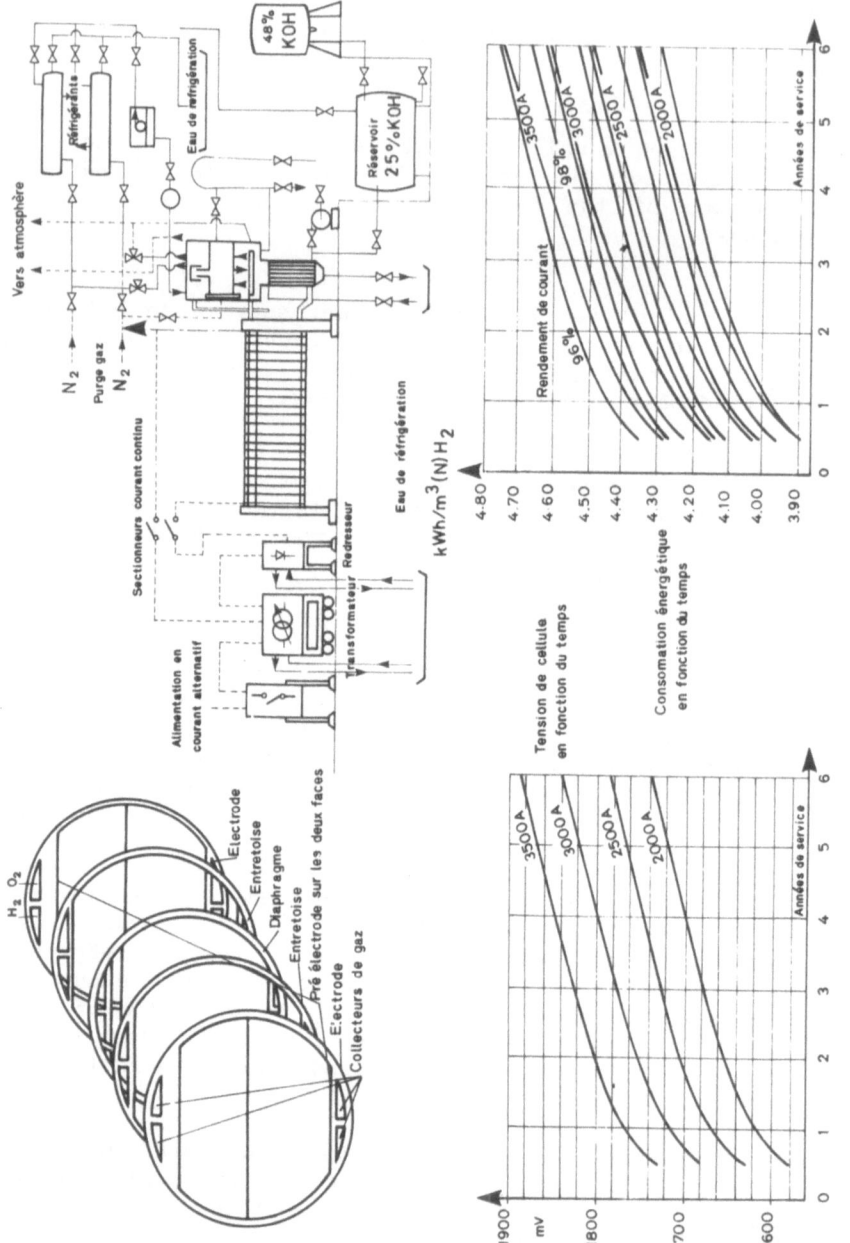

Fig.2 NORSK HYDRO ELECTROLYSER

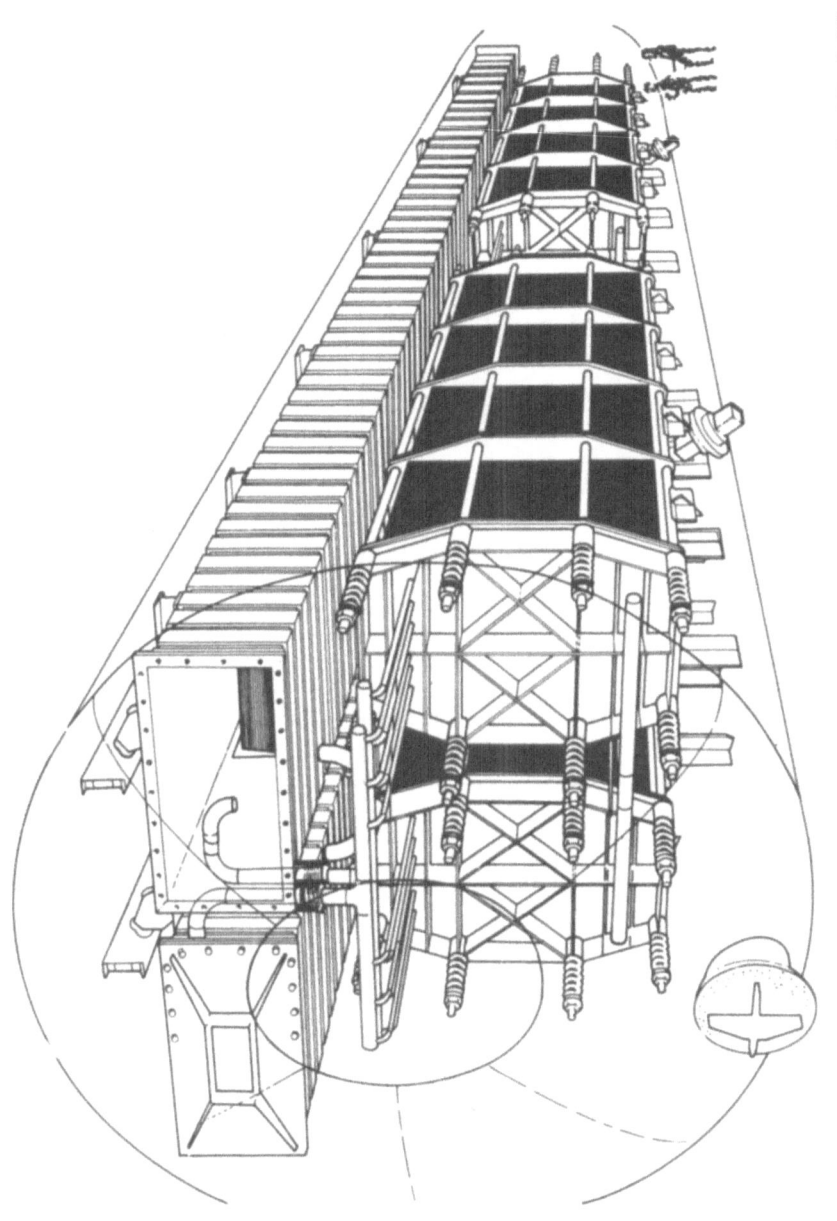

Fig. 2 bis 74 MW EDF-GDF PROJECT

4,5 kwh/m^3 N of hydrogen - Electrolyte shows potash content 25/30%.

Main industrial electrolyzers available today are the following, made by a stack of electrolytic cells.

Electrolyseur	NORSK-HYDRO	B.B.C.	DE NORA	DEMAG	LURGI
Production (Nm^3/ H$_2$/h)	30-320	2,5-210	18	150	145
Intensité (kA) (exemple)	2-3,5	8	2,5	7,6	3
Tension (V) (exemple)	570	170	34	87-90	217
Electrolyte Solution de Potasse	25%	25%	29%	28-30%	25%
Temperature (°C)	80	75	75	80	70-90
Consommation spécifique (kWh/Nm^3 H^2)	4,5	4,3-4,4	4,6	4,3-4,5	4,5

Atmospheric pressure + 500 mm water   30 bars

Each cell (see figure 2) is made of two electrodes separated by one porous separator impervious to gas, permeable to electrolyte. From each cell gas are collected as electrolyte. This one is driven into a cooler which achieve separation of small gas bubbles from electrolyte (see figure 2 bis)

The electrolysers are machines which transform electrical energy into chemical energy, with a certain efficiency and at a certain cost. The theoretical voltage for decomposing water at 198 K is 1.23 V, requiring the delivery of 48.7 kJ of heat per mole of water split; if all the energy has to be provided by the electricity, the minimum voltage for water electrolysis rises to 1.47 V.

CURRENT DENSITY INFLUENCY

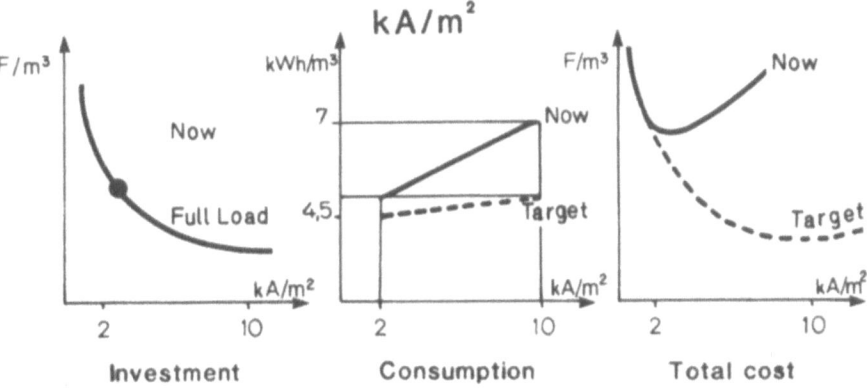

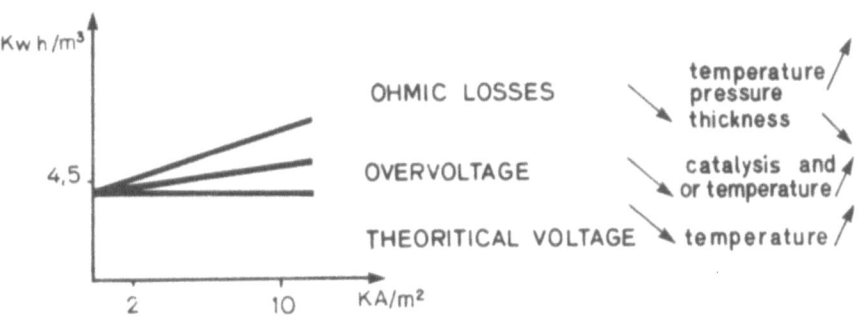

Parts of total consumption

Fig.3 DEVELOPMENT

Supposing a practical voltage of 2 V, the exergy efficiency is 61.5 %, and the thermal efficiency would be 73.5 %. The cost of electrolysers is also high because energy must flow through a composite surface (electrodes, electrolyte and separator) at quite a low power density; operating at 2 V and at 2 kA/m^2 the power density is only 0.4 W/cm^2.

In order to obtain cheeper service from a given electrolyser one wishes to increase the power density. This induces an increase of energetic losses which are overvoltage at the electrodes and ohmic losses in electrolyte and separator.

The possibilities for improvement in the performance of electrolysers are :

a) reduction of theoretical voltage for decomposition of water by working at much higher temperatures,

b) reduction of overvoltage at the electrodes, by more effective electrocatalysts and by increasing the working temperature,

c) reduction of ohmic losses in the electrolyte and in the separator by proper design of individual cells or by new separator materials.

Many studies are in progress in the various fields mentioned here, fig.3.

a) High Temperature Electrolysis

In order to obtain a substantial reduction of the theoretical voltage for water splitting, it is necessary to reach quite high temperatures. At 1200 K the reduction is about 25%. In these conditions pure steam is electrolysed; the cell is usually constituted by the electrolyte, an oxygen-ion-conducting ceramic oxide, with the electrodes plated on both sides. The high temperature is also necessary to obtain sufficient ionic conductivity of the ceramic electrolyte (doped ZrO$_2$).

b) Improvement of the Electrocatalyst Effectiveness

The purpose of the studies in this field is twofold : to reduce the working voltage and to increase the current density at the same time. This double result can only be obtained by a great improvement in the electrocatalyst performance. The active part of the electrodes of actual industrial electrolyzers is made mostly of nickel; the highest overvoltage is present at the anode, and most studies are now concentrated on the anodic electrocatalysts.

c) Reduction of Ohmic Losses

This improvement is achieved for instance by increasing the working pressure, so reducing the volume of the gas bubbles and the effect of the bubble curtain between the electrodes and the separator, or by reducing the thickness of the elect- rolyte placing the perforated electrodes directly against the separator, or by using separators other than asbestos, such as potassium titanate or polyantimonic acid or porous metal- lic structures electrically isolated by oxide coatings.

The electrolytic production of hydrogen is clean, can produce hydrogen directly under pressure with a minimal energy con- sumption.

The construction is modular, so that electrolyser's size can match the power availability without economic penalties. This one has an easy start-up and shut-down procedure, so that it can also be used for intermittent operation.

3 HYDROGEN STORAGE [4]

All the examples would be choosen in natural gas industry and very after comparison would be made between hydrogen and natural gas storage. This last is widely used. This massive storage is achieved in underground storage.

3.1. General principle

The creation of an underground storage corresponds with the use, as a reservoir capacity, of vacuums in a geological site. Va- cuums which can be natural or man made, isolated from the sur- face by one or more tight layers. The reservoir is linked to the surface installations by one or more wells or shafts properly equiped.

These vacuums may be microscopic (it is the case for porous and permeable rocks : reservoirs in aquifer layer) or gigantic (old mines, karsts). These two techniques of underground storage are used now for natural gas and could be used for hydrogen.They can give reserves up to several hundred millions cubic meters (n).

3.2. Storage in cavities

There are four possibilities for the use of subterranean cavities for large-scale storage of gas under pressure :
-two consist of using cavities specially for this purpose, either by controlled dissolution of salt formations or by underground nuclear explosion,

- two depend on the use of existing cavities whether natural, like karstic caverns, or artificial, like abandoned mines.

Storage cavities formed by nuclear explosion and the use of abandoned mines have, so far, only been tried for methane in the USA, in the USSR, and in Belgium (in old coal mines).

The use of karstic cavities is presently being studied in France with a view to storing natural gas. The preliminary results of these studies reveal one of the principal problems in the use of this type of cavity, namely the possibility that the ratio of the volume of gas stored to the total available volume is too low, which would have lesser economic consequences for hydrogen than for methane because of the lower calorific value of hydrogen. In all other respects there is no reason to think that this type of storage could not eventually be used for hydrogen as well as methane.

Salt (sodium chloride) is a substance found abundantly underground ; it is a substance with a fine and low porosity and it is practically impermeable. It can be found in substantial quantities at various depths in evaporation basins or diapirs. Large cavities can be made in such salt formations by excavation with water, salt being highly soluble. These cavities, for gas storage, mut have a regular shape with a conical vault, with a small aperture angle at the top. They are operated under conditions of variable pressure drop. The volume of each cavity depends on the thickness of the salt formation and could reach several hundred thousands of cubic meters.

Each cavity is created by a washing process that is carried out through a drill-hole of the type used for oil wells, the casing being cemented to the surface of the ground ; this washing is done at a level below the roof of the salt bed. Figure 4.1 shows this operatin schematically. The drill-holes are fitted with double tubing, and fresh water is injected through the inner tubing wheras the brine resulting from the washing is extracted through the intermediate tubing. A blanket of inert material, vis-a-vis the salt, is placed in the outer tubing from ground level down to a few metres above the end of the tubing of greatest diameter ; this is to control the rate of dissolution of the upper part of the cavity being created. The positions of

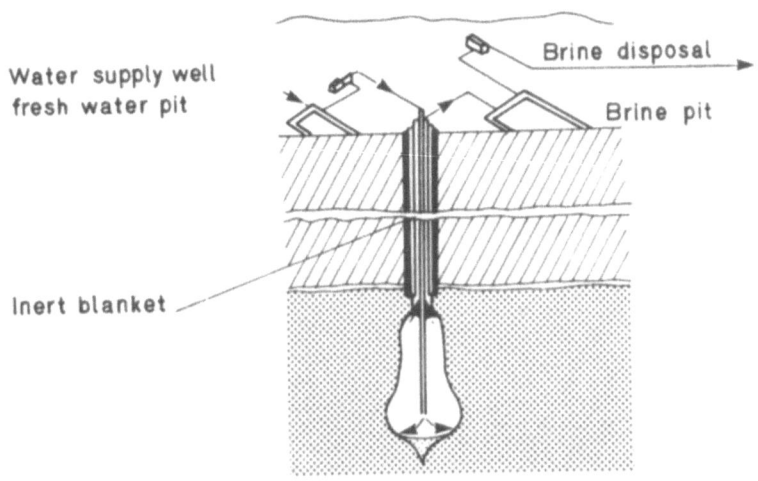

1. CAVITY FORMATION BY WASHING OPERATION

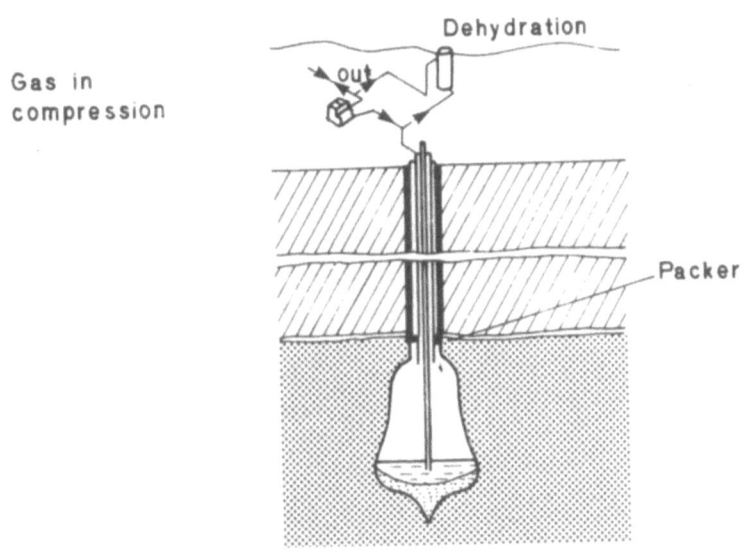

2. CAVITY USED BY VARIATIONS OF GAS PRESSURE

Fig.4 GAS STORAGE IN CAVITY CREATED BY DISSOLUTION IN MASSIVE SALT

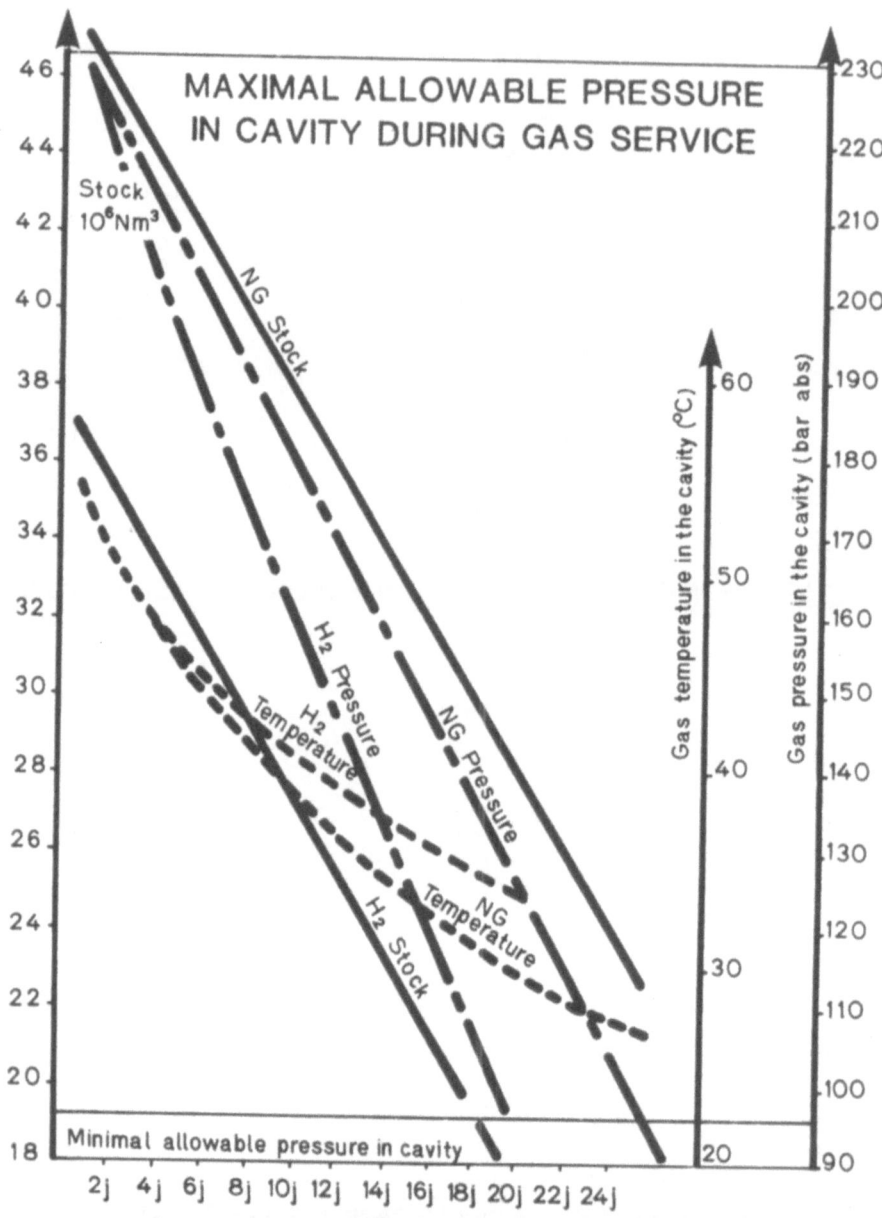

Fig.5 CAVITY CONTAINING
EITHER HYDROGEN : H_2
OR NATURAL GAS : NG

the bases of the two columns of tubing are altered according to the growth in the size of the cavity.

When the cavity has reached its final size and after the operating machinery has been installed, the gas is injected through the intermediate tubing and the brine extracted through the central tubing. After the brine is completely extracted, the gas is injected until a pressure is reached in the cavity that has been calculated to ensure that the cement of the lining of the drill--hole will not be damaged. Gas is then withdrawn from the cavity by simple decompression until a predetermined minimum acceptable working pressure has been reached (figure 4.II).

Reasonable values for the operating pressure range are between 0.6/0.7 and 1.6 times the hydrostatic pressure at the top of the cavity, which, for a cavity at the depth of 1 000 m, would be between 60 bar, at the lower limit, and 160 bar, at the upper limit.

Storage of this type has been used for natural gas in the USA, USSR and France, and several such storage cavities are being made in these countries as well as in the Federal Republic of Germany and in Great Britain. We have had several years of experience in the storage of natural gas in underground cavities, and numerical models that allow us to estimate their operating characteristics (total storage capacity, operating volume, residual volume after gas withdrawal, supply and available stock at any given moment, etc.) have been developed by Gaz de France, for example.

It is possible to compare the energy reserves constituted with natural gas on one side and hydrogen on the other, in a 220 000 cubic metres cavity, located at 1130 m deep, in which 10^6 $m^3(n)$/day quantity of gas would be extracted. The expected differences would be due to the differences between thermodynamic behaviours of these gases which have not the same state equations (cf fig. 5) :

- for the same maximum admissible pressure in the cavity and the same temperature (55°C) of the gas the cavity holds about 47 x 10^6 Nm3 of natural gas but only about 37 x 10^6 Nm^3 of hydrogen,

- the decrease in cavity pressure is much more rapid for hydrogen than for methane, so that the speed of variation of mechanical stresses in the salt mass is greater but in the present state of knowledge it is not possible to evaluate the influence on the visco-plastic behaviour of the salt,

- the cooling of the gaseous phase is less rapid and the observed temperature differences between natural gas and hydrogen
after complete withdrawal (determined by a cavity pressure equal
to the minimum authorized pressure) is about 7°C,

- the respiration, or difference between the maximum gas content
of the cavity and the content upon complete withdrawal, is, in
terms of volume, lower for hydrogen than for natural gas ; the
same is true for the residual stock after complete withdrawal,

- the pressures at the well-head are about 24 bars higher for
hydrogen (because of lower pressure drops but especially because
of the lower weight of the gas column).

If one no longer argues in terms of a steady volumetric rate of
withdrawal but tries, for a given gas stock, to obtain the same
respiration and the same steady rate of withdrawal in terms of
energy, one observes that :

- it is necessary to have four cavities of hydrogen for each one
of methane,

- the retained residual volume is about 3.1 times greater for
hydrogen so that the number of thermies retained is approximately the same in both cases.

The compressor stations, the treatment units and the distribution network hardly pose any more problems for hydrogen
than for natural gas. There is always the possibility with these
installations, which usually function at pressures in excess of
70 bar and sometimes even in excess of 200 bar, that problems of
metal brittleness might give cause for concern. On the other
hand the risk of formation of hydrates in the service pipes of
the drill-holes and in the ground installations, though representing real constraints with respect to natural gas does not
exist with respect to hydrogen.

3.3. Storage in porous and permeable geological formations

This type of natural gas storage was first tried in the USA
where petroleum extraction was started in the last century. Exhausted oil fields and especially exhausted gas fields are very
numerous there, and the idea of using them to store natural gas
from geographically less favourably situated oil fields, or obtained during oil extraction, naturally arose. Their use for
storing hydrogen could also be envisaged but the hydrogen at the
time of withdrawal would necessarily contain hydrocarbons and
would have to be treated if it was to be used in processes

①   **Production well (injection.withdrawal)**

②   **Log survey well**

③   **Peripheral survey well**

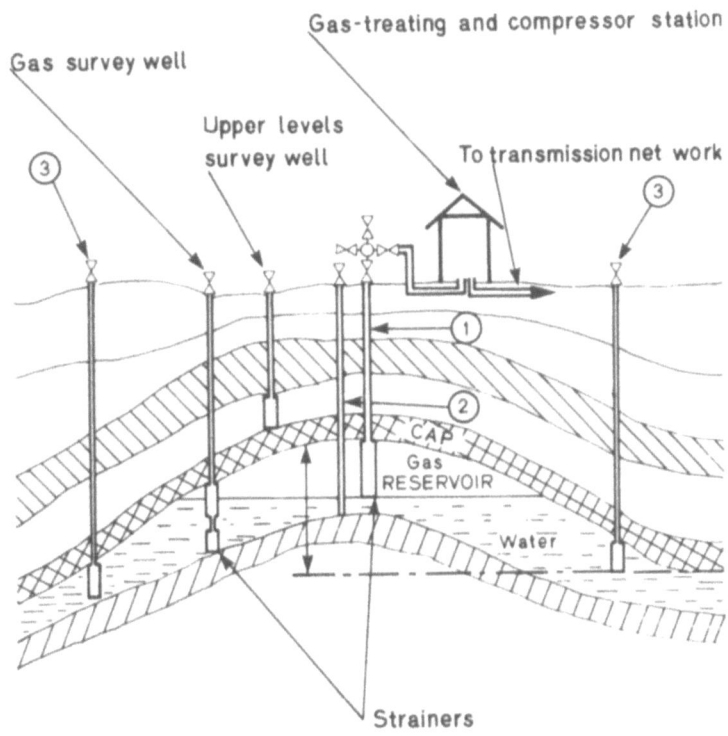

Fig.6 1 SCHEMATIC CUTTING VIEW OF UNDERGROUND AQUIFER STORAGE

requiring hydrogen with a high purity. For this reason it would be better to use aquifers that have never contained hydrocarbons.

The use of this type of structure has moreover already been widely developed for natural gas.

In France where there are practically no exhausted oil fields, aquifers have been used from the start for storing natural or manufactured gas.

The principle of underground storage in aquiferous layers is to create an artificial gas field by injecting gas into the intersticial spaces in the rock that were occupied by water. However, the use of this technique requires special geological conditions.

Underground storage requires, from the sedimentary standpoint, the superposition of two kinds of rock ; the reservoir, consisting of rock that is porous and permeable, the pores of which, after the displacement of the intersticial water, can be used to store the gas, and the overlying cover, consisting of rock capable of maintaining the gas in the reservoir either because it is impermeable sensu stricto or because its texture would require, for gas to enter it, the application of capillary pressures exceeding the pressure of the gas in the reservoir. The gas injected into the reservoir should be able to accumulate around the drill holes, and this accumulation or "balloon" should not disperse with time, for which purpose it is advantageous to have the above-mentioned rock formations in the form of a trap, usually what is called an anticline, in which the reservoir-cover combination is an inverted concavity in which the gas is trapped.

Figure 6.1 shows a vertical section of such a structure ; it is a trap with a characteristic enclosure. Upon withdrawal of the stored gas, the height of the zone occupied by the gas diminishes until the desired supply of gas cannot be maintained because the water enters the bottom of the drill holes ; the quantity of gas that one can inject and then withdraw throughout the annual operating cycle is called, as in the case of cavities, usable gas or reservoir respiration. This quantity is about half the total in storage.

The introduction of the gas is accompanied by the partial displacement of the water in the pores of the rock. This displacement only takes place if the gas is compressed at a pressure exceeding that observed in the water-bearing layer. This over-pressure should be limited so as not to exceed the value of the capillary pressure in the covering rock. The pressure in

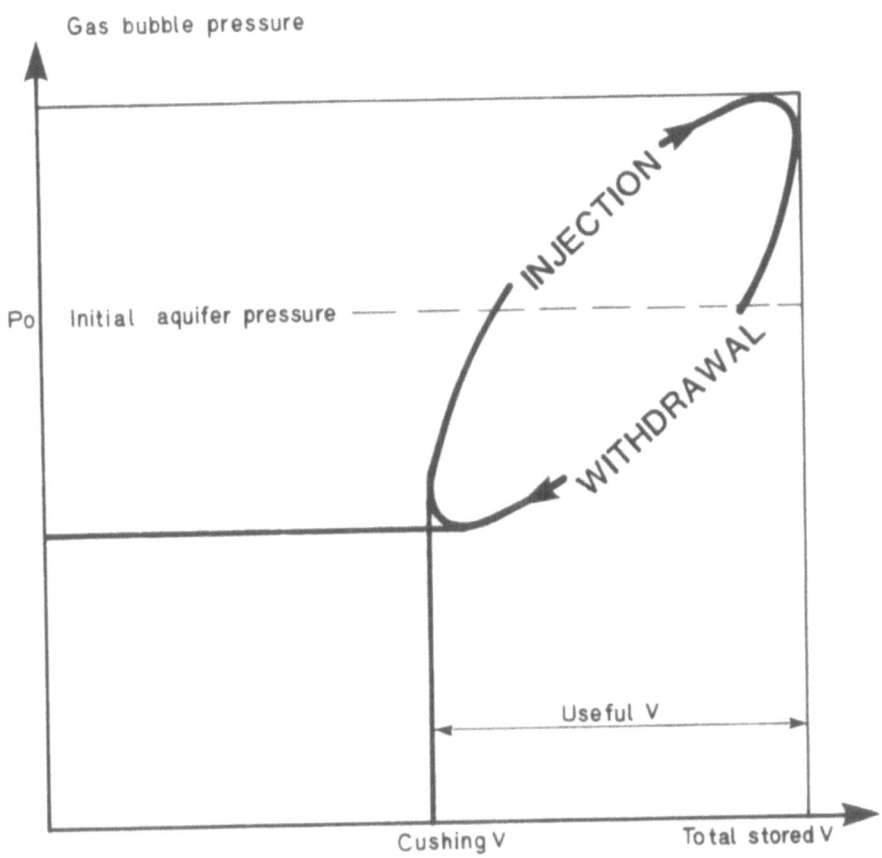

Fig.6 2 RELATION BETWEEN PRESSURE AND GAS QUANTITY IN STORAGE

the aquifer is generally close to the hydrostatic pressure (1 bar per 10 m. of depht). During the injection of the gas this pressure should be kept to between 1.2 an 1.5 bars for each 10 m of depth.

Inversely, during withdrawal, it becomes less than the initial equilibrium pressure, so an annual pressure cycle is established (see fig. 6.2).

It is necessary to be able to forecast pressure changes in the reservoir as a function of gas movements, injection and withdrawal.

Generally speaking, forecasting of the pressure distribution in the reservoir is, in the various models now in use, based on a radial circular flow in a flat homogeneous layer away from the walls of a cyclindrical well. The injections or withdrawals of gas are equated with injections or withdrawals of water, volume for volume, in the imaginary well, and the resulting pressure drops at the rock face of the well determine the pressure changes in the "balloon" of gas in the middle of which pressure drops are generally considered to be negligible.

These considerations which have been applied regularly for many years to the storage of natural gas, may also be applied to hydrogen. The greater diffusivity of hydrogen leads one, however, to ask whether there could be a transfer of this gas through the covering rock.

In this regard it should be noted that the penetration of gaseous hydrogen into the covering rock is subject to capillary forces and that the threshold capillary pressures which, for a given pore morphology, depend only on the interfacial tension, are comparable for the two combinations : water-methane and water-hydrogen.

Regarding the spread of gaseous hydrogen once it has penetrated the convering rock, Darcy's law applies and the rates of spreading are of the same order as those for natural gas (the viscosity of methane being only about two times greater than that of hydrogen, under the conditions normally found in storage aquifers).

An examination of the problems posed by the diffusion of hydrogen in solution in the interstitial water of the covering rock leads to the following observations :

a) from the theoretical point of view the phenomenon has been insufficiently studied in rocks with a low permeability, and especially in clays, and a complete study should be made, not only of the pore morphology but also the effects of pressure and temperature gradients which form part of the background conditions, and the various phenomena due to the very small intersticial dimensions (osmosis, ultra-filtration, ionic effects in clays) ; the combination of these factors can modify considerably the conclusions drawn,

b) an examination of laboratory studies carried out in the USA and the USSR shows that although GRAHAM'S law* cannot be applied to the diffusion of gases in solution, this diffusion does not differ very much, in rock of low permeability saturated with water, as regards the ratio of the diffusion of methane to that of helium from which we may make an assumption regarding the diffusion of hydrogen compared to that of natural gas : hydrogen would diffuse four times as fast as natural gas through a water--saturated rock.

c) from the industrial standpoint it should be noted :

- that the operation of a town gas reservoir with a capacity of 330×10^6 Nm³, at Beynes, France, between 1957 to 1974, a reservoir in which the gas was about 50 % hydrogen, revealed something of special interest for the study of hydrogen diffusion : because of the exceptional thinness (7 m) of the covering rock a very strong concentration gradient existed between the reservoir and another aquifer overlying the cover rock, but no trace of hydrogen has ever been found in the gases extracted under vacuum from water taken from the upper aquifer,

- that a helium reservoir of more than 10^9 Nm³ under a calcareous covering rock, near Amarillo, Texas, USA, has been exploited without detectable losses since 1963 ; calculations made on the basis of the above-mentioned laboratory data have shown a theoretical diffusion rate of 3 Nm³ per year and per km² through a covering rock 100 m thick, a result which would be at least doubled for hydrogen and but which, in spite of the theoretical interest in a much deeper study of this diffusion, would be of limited industrial importance.

The other problems refer to the technology and economics of exploitation which are different between hydrogen and natural gas owing to their different physical characteristics regarding compressibility, viscosity, density and calorific value.

---

*GRAHAM'S Law : The diffusion of a gas through a porous layer is, for a given temperature and pressure, inversely proportional to the square root of the gas density.

Consider a site of given geometric characteristics in which 10^9 Nm^3 of natural gas can be stored at a maximum pressure of 70 bar, with a respiration of 0.5 x 10^9 Nm^3 and a final pressure of 40 bar upon completion of operational discharge, and with a ground temperature of 30°C. This would allow the following possibilities for the storage of hydrogen :

total stock	:	860 x 10^6 Nm^3
residual stock	:	460 x 10^6 Nm^3
respiration	:	400 x 10^6 Nm^3

The maximum heating values being 3.05 thermies per cubic metre for hydrogen, and 9.85 thermies per cubic metre for natural gas, a geological structure of the type just described will permit the storage of 4 925 x 10^9 thermies of natural gas or 1 200 x 10^9 thermies of hydrogen. So, for any given quantity of energy the storage capacity should be about four times greater for hydrogen than for natural gas.

As regards the movements of the interface between the gas and the water, and the changes in the shape of the "balloon" of gas, it does not seem likely that hydrogen would behave very differently from natural gas. In effect, the relation between the mobility fo the water and that of the gas will be enhanced by the lower viscosity of hydrogen, which would tend to accentuate the effects of horizontal deformations such as interdigitation of gas water. On the other hand the less important pressure drops occurring inside the "balloon" of gas will quickly reach the surface of equipotential and should reduce certain deformations in the vertical direction during withdrawals, as for example coning effects due to water ascending into the gas supply zone. It should also be noted that the tilting of the gas-water interface when injections are taking place depends basically, for a given displacement speed, on the difference between the viscosities of the water and the gas, which difference is of the same order of magnitude for hydrogen as for natural gas.

An important consequence, from the technical and economic standpoints, of the low values of viscosity and specific gravity of hydrogen is the lower number of wells required for a given volumetric output of hydrogen. Indeed, for the same differential pressure applied to the reservoir, the output of a hydrogen well can be about twice as high as that for natural gas. This property leads in practice to the establishment, for a given volumetric production, of a number of supply wells that are twice as small for a hydrogen reservoir as for a natural gas reservoir.

GAZ de GRONINGUE

GNL

Lille
Taisnières
Amiens
Gournay
Le Havre
Rouen
Reims
Caen
St Illiers
Metz
Paris
Velaine
Beynes
Nancy
Strasbourg
Brest
Mulhouse
Rennes
Le Mans
Orléans
Nantes
Chemery
Dijon
Besançon
Angers
Tours
GNL
Etrez
Lyon
Limoges
Angoulème
Clermont-Ferrand
St Etienne
Grenoble
Tersanne
Bordeaux
GAZ de LACQ
Nice
Lussagnet
Nîmes
Montpellier
Fos
Toulouse
Marseille
LACQ
Toulon
Perpignan
GNL

LEGEND

───── ⎱ GAS Lines
- - - - - ⎰

▽ Lacq
★ GNL Terminal
○ Aquifer   ⎱ Underground
⊕ Salt cavities ⎰ Reservoir

Fig.7 GAS TRANSMISSION GRID IN FRANCE

STOCKAGES	BEYNES (Yvelines)	LUSSAGNET (Landes)	S¹ ILLIERS (Yvelines)	CHEMERY (Loir et Cher)	TERSANNE (Drôme)
Operating society	GDF	SNEA (P)	GDF	GDF	GDF
Type	Aquifer	Aquifer	Aquifer	Aquifer	Salt
First year of service	1956	1957	1965	1968	1970
Top depth (m)	405	600	470	1120	1400
Maximal storing capacity $10^6 m^3$(n)	480	1300à3000	1400	2800	540
Compressing station (power)	10600 kW	4800 kW	18270 kW	12800 kW	4660 kW
Number of wells — exploitation	14	34	28	43	5
Number of wells — survey	14	13	16	17	-

STOCKAGES		VELAINE (Meurthe - et Moselle)	BEYNES PROFOND (Yvelines)	GOURNAY SUR ARONDE (Oise)	ETREZ (Ain)
Operating society		GDF	GDF	GDF	GDF
Type		Aquifer	Aquifer	Aquifer	Salt
First year of service		1970	1975	1976	1979
Top depth (m)		470	740	750	900 à 1400
Maximal storing capacity $10^6 m^3$(n)		1250	500	1300	indéterminé
Compressing station (power)		11300 kW	3000 kW	3000 kW	
Number of wells — exploitation		36	7	10	-
Number of wells — survey		14	10	10	-

Fig.8 UNDERGROUND STORAGES OPERATED IN FRANCE

(Situation 1st January 1978)

Storage	Maximun volume start of winter $10^6 m^3$ (n)	Withdrawable volume during winter		Possible peak flow rate	
		$10^6 m^3$(n)	10^9 kWh	$10^6 m^3$(n)/j	10^6 kWh/j
BEYNES	347	160	1,8	4	44,8
SAINT-ILLIERS	1 232	550	6,2	15	168
CHEMERY	2 515	1150	12,4	22	246
TERSANNE	120	75	0,9	5	56
VELAINE	1 189	530	4,8	12	134
BEYNES PROFOND	238	65	0,7	1,5	16,8
GOURNAY-sur-ARONDE	395	80	0,9	3	33,6
TOTAL	6 036	2 610	27,7	62,5	699

Fig.9 WINTER 1977–1978

	BEYNES	SAINT ILLIERS	CHEMERY	VELAINE	BEYNES PROFOND	GOURNAY SUR ARONDE
Total porosity (%)	25 à 30	30	25	17	25 à 30	15 à 25
Permeability K (darcy)	3 à 5	1	1	0,76	1	0,3
Thickness (m) *	10	30	30	60	20	70
Initial aquifer pressure Top of structure (bars abs)	36,8	46,8	114,2	43,5	78	78
Maximal Allowed pressure (bars abs)	48,7	69,5	140	61,2	97	97
Cap thickness (m)	8	180	120	80	150	70
Proved closure (m)	27	130	80	85	27	65
Closed area (km²)	8	7	18	21	8	27

*It is either a real thickness (Beynes, Saint Illiers and Gournay) or a similar thickness for the fluids move (Chemery, Velaine).

Fig.10 CHARACTERISTICS OF SAME FRENCH UNDERGROUND AQUIFER STORAGES

3. 4. EXAMPLES OF ACHIEVEMENTS

As it is noted in the first item, the examples come from the achievements realized in the natural gas industry in France.

The figure 7 gives the location of storage compared with the main consumption centers and the feeding points of the transport networks.

The tables 8a and 8b give some indications on the different underground storage exploited in France. The table 9 collects the exploitation characteristics of the underground storage of GAZ de FRANCE (GDF). The table 10 shows some characteristics of the storage in aquifer sheet exploited by GDF.

4 CONCLUSION

Hydrogen as energy carrier seems to be connected to the development of nuclear power used for electric power production. Then, produced through an easy to operate way like electrolysis, efficiency improvement of which is under study, this gaseous product can be easily stored in very large reservoirs, such as those well known through a long operation time with natural gas.

Thanks these two auxiliary means heat could be stored through hydrogen production and storage.

REFERENCES

[1] Hydrogen as chemical storage : methods of production from water

Gianfranco de Beni - CEC Joint Research Center - Ispra Establishment.

[2] Thermochemical production of hydrogen - Myth or reality ?
G. DONAT, B. ESTEVE and J.P. RONCATO
Revue de l'Energie N° 293   Avril 1977

[3] Exposés Electricité de France et Gaz de France sur l'Electrolyse de l'eau.

Deuxième forum Electro-industriel national - Atelier Electrochimie
Comité Français d'Electrothermie - Lyon 6-7 NOV 1980

[4] Underground storage of gaseous energy carriers

Michel Pottier - Direction des Etudes et Techniques Nouvelles - Gaz de France - Paris

ISPRA COURSE : ENERGY STORAGE AND TRANSPORTATION

STORAGE IN THE GROUND

Ghislain de MARSILY

Centre d'Informatique Géologique, Ecole Nationale
Supérieure des Mines de Paris, Fontainebleau, France.

ABSTRACT

After a brief examination of why and how energy is stored,
the problem of heat storage in the ground in impervious and porous
media saturated with water is described. The equations for heat
transfer in natural medial, and a summary of the ideas, experiments and projects in this field, are given. The technique of
aquifer heat storage appears promising for the storage of around
a million megacalories, but there are still some problems which
are not yet completely solved.

1. WHY STORE ENERGY ?

It is hardly necessary to point out that energy must be
stored in order to adapt the capacity to the needs. In all our
present energy producing systems, storage is used, as for
example:
- storage of hydraulic energy which can be called upon
immediately to meet demands of electrical power;
- storage of gas for distribution;
- heat storage for space heating (thermal inertia of
buildings or accumulation systems);
- finally, and above all, innumerable stores of liquid
or solid fueld which can provide energy at very short
notice.

Until the recent energy crisis, these more or less natural
stores were sufficient to balance supply and demand[1]. The only
instances of artificial storage of energy excess were found in

certain hydro-electric power plants, which filled their reservoirs by pumping during slack hours or, elsewhere, by storage of steam under pressure.

This situation was basically due to the great flexibility of fossil fuel energy sources.

The replacement of these fuels by new energy sources, as well as the growing cost of energy, makes storage increasingly necessary:

- it is very difficult to regulate the production of nuclear energy and, furthermore, little is gained by doing so, as the cost of the fuel plays a very small part in the cost of the energy produced;
- solar energy and wind energy are practically inconceivable without storage, as the supply is independent of, ans sometimes opposed to, demand;
- utilization of wasted energy, often presented as a new energy source by the Agencies of Energy Saving, often requires storage;
- geothermal energy could also include storage if maximum flow is maintained throughout the year, in order to maximize the return on investiments in deep wells.

Eventually, a society based on hydrogen energy produced by nuclear reactors, electrolysis or chemical reactions, which methods are currently being studied, could maybe solve the problem by storing hydrogen. However, this source of energy will not be available in the near future.

2. HOW TO STORE ENERGY

One may consider storing (1):
(i) electrical energy directly by electromagnetic storage in supra-conductors at a temperature of a few degrees Kelvin. This process is being studied for large storage capacities, but at very high costs;
(ii) chemical energy (batteries, accumulators). Prices are high and efficiencies low;
(iii) mechanical energy: a large number of methods are conceivable, many of which concern the underground. However, these methods of storage require a superior type of energy which excludes any recovery of degraded energy. For example:

- fly wheels (short term storage),
- storage of steam under pressure in tanks,
- compressed air to feed gas turbines: the storage will be located in sealed subterranean cavities using, for example, an abandoned mine- projects like this

exist in Sweden and in Germany,
storage of water in reservoirs with 2 different eleva-
tions as already mentioned. Since there are few natural
sites available close to the places where the energy
is needed, it has been proposed that the second reservoir
be buried approximately 1000 m underground (1000 MW
project at Mount Hope, New Jersy (2)) - the efficiency
is generally rather high.

iv) thermal energy: this seems the most promising. One may store
sensible heat or latent heat. For sensible heat storage, a
substance with high heat capacity is needed (preferably water
or other heat-carrying organic fluids or even solid materials).
For the storage of latent heat, recent research has concentrated on eutectic mixtures of salts and water with high latent
heats of fusion. For low temperatures, however, especially
when large quantities of heat are to be stored, such artificial storage becomes expensive and the use of natural storage
media becomes attractive.

In this article, we shall review the state of the art and
the various projects which are underway in this field.

To conclude this brief review, we can also mention the
storage of natural gas underground (in aquifers, or cavities in
salt domes). This type of storage could become very important if
a hydrogen-based economy were to develop.

3. HEAT STORAGE IN THE GROUND

3.1. If there is no water

The idea of storing heat in the ground is far from new.
We need only recall the Moroccan way of cooking a sheep in a hole
in the ground, where the fire's heat is stored in the earth to be
used later to cook the meat, away from the flames; or, on the
contrary, the accumulation during the winter of ice and snow in
underground cellars for the ices and cold drinks of the following
summer, an age old practice. This ability of the ground to store
heat is due to the very low thermal conductivity of dry rocks
(generally less than 10^{-3} kcal/m.s.°C), i.e., of the order of
magnitude of that of refractory brick walls or even of asbestos,
which is an excellent natural insulator. The thermal capacity per
unit volume of rocks is relatively high, about half that of water.

Thus, if one can introduce heat into ground, this heat leaks
away so slowly, that most of it can be recovered, while only a
small amount is lost at the boundaries of the storage.

To illustrate this point, Fig.1 shows the evolution of the temperature as a function of time and distance for 1-dimensional heat flow in a rock mass, one face of which has suddently been brought to and maintained at a constant temperature. The thermal diffusivity[2] used is that of a clay (10^{-6} m^2/s) (see Table I).

When heat is stored in the ground, three problems arise:
a) how to introduce the heat into the ground ?
b) how to determine the size of the storage so that the quantity of heat stored (which depends on the volume of heated rock) is large in comparison to the heat losses at the boundaries (which depend on the size of the surface enclosing the heated volume) ? Because of this volume/surface ratio, heat storage in the ground necessarily implies large volumes:

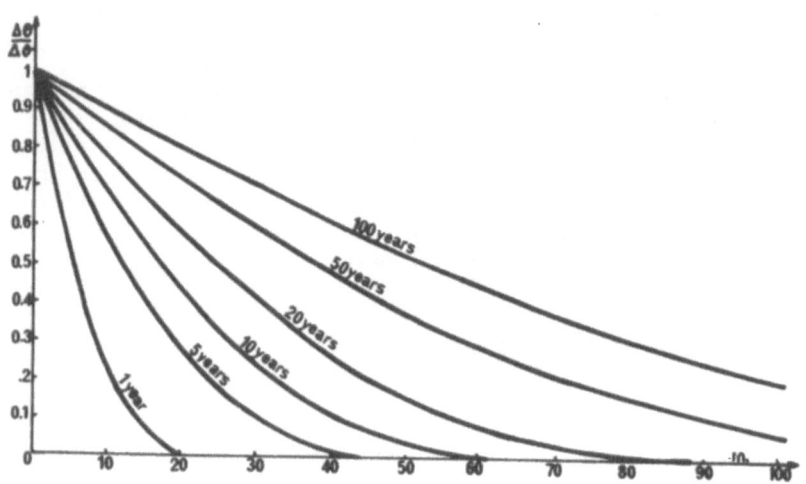

Fig.1. Temperature increase as a function of distance and time in a semi-infinite impermeable mass, one face of which is brought to a constant temperature $\theta_o + \Delta\theta_o$

Thermal diffusivity $\dfrac{\lambda}{\gamma} = 10^{-6}$ m^2/s

c) how to find a suitably situated rock ? It is particularly important to consider the ecological consequences of heating top soil, if the storage is not deeply buried.

Table 1. Thermal conductivity of common rocks in 10^{-3} kcal/m.s.°C
at room temperature

Granites	0.5 - 0.9	Marble	0.50 - 0.70
Basalts	0.34 - 0.40	Dry clay	0.20 - 0.50
Gneiss	0.46 - 1.14	Wet clay	0.60 - 0.80
Quartztites	0.87 - 1.92	Consolidated	
Schists	0.40 - 0.90	clays (shale)	0.30 - 0.70
Sandstone	0.30 - 1.	Dry sand	0.10 - 0.20
Limestone	0.47 - 0.80	Wet sand	0.60 - 0.80
Calcite	0.84 - 0.94	Rock salt	1.27 - 1.72

For comparison:

Concrete	0.20 - 0.30
Plaster	0.07 - 0.17
Glass	0.17 - 0.25
Copper	95.
Aluminium	56.
Iron	19.
Water	0.14

References: - D. Daly, Igneous rocks and the depth of the earth,
Mc Graw Hill, 1933,
- F. Grout, Petrography and Petrology, Mc Graw Hill,
1932,
- Cazal, op. cit. (13),
- BURGEAP, op. cit. (14),
- K. Schneider and A. Platt, High level waste management
alternatives, report Battelle Institute BNWL 1800,
May 1974,
- R. Weast, Handbook of Chemistry and Physics, The
Chemical Rubber Co., 1968.

3.2. If there is water

At first sight, the presence of water seems favourable: heat
can easily be introduced into the ground by circulation of hot
water. Furthermore, the thermal conductivity of rocks containing
water is of the same order of magnitude as that of dry rocks:
the ground's capacity to confine heat remains the same. But there
are obvious disadvantages connected with the convective phenomena
of water circulation:
- horizontal carrying of heat by the circulating water, if
we use a flowing aquifer,
- possible appearance of natural vertical convective motion
due to the lower density of hot water (2).

To illustrate this point, Fig.2 gives the evolution of the temperature as a function of time and distance, for one-dimensional flow of heat in a mass of porous rock in the direction of the flow of the aquifer, if one of the surfaces of the rock is suddenly brought to and kept at a constant temperature.

We used a permeability of 10^{-3} m/s, a hydraulic gradient of 0.2% (average flow conditions for an aquifer), a heat capacity of the porous medium of 600 kcal/m^3°C (see below). It is interesting to compare this graph with that of Fig.1 and to understand that convection changes the order of magnitude of the phenomena.

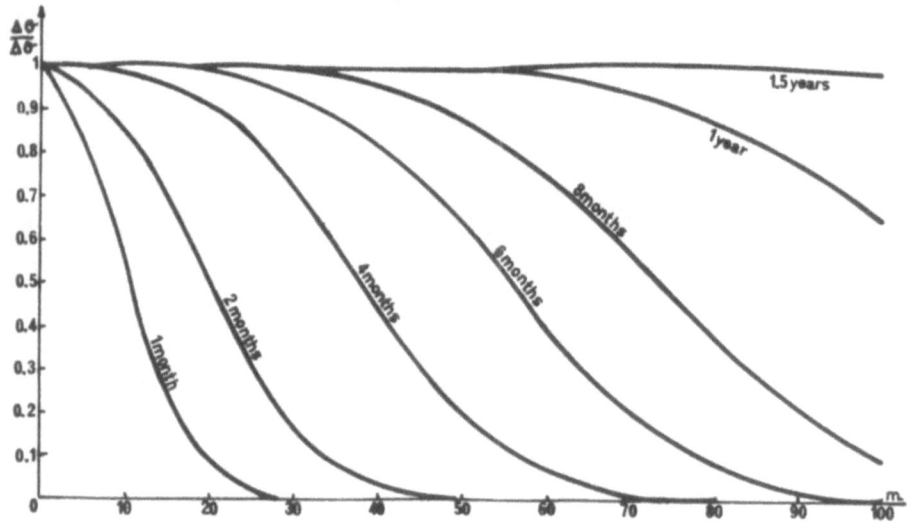

Fig.2. Temperature increase as a function of distance and time in a flowing aquifer (in the direction of flow) when one face is brought to a constant temperature of $\Theta_o + \Delta\Theta_o$.

Permeability: 10^{-3} m/s; flow gradient: 0,2 %; "equivalent" thermal conductivity: $\lambda' = 0.6 \ 10^{-2}$ kcal/m.s.°C; thermal capacity: $\gamma = 600$ kcal/m^3°C.

Water also presents another problem: if the temperature of the storage has to be higher than 100°C, the storage must be situated at a depth, where the pressure can be maintained[3]. Moreover, the chemical and biochemical balance between the water and the soil is a function of the temperature and, thus, if the temperature is changed, there may be problems of clogging or corrosion.

Thus, the problems in this case are different from those of dry rocks:

a - the introduction of the heat is easy, the water acts as a heat carrying fluid;

b - the problem of storage size remains unchanged with the same conclusion;

c - the choice of the rock is easier, as there is a relatively large number of aquifers, at least in Europe, but there may be conflicts concerning the uses of the water; the ecological problem on the surface remains;

d - but one must limit, or stop, losses by horizontal convection (natural flow of the aquifer) and vertical convection (rising of hot water);

e - finally, the geochemical and biochemical changes in the water due to the rise in temperature must be brought under control.

Before examining the technical solutions to these problems, it is useful to point out briefly how one can model the flow of heat in the ground in order to predict, by computation, the theoretical behaviour of a storage project.

4. HEAT TRANSFER IN NATURAL MEDIA

4.1. Dry rocks, i.e. where water does not circulate

The problem has been completely solved: the general heat equation is applied to these media; it writes:

$$\text{div } (\lambda \text{ grad } \Theta) = \gamma \frac{\partial \Theta}{\partial t}$$

with λ: thermal conductivity,
   Θ: temperature
   γ: heat capacity.

Analytical solutions of this equation are well known (4,5), and there are numerous computer programs to calculate solutions for complex geometries.

It is therefore sufficient to determine, on a sample or from tables of properties (tab. I), the thermal conductivity and the heat capacity of the ground.

4.2. Rocks containing circulating water

The convection phenomena in principle are described by substracting a convective term from the left hand side of the heat equation $\gamma_f \text{ div } (\vec{u}\Theta)$:

$$\text{div } (\lambda' \text{ grad } \Theta) - \gamma_f \text{ div } (\vec{u}\Theta) = \gamma \frac{\partial \Theta}{\partial t}$$

with λ': "equivalent thermal conductivity" (see below)
  γ_f: heat capacity of the fluid,
  $\vec{u}$: Darcy's velocity of the flow.

In fact, to some extent, the grains of the porous matrix play the role of heat exchanger and absorb the heat brought by the water. We assume that, at every instant, an instantaneous thermal equilibrium is established between the temperature of the fluid and that of the matrix[4], which means that we need only add to the heat equation the thermal flow brought by the circulating fluid, i.e. the difference in water flow × temperature between the entry and the exit of an elementary volume, multiplied by γ_f.

There is, however, a problem concerning the conductivity λ' which thus becomes the "equivalent thermal conductivity" of the medium. To understand this problem, we will examine in detail the physical phenomena involved.

If the aquiferous layer were homogeneous throughout its thickness, there would be no problem: the vertical thermal front would advance slowly under the influence of horizontal convection, if we ignore, for the moment, the vertical convection linked to the density differences.

The role of the thermal conductivity of the (rock + water) medium would be to propagate the heat at the passage of the front by pure conduction: instead of a sharp front, there would be a zone of temperature variation between hot water and cold water (Fig.3).

In view of the mean thermal conductivity values measured in laboratories, the transition zone would not be very wide and the phenomenon would not be significant.

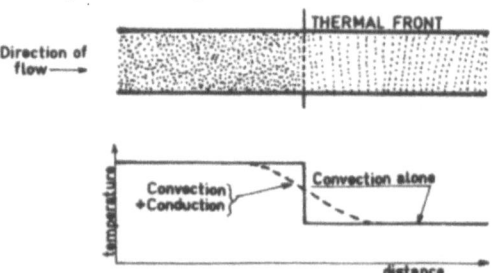

Fig.3. In an homogeneous layer, if there were only convection, the thermal front would be absolutely sharp. As there are conduction phenomena, the temperature varies progressively from the hot to the cold zone and the higher the thermal conductivities, the larger the transition zone.

On the other hand, let us consider a formation with vertical heterogeneity: e.g. a less permeable layer separating two more permeable and different layers. In each layer, the water - and therefore the thermal front - advances at different speeds. At any given time and without conductivity, the mean temperature of the layer will have the staircase form shown by the continuous line in the lower part of Fig.4:

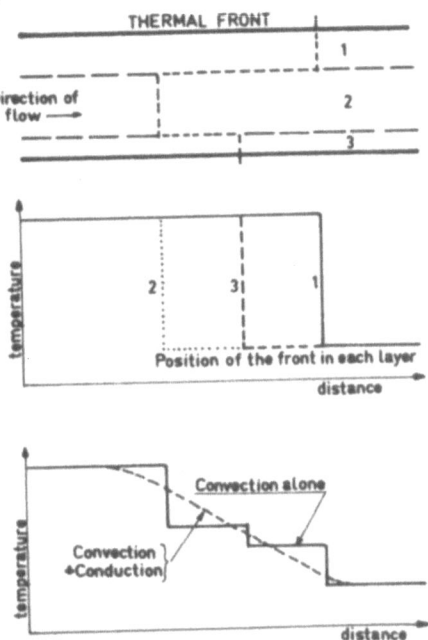

Fig.4. In an heterogeneous layer, each bed behaves accordingly to its own permeability and, without conduction, the mean temperature curve of the layer would have a staircase form. But because of the conduction phenomena, both horizontal and vertical, the mean temperature develops progressively (dashed curve).

We now add thermal conductivity:
- horizontal: each front in each layer will be blunted as in the first example;
- vertical: vertical heat exchange between layers will tend to attenuate the phenomena (heating the cold layers and cooling the hot layers) again smoothing the appearance of the mean temperature curve of the layer (dashed curve in the lower part of Fig.4).

If we now look at the dashed curves of Fig. 3 and 4, we see that, in reality, the heterogeneity of the aquifer gives rise to a development curve of the mean temperatures which can be represented by a much higher "equivalent thermal conductivity" than that, stricto sensu, measured in a laboratory.

The role of the heterogeneities, or of the thermal equivalent conductivity, is therefore to progressively enlarge the transition zone between hot and cold water as the thermal front spreads.

A plane heterogeneity (more permeable zone, fissure directly joining the injection wells to the pumping wells) would play exactly the same role.

As one does not know, or would have great difficulty in evaluating, the heterogeneity of the underground, one may have to base the predictions of the velocity of the fronts on relatively high thermal conductivity coefficients, assuming, furthermore, that the medium is homogeneous. One therefore tries to measure, in situ, this equivalent thermal conductivity, by heat tracing experiments. It seems indeed that this conductivity is a linear function, as a first approximation, of Darcy's velocity as is accepted in the theory of dispersion.

A last problem concerns the calculation of Darcy's velocity, $\vec{u}$, of the flow: whether or not to take into account the influence of the temperature variation of the water on its viscosity and density. The fact that the density of the injected water is lower modifies the aspect of the flow network. In fact, a vertical component due to gravity, following the generalized Darcy's law:

$$\vec{u} = - [\frac{K}{\mu} \, \vec{grad} \, p + \rho g \, \vec{grad} \, z]$$

(see notation below)
is superimposed on the gradient of the horizontal flow.

The hot water, therefore, has a tendency to rise to the top of the layer, which means that all the cold water in the area will not be forced out by the injected hot water. Some colder water will stay at the bottom of the layer, and may be repumped when the storage is emptied.

This parameter seems, however, to be less important than accounting for the heterogeneities. Indeed, the vertical movement of the hot water will at first be greatly reduced by the vertical/horizontal anisotropy of the permeability, which one is quite likely to find when storing in sedimentary layers. Morevoer, once this segregation has occured, the transversal conductivity will also contribute to equalizing the temperatures and therefore the densities.

A priori, the viscosity has an unfavourable effect on the injection and a favourable one on the recovery. Indeed, at the injection, the hot water, which is less viscous, circulates more easily than the cold water and may create fingering phenomena at the contact with the cold front, fruther increasing the longitudinal dispersion across the front. On the other hand, when recovered, the hot water will circulate towards the well more easily than the cold water: the heat recovery will be larger than that predicted without taking the variation in viscosity into account.

It is possible to account for these two effects in the calculations: one simply solves the continuity equation in porous media, known as the diffusion equation, which, at every moment, provides the velocity, $\vec{u}$, which must be introduced into the convective term of the heat equation:

$$\text{div } (\rho \vec{u}) + \frac{\partial (\rho \omega)}{\partial t} = 0$$

$$\vec{u} = - \frac{\overline{\overline{K}}}{\mu} \, (\text{grad } p + \rho g \, \vec{\text{grad }} z)$$

or again:

$$\text{div } \frac{\overline{\overline{K}} \rho}{\mu} \, (\vec{\text{grad }} p + \rho g \, \vec{\text{grad }} z) = \rho \omega \, (\beta_\ell - \beta_s + \frac{\alpha}{\omega}) \, \frac{\partial p}{\partial t}$$

where ρ: mass per unit volume of fluid,
  ω: porosity,
  $\vec{u}$: Darcy's velocity,
  $\overline{\overline{K}}$: intrinsic permeability tensor,
  p: pressure,
  g: acceleration due to gravity,
 $\vec{\text{grad }} z$: vector of component 1 on the vertical direction, directed upwards,
  β_ℓ: compressibility of the fluid,
  β_s: compressibility of the solid grains,
  α: compressibility of the porous matrix.

The complete solution of the problem of heat transfer therefore requires the solution, in time, of two series of equations:
- the diffusivity equation which gives the velocity $\vec{u}$ at time t;
- the convection-conduction heat equation, which gives the temperature Θ, from which the new values of ρ and μ are deduced, to solve again the equation of diffusivity, for time $t+\Delta t$.

However, because of the unwieldiness of these calculations, one often prefers to ignore the effect of density and viscosity variations and solve the diffusion equation once and for all. This generally amounts to solving a steady state equation if the injection rate is constant.

5. THE IDEAS, THE EXPERIMENTS, THE PROJECTS

5.1. The ideas

Already in 1958, P. Laffitte (7), studying heat propagation
in rocks, proposed that thermal energy be recovered from a nuclear
reaction set off underground. Water would be injected into boreholes and steam leaving the heated zone would be recovered. This
was a sort of artificial geothermal project.

In 1965, G. Brun (8) proposed that energy from a thermal
solar power plant be stored in a limestone or dry granite massif
at 500°C. The solar station, made up of cylindroparabolic mirrors,
produced steam at 230 atm. and 500°C. This steam was sent into
wells with two concentric steel casings, which made closed circuit
operation possible (Fig.5).

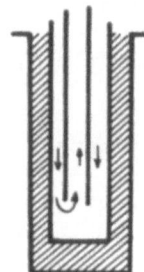

Fig.5. Sketch of the injection of steam using two concentric
steel casings proposed by G. Brun. To ensure a better
transmission of the heat in the ground, the space
surrounding the casings is packed with rock-salt, or with
cement with iron filings, or with compressed metallic
shavings.

For a 250 MW station, he planned to heat a massif 200 m high,
300 m in diameter, equipped with 9000 wells 3 m apart. The steam
would circulate in series from the centre towards the periphery
for storage, and in the opposite direction for reextraction. He
arrived at a storage cost of $ 0.001 per kWh and an insignificant
thermal loss at the boundaries. The economic calculations would
obviously need revising today.

The great problem with this project, as pointed out by the
author himself, was the appearance of thermal constraints in the
rock, linked to the increase in temperature of the massif. These
constraints could cause casings to broke due to expansion. We
will see this problem of thermal constraints in the medium reappear
later.

Finally, the idea of injecting hot water into an aquifer and

repumping is very old; a patent in this field was taken out by
M. Georig in Switzerland as early as 1946, or more recently by
Harris et al. (9) and often reproposed, in particular by D. Todd
in the United States. He proposed, with C. Meyer (10), in 1973,
the injection of water under pressure at about 150°C into a modera-
tely deep confined aquifer, e.g. between 100 and 200 m deep. The
hot water was produced by an ordinaty thermal power plant operating with a cold source at 150°C. When repumped, the hot water
provided space heating during the winter. At the same time, this
solved the problem of thermal pollution by the plant. For a storage
of 340,000 m³ of water, the author predicted, after 90 days storage,
a recovery rate of 70 to 80 % of the heat injected, following a
period of 5 years during which the storage aquifer was brought to
the proper temperature. The thermal calculations were, however, not
very sophisticated and the aquifer was supposed to be immobile.
As the aquifer was confined, the hot water was enclosed by impervious upper and lower layers.

Let us, finally, mention the exploratory study carried out
for the French Agency for Energy Savings by M. Latil at the
French Petroleum Institute in 1976 (11) on the same type of device
as that described by Todd; immobile confined aquifer, single
injection and recovery well. Latil uses a more elaborate digital
model than Todd, but ignores the influence of the variation of
density and viscosity in the water. He calculates the loss of
heat into the confining layers according to the conduction of heat
in impermeable media. Three injection temperatures have been tested
(90°C, 130°C, 180°C) with a minimum withdrawal temperature. As it
is intended for space heating, the storage receives calories during
150 days in summer and restores them during 125 days in winter
with two 45-day periods of inactivity in midseason. The efficiencies
are calculated as the ratio of the heat injected to the heat withdrawn, with a chosen reference temperature of 60°C. These efficiencies are between 40 % and 50 %, once the storage has reached the
proper temperature (5 to 10 years). The efficiency increases, of
course, with the volume stored and the thickness of the aquifer
(10 or 20 m, which reduces the contact surface for an equal
volume). The author considers storages of 200,000 to 500,000 m³
of hot water, i.e. several tens of millions of megacalories.

5.2. The experiments

As far as dry rock is concerned, few experiments have been
carried out so far. The only large project which can be mentioned
in this context is the geothermal test in hot dry rocks carried
out at Los Alamos (New Mexico) by Doe (12). Strictly speaking,
this is not a heat storage, but rather a geothermal experiment, as
the rock is naturally hot. But the problem of extracting the heat
is similar. The idea tested was to drill two slanted deep wells,
30 m apart, with controlled direction 3 km down into a hot

crystalline rock (granite) and to try to join the two wells by
hydraulic fracturing (Fig.6).

It is hoped that once the hydraulic fracturing has been set
off, it will be possible to force water,heated by contact with
the rock, to circulate from one well to the other. Furthermore,
one assumes that the thermal contraction shock of the rock caused
by the contact with the cold water will generate secondary fissu-
ring, thus extending the operation zone of the device.

Considerable sums have been invested in this project. Finally,
after much difficulty, a fissure made by fracturing from the
first well was reached by deepening the second well, thus establishing communication between the two holes. A thermal production of
5 MW at a temperature of 132°C was obtained for 3 months in 1978
with a constant improvement in power. But the existence of secondary thermal fracturing has still to be proved. The method of heat
recovery from a hot rock, which is not naturally fissured, has
still to be developed.

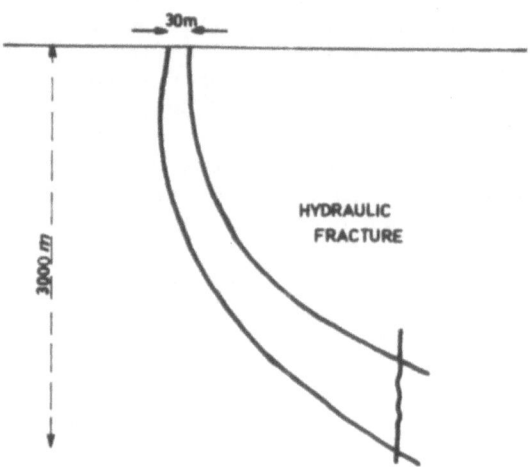

Fig.6. Diagram of the device put into operation in New Mexico
       to extract heat from dry rocks: two 3,000 m deep oblique
       wells were bored 30 m apart, fracturing of the rock was
       induced and water is heated while circulating between the
       two wells.

In 1980, hydraulic fracturing between two wells was obtained
in granite in the "Mayet de Montagne", France, by Institut National d'Astrophysique et de Géophysique and Institut de Physique du
Globe de Paris (M. Cornet), at 200 m depth, and 30 m distance. In
an experiment of heat storage in this single fracture performed
by Ecole des Mines, it took two months for the heat injected in

one well to be detected at the other, at an average inflow of 1.5 m^3/h, in this single fracture.

As to the injection of hot water into aquifers, at least half a dozen experiments have already been completed. In 1959 and 1960, A. Cazal (13) and H. Schoeller at St-Girons made a first experiment injecting hot water into the aquifer in the sands of les Landes for 7 months. Actually, the problem was rather that of disposing of the hot cooling waters from an industrial plant while protecting the aquifer. Before 1973, many studies and projects concerning the reinjection of cooling water were carried out without any storage objective.

At S-Girons, 11 m^3/h of water at 43°C were injected for 7 months into an unconfined aquifer at a depth of 16 m with a natural temperature of 13°C. The injection was not followed by repumping, but Cazal noted and explained by calculation the differences in behaviour between the wells (20 m from the injection point) in the flow direction of the aquifer and in the orthogonal direction (27° and 22° respectively). He also drew attention to the importance of the geochemical studies, as an increase in temperature of the water could lead to precipitation and clogging.

In 1973, BURGEAP carried out a first experiment of heat storage[5] by injecting into and pumping out 8,000 m^3 of water from the Albian aquifer at a depth of 600 m, at Noisy. Although originally not planned for storage, the experiment was interpreted by Bourguet, Clouet d'Orval (BURGEAP) and Ledoux (School of Mines) (14,15), who obtained very important information on the "equivalent conductivity" of porous media which we have discussed above.

Let us mention the experiments conducted by B. Mathey (16) in 1974 at Neuchâtel, Switzerland, where 500 m^3 of water at 51°C were injected for 9 days into an unconfined aquifer in a 10 m thick alluvial deposit. After 4 months, 16,000 m^3 of water were repumped, recovering 40 % of the heat injected with a reference temperature of 11°C. But the temperature at the pumping was very low, of the order of 14°C; for the first 5,000 m^3 of recovered water, the efficiency was less than 7 %. Mathey indicated a very important density effect in his experiment: the hot water had spread over the surface of the aquifer without penetrating it. Furthermore, a significant part of the heat escaped to the ground surface through the unsaturated layer. Finally, the "bubble" was moved laterally downstream towards Lake Neuchâtel, a dozen meters in 6 months.

Apart from this experiment, there are now several heat storage projects in Switzerland, in connection with releases from power plants.

In Germany, Key and Nieskens (17) carried out a similar experiment at Krefeld, in 1974, this time in a superficial aquifer 3.5 m thick, but confined by 0.75 m of silt. In two months, they injected 430 m^3 at 45°C, and observed the temperature changes without repumping. No efficiencies were given, but the authors did not find the important density effects cited by Mathey, their "bubble" of hot water remained almost cylindrical.

A similar experiment was carried out in France as part of a joint action by the General Delegation for Scientific and Technical Research (DGRST), at Bonnaud in the Jura, by BRGM, BURGEAP, CEA-CENG and the School of Mines, Paris (18) to measure the thermal parameters of an aquifer.

They also worked on a small superficial aquifer, 3 m thick, in sand, confined by 4 m of clay. As a last experiment, 1,400 m^3 of water at 40°C were injected for 20 days, then repumped after an interval of 4 months. In this aquifer, where flow rate is very low, there was no significant lateral displacement of the hot water or segregation of the heat towards the top of the aquifer. However, in 4 months, there was considerable heat loss, as the mean temperature of the repumped water was 16°C for a mean water temperature of 12°C. With a reference temperature of 12°C, the quantity of heat recovered was 30% for a pumped volume of 3,000 m^3.

To conclude, les us mention the experiment of Auburn in the United States, conducted in 1976 by the US Geological Survey and the University of Alabama, where cooling water from a power plant was injected into a confined aquifer at a depth of about 30 m. Unfortunately, this river water contained fine clay particles in suspension which clogged the well very rapidly, thus stopping the injection. There again, the effect of density was mentioned in the interpretation of the temperature measurements, but it was attenuated by a 1/10 anisotropy in the permeability.

In 1978, the experiment was resumed, this time by injecting high quality water pumped in a shallow aquifer (22). 55,000 m^3 of water at 55°C were injected in 79 days, followed by 51 days of storage, and then by a production of 55,000 m^3 of water, in 41 days. The temperature of the produced water decreased from 55°C to 33°C, and 65% of the injected thermal energy was recovered. This time, vertical thermal convection was not significant, but some clogging of the injection well occurred again, as the clay particles of the aquifer swelled because of the different chemical composition of the injected water.

At last, in 1981, the experiment was performed again, but with a doublet of wells pumping and injecting into the same aquifer. This time, no problems of clogging occurred.

All these experiments lead us to three conclusions:

1) For storage in an aquifer, a minimum size must be attained, below which heat losses are too great. This minimum volume amounts to, at least, thousands of m^3.

2) The injected water must be very clean to avoid clogging of the wells, and preferably from the same aquifer, to avoid geochemical clogging.

3) The influence of water density must be reckoned with, since it can spread the "bubble" of hot water over the surface, as well as the natural circulation of the aquifer which may transport the bubble downstream.

Concerning dry rock, the only existing experiment known at the moment is that of J. Guimbal at St-Etienne. It dealt with heat storage for space heating by circulation of water in wells, quite similar to the idea of G. Brun. Here, the storage is made up only of the heat introduced by conduction into the impervious rock.

The selected site was on a hill, with crystalline rocks. Unfortunately, when the boreholes were drilled, it was discovered that the rock was not as impervious as expected, because of numerous fractures. This should indeed be considered as the normal situation in any compact rock, especially near the surface. Grouting was attempted, but it drastically increased the cost of the project and was not 100% effective. As a result, most of the heat stored was leached away by water flowing through the fractures.

As examples of storage in aquifers, we will present four French studies: the project of the Rhône-Progil building, carried out by the Serete and the School of Mines (19), the projects of storage in captive aquifers of the CEA and the SNEA (20) at Provins, and of the BRGM, COFRETH and Trépaud, at Blois (29) and finally the experiment on heliogeothermy carried out at Nîmes by Electricité de France and the School of Mines of Paris, for the Plan Construction (21). We will finish with examples of natural storage of energy in the ground.

a) Storage in a water-tight enclosure: The Rhône-Progil building, built at Neuilly on the alluvial deposits of the Seine, is heated and cooled mainly by the recovery of energy from artificial lighting, using refrigerating units functioning as heat pumps. An electric boiler operating during slack hours provides extra energy, if needed. Devised by the Serete, this heating system already uses a reservoir storage system by day, but would be more efficient if it could store excess calories from summer air-conditioning for use in winter[6]. The temperatures of the cooling and warming water are very low (25-40°C) and thus especially well suited for storage.

Furthermore, the foundations of the building consist of a dug-in concrete wall, which goes through the alluvial deposits of the Seine down to the clays of the Sparnacian, thus hydraulically isolating an aquifer layer of 20 m, composed of Cuise sands and Lutetian limestone. The thermal study of the heat storage in this isolated layer was made by M. Lebrun in 1975 (19). The obvious advantage of the arrangement was the absence of horizontal convective displacement of the bubble because of the concrete walls. The thermal and economic evaluation of the storage was soon found to be very positive: on a storage of about a million megacalories, there was a recovery rate higher than 50% already in the first year, and it could be increased later. The savings in fuel provided by the storage was 5% of the present consumption, for a building which is already thermally efficient. Finally, the cost of the installations could be paid off in about 5 years.

Unfortunately, the project could not be completed as there were major problems with expansion of the concrete wall: as it was not designed for heat storage, the concrete wall did not have expansion joints (which it would have been technically possible to put in when the wall was constructed) and the increase in temperature could have led to expansion stresses of the order of magnitude of the traction strength of the concrete: the project thus threatened the stability of the building and, despite its favourable thermal aspect, could not be pursued further.

Proposals have been made to build similar storages in aquifers with some kind of hydraulic barriers to stop the natural circulation of the aquifer and the transport of the hot bubble downstream. So far, these projects (dug-in concrete walls, grouted curtains,....) are too expensive if these barriers are built only for the benefit of the store.

b) Storage of overheated water in a confined aquifer: This project, developed by the CEA and the SNEA, was presented by J. Despoix and F. Nougarede (20). The heat source could be, e.g. a purely heat-generating nuclear reactor, intended for urban heating which would be operated all year, if possible, storing the heat in summer in a confined aquifer 30 to 40 m thick, at a depth of 600 m. The storage volume quoted in the example is already large: 500 millions megacalories, i.e. 5 millions m^3 with·a $\Delta\theta$ of 100°C (between the input and the return of the network) and a maximum recovery flow of 2,000 m^3/h. Three hot wells are necessary, with six cold peripheral wells and six small control wells. The diameter of the storage is said to be about 500 m. The cost of the project is estimated at $ 7,000,000 and the thermal efficiency, after the project has reached its projected rate of operation, in a few years, would be 70 to 80%.

The project considers the existence of a "tepid" crown

surrounding the storage, created by the return water, which would act as an insulator between the central storage and the cold exterior layer. The authors have submitted a patent application for the technical devices, which make it possible to control the pressure and the flow of hot water in the system without causing water vaporisation.

The authors justly insist on the need for a large storage, without which the thermal losses would be too large, and on the need to take into account the variation in density and viscosity of the water in the calculations, even though these calculations are very lengthy. Such a model has been developed at the ELF-ERAP computer center, but will only be used for designing the final project. The implementation of this project is now being considered near Provins (Seine & Marne). The heat would be generated by a large garbage incinerator.

c) Injection of hot water in a tepid aquifer: At Blois, the Bureau de Recherches Géologiques et Minières, the COFRETH and the Sté des Echangeurs TREPAUD (29), are carrying out a feasibility study of storing hot water in a deep aquifer, namely the Lusitanian, at 500-600 m. The normal temperature of this aquifer is already in the order of 30°C. Hot water (e.g. 60°C) will be produced in summer initially by a garbage incinerator, later also by solar collectors.

The water is pumped at, say 30°C from the same aquifer, heated at 65°C, and reinjected, with a doublet of wells. In winter, the same doublet is used to provide a cold source for a heat pump for space heating, while the garbage incinerator provides heat directly to the same housing units. But the originality of this system is that the operation of the doublet is not reversed: water at 30°C is extracted from the same well as in summer, cooled at e.g. 10°C by the heat pump, and reinjected into the well where hot water was injected in summer. The aquifer serves as a "mixing well"; it is said that by careful selection of the distance between the wells, the "waves" of hot and cold water will mix, because of thermal conduction and hydrodynamic dispersion in the medium. If the energy injected and extracted annually is balanced, the temperature of the producing well should be kept almost constant indefinitely.

In such a project, the water injected at 65°C is recovered at 30°C, but globally, there should be no heat losses, and furthermore, the operation of a doublet always in the same direction makes the installation much simpler.

d) Heliogeothermal experiment in Nîmes: Strictly speaking, this is the first experiment in the world, where a large quantity of heat is stored and used in an aquifer. This project is managed jointly by Electricité de France (EDF), Direction des Etudes et Recherches, Département Application de l'Electricité and the Centre d'Informatique Géologique of the School of Mines of Paris, for the Ministère de l'Equipement, Plan Construction (21).

The Plan Construction is interested in the problem of heat storage in the ground for space heating purposes, with the particular objective of using solar energy.

EDF joined this project because it offers a means of paying off, over long periods, and with the best rate for the consumer, the heat pumps (and the power lines) installed for heating (horticulture greenhouses, floors heated by low temperature water).

A heliogeothermal project, therefore, consists in using solar collectors to provide hot water in summer, which is injected into an aquifer, thus creating an artificial geothermal storage for use during the following winter.

This experiment began on the 4th of July 1977, in the unconfined aquifer of the Costières du Gard, near Nîmes. The aquifer is situated in the large alluvial deposits at a depth of between 2 and 10 m with a natural temperature of 14°C. From the 4th of July to the 30th of September 1977, 20,000 m^3 of water at about 33°C were injected into the aquifer at a rate of about 200 m^3 per day, creating a storage of about half a million megacalories. Six weeks later, the first production began and lasted from 8th of November to the 19th of December 1977, at a rate of 150 m^3 per day.

The temperature of the recovered water decreased from 33°C to 22°C, with a ratio of recovery of 13.5% of the injected heat. The heat was used for heating greenhouses.

A second waiting period lasted from the 19th of December to the 15th of January, as the greenhouses were not used.

The second production period lasted from the 15th of January to the 15th of March, at a rate of 200 m^3/day. The temperature decreased from 18°C to 15°C, with a heat recovery ratio of 5%.

On the whole, the heat recovered amounts to approximately 20% of the injected heat; this somewhat low efficiency was mainly caused by the proximity of the water table to the ground surface (2 m in summer, 1 m in winter after infiltration of rainfall), created by very heavy rainfalls in 1977 and 1978.

The hot water injected was pumped out of the same aquifer through a well 200 m away from the injection wells; the heat came partly from the use of an electric heat pump during slack hours at night, using, for its cold source, a second well in the same aquifer at some distance.

In winter, the recovered water was used to heat 3,800 m^2 of greenhouses, by circulating it through plastic sheeting placed on the ground[7]. This sheeting was used in summer as rudimentary solar collectors inside the greenhouses.

Depending on the thermal needs of greenhouses and their temperature, there seems to be three possible uses of water from the storage:
- direct use by heating the sheeting in the greenhouses,
- use after further heating of the water by heat pumps,
- indirect use on the evaporator of the heat pump, to increase its efficiency by raising the temperature of the cold source.

Calculations showed a reduction of about 1/3 in the cost of heating the greenhouse by this process as compared to warm air blowers using ordinary fuels, for a thermal efficiency of 1/2 of the storage. The cost of solar calories is in fact very low in this device.

Interpretation of the observed temperatures in a dozen observation wells was made on a digital model of the system. The following figures were obtained, explaining the heat losses:

heat loss at the boundary of the storage:	35%
heat loss towards the atmosphere:	33%
heat loss by convective transport by the aquifer:	6%
heat loss towards the substratum of the aquifer:	3%
heat loss by infiltration of the cold rainfall:	3%
heat recovered:	20%
	100%

By reducing the heat loss towards the atmosphere, an efficiency of 50% could easily be achieved. This can be done by using a deeper aquifer, or even by artificially insulating the ground over an area of 8,000 m^2 above the storage.

There was neither drastic influence of vertical or horizontal thermal convection, nor clogging. However, the injected hot water had a very high bacterial content, and had to be treated by chlorination.

Fig. 7 and 8 show typical temperature profiles in the aquifer at various times during storage and recovery, at 20 m from the injection center. Fig. 9 shows the measured and computed

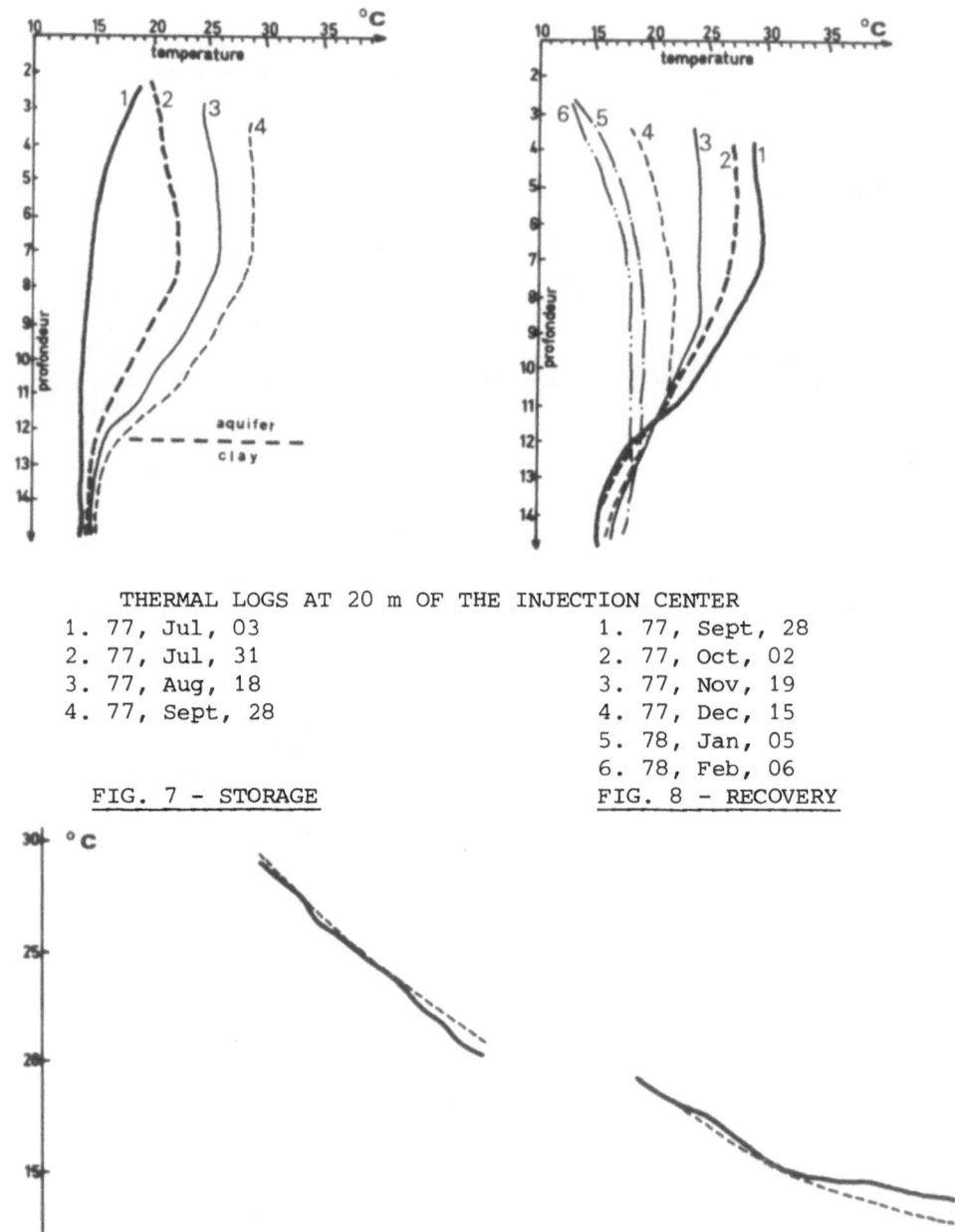

THERMAL LOGS AT 20 m OF THE INJECTION CENTER

1. 77, Jul, 03
2. 77, Jul, 31
3. 77, Aug, 18
4. 77, Sept, 28

1. 77, Sept, 28
2. 77, Oct, 02
3. 77, Nov, 19
4. 77, Dec, 15
5. 78, Jan, 05
6. 78, Feb, 06

FIG. 7 - STORAGE

FIG. 8 - RECOVERY

———— measured in situ

--- computed with correction of the losses by natural convection

FIG. 9 - RESTITUTION CURVES

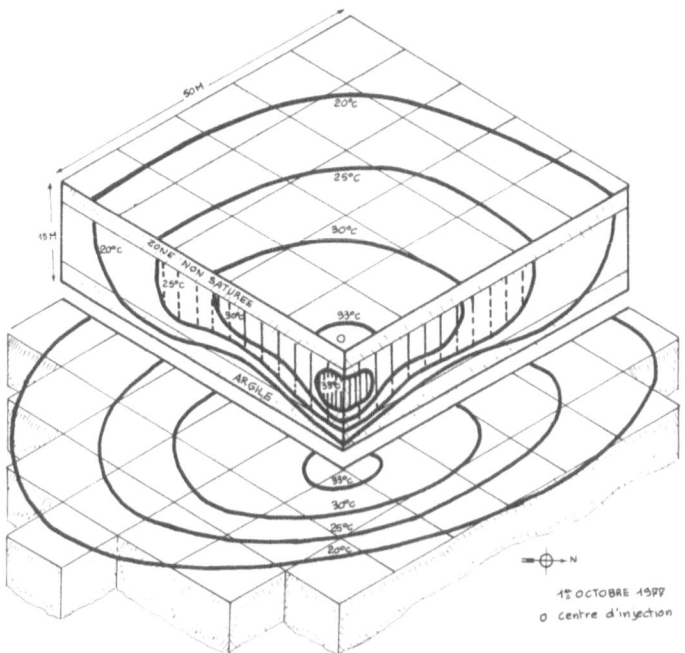

Fig. 10 - Spatial distribution of the temperature in the
         aquifer, Oct 1, 1977.
         O is the injection center.

Fig. 11 - Storage area.

Fig. 12 - Inside a greenhouse, with the plastic sheeting on the ground (heaters and also solar collectors).

Fig. 13 - The diesel heat pump used.

Fig. 14 - Observation well and equipment for temperature logging.

temperature of the pumped water during the recovery phase. Fig. 10 shows a 3-D representation of the spatial distribution of the temperature in the aquifer at the end of the injection period (September 1977). Fig. 11 through 14 give pictures of the site, the greenhouses, the heat pump, and an observation well with measuring equipment.

e) An alternative for using solar heat: the heliogeothermal doublet of interseasonal recharge: We have seen that one of the difficulties of heat storage is that the temperature of the recovered water drops during the pumping period. It is also evident that:
- if the water is stored at high temperature (e.g. 80°C), the solar collectors will have to be very sophisticated and expensive;
- if the water is stored at low temperatures (e.g. 40°C), heat pumps are necessary to make use of this heat, especially at the end of the recovery period. Heat storage merely serves to increase the temperature of the cold source of the heat pump, thus increasing its efficiency.

On the other hand, the water contained in shallow aquifers at normal temperatures (i.e. 10-15°C) represents a very valuable energy source that can be utilized with heat pumps (23). Several modes of utilization are possible:
i) withdrawal of the water from the aquifer, and disposal of the water cooled by the heat pump in the drainage network; this solution is only acceptable in areas where the water resource is large, and the users far apart, otherwise the water resource will soon be depleted;
ii) reinjection of the cooled water into a second well, creating a "geothermal doublet"; the water resource is conserved, but the aquifer is progressively cooled. The two wells have to be placed far apart (e.g. several hundred meters), or the cool water will soon by recycled, and the efficiency of the system will drop (26,27). Fig. 15 shows the drop of efficiency of such a system, as a function of time and of the distance between wells, when there is no natural circulation (zero hydraulic gradient) in the aquifer. Calculations were made for a 20 m thick aquifer, with an annual extracted volume of 120,000 m^3. Fig. 16 shows the temperature distribution in the aquifer after 10 years.

In cases where there is a strong hydraulic gradient (e.g. a water velocity of 10 cm/day, and where the producing well is placed upstream of the injection well) the cold water will not be recycled (Fig. 17), but a very large portion of the aquifer will be cooled (10 ha in 30 years for heating 200 housing units). Such a system is thus totally unacceptable in an urban area.

iii) heliogeothermal doublet of interseasonal recharge. The method that we propose (patent by P. Iris, 25) works as shown on Fig. 18 (24,26,27).

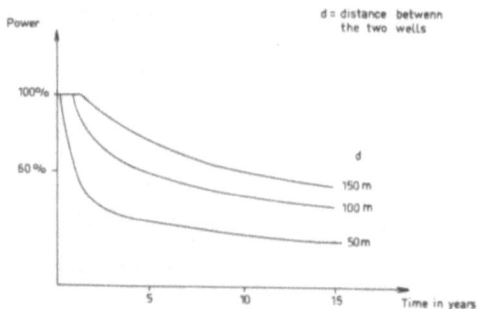

Fig. 15 - Evolution of the power at the producing well.

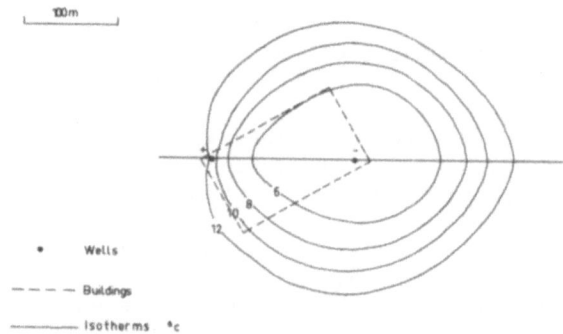

Fig. 16 - Thermal field in the aquifer after 10 years, with no
regional circulation.

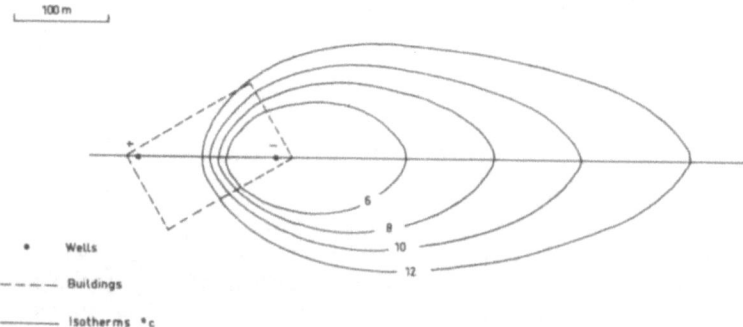

Fig. 17 - Thermal field in the aquifer after 10 years, with a
10 cm/day natural water velocity in the aquifer.

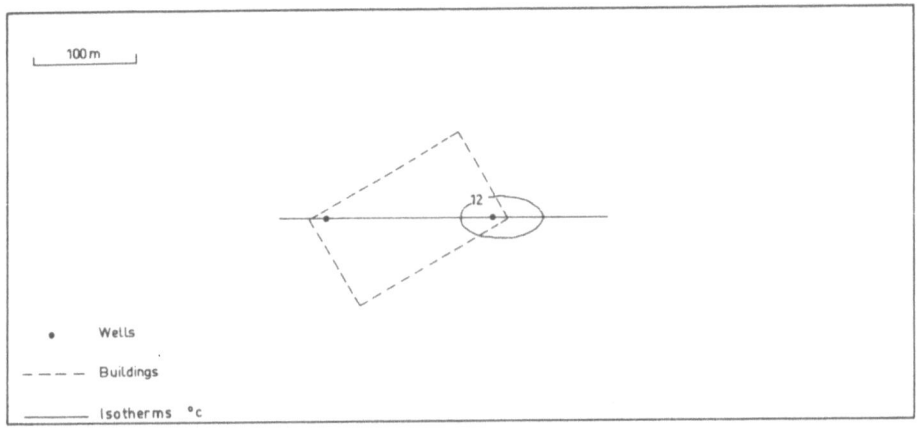

Fig. 18 - The heliogeothermal doublet, winter and summer.

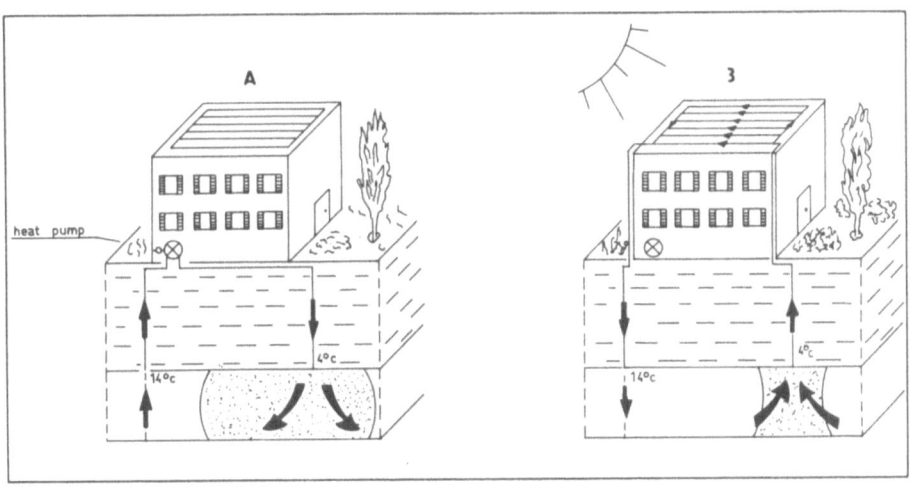

Fig. 19 - Thermal field in the aquifer after 10 years,
heliogeothermal doublet with no regional circulation.

- In winter (A), the water is pumped at the natural temperature of the aquifer (10-15°C, according to depth and location), and reinjected after cooling by the heat pump (4°C).

- In summer (B), the "cold bubble" is pumped back, heated by aero-solar collectors working at a very low temperature, and injected into the producing well at the initial temperature of the aquifer (10-15°C).

This method has the following advantages:
- Conservation of both the water resource and the energy resource of the aquifer, indefinitely. Fig. 19 shows the size of the cooled area after 10 years in the case of negligible hydraulic gradient.
- Small distance needed between the two wells, compatible with land owernship in urban areas.
- Behaviour of the energy source at a constant temperature throughout winter, making it possible to design a simple heating system with heat pumps.
- Collection of solar heat at very low average temperature (10°C), below the air temperature, using very simple collectors (e.g. array of pipes on a roof without any glass cover) with a very high efficiency. Thus a large amount of heat is extracted simply from the air.
- Feasibility exists at any scale, from individual houses to large-scale programs (whereas heat storage is only feasible on a large scale).

This method is being implemented in 1981 in France, at Aulnay-sous-Bois, near Paris, in an aquifer lying at 60 m below ground. This project is funded jointly by the Commission of the European Communities, the Commissariat à l'Energie Solaire and the Plan Construction, Ministère de l'Environnement, in France, and involves the Ecole des Mines, the Cabinet Antonelli and the Housing Company FFF. 220 housing units will be heated in this way, starting in the fall of 1982.

Estimated efficiency of the systems shows a primary energy saving of 70% as compared to electric direct heating (50% as compared to oil heating) at a global cost of the same order of magnitude as usual solutions (oil, electric heating). A detailed energy balance and economic evaluation of the system will be published after the first year of operation.

f) A natural storage: heat pumps with the ground as a cold source: By burying a simple plastic pipe about 1 m underground, through which a heat transporting fluid is circulated, a cold source can be created for a heat pump. Of course, this buried pipe must be of a certain length to provide a certain amount of power, e.g. a few hundred meters for the heating of a single family residence. Such a device would also act as a thermal storage of very low temperature: in winter, one extracts (at 0°-15°C) the solar heat naturally stored in the ground during the summer.

The AGA company in Sweden has been marketing such a device for some time and about 1000 of them were said to be in operation in 1977. EDF in France is also studying this type of heating.

However, with a cold source at such a low temperature (the ground may even freeze in winter), the Carnot efficiency of the heat pump is rather low (e. g. a coefficient of performance of 2 or lower), and the energy gains are small if not zero. This is why artificial storage of heat in the ground by one of the methods described above seems preferable in order to noticeably increase the thermal level of the storage, even when it serves as a cold source for a heat pump.

Recently, ELF-France (28) has built and experimented a solar house in Solaize (near Lyon), using the same concept. But the heat pump and pipes are associated with low temperature solar collectors placed under the tiles of the roof, using the soil as a daily (or weekly) storage. The solar collectors in the roof are used directly as a cold source for the heat pump in the daytime, whereas the soil is the cold source by night. But the extra heat collected by the roof is also injected into the soil through the pipes, as soon as the temperature in the solar collectors exceeds that of the soil. In this way, it is possible to avoid freezing of the soil, and to keep the global coefficient of performance of the system between 3 and 4.

CONCLUSION

The storage of heat in the ground appears promising, because it provides:
- storage of very large quantities of heat at reduced costs (about ∮ 5/useful gigacalories in capital return and operation cost in the Nîmes experiment);
- a possibility of combining it, very usefully, with solar energy which can hardly exist without storage and which, at present, can only with difficulty pay for the cost of artificial storage;
- the storage of continuous or intermittent recovery energy thus avoiding the release into the environment of harmful quantities of heat.

The problems posed by this type of storage are: first, the determination of the critical size below which the heat losses are too great; this size is certainly in the order of millions of megacalories. Then, there are the limitations of the convective shifting of the "bubble" of hot water, if storage is made in an aquifer with non-negligible natural movements, and the biochemical reactions of the heated water in the ground. Finally, if the storage is shallow the tolerance of the ground surface to the increase of temperature must be examined, as well as the compatibility between the local use of aquifers for thermal needs and other traditional uses.

Storage of energy at present is economically appealing for "lost" energy sources in summer (e.g. combustion of urban refuse). For purely solar energy, the economics depend heavily on the other components of the system (collectors, heat pumps, method of heating). Thus, it seems that heliogeothermy is often more appealing than storage of solar heat: calories are "borrowed" from the ground (aquifer, soil,....) at low temperature by means of heat pumps, and solar energy is only used to return to the system, at low temperature, the quantity of energy necessary to prevent it from deteriorating, as the ground is deprived of its natural heat. Such a system has no heat losses, but on the contrary, heat "gains" from the natural medium.

NOTES

[1] not always: certain famous electricity blackouts in New-York or in Paris showed how the difficulty of storing energy makes our distribution systems precarious.

[2] the thermal diffusivity is the ratio between the thermal conductivity and the heat capacity of the medium.

[3] an increase of temperature in the ground creating an overheating of the water in a zone, where the permeability is too low, may, under suitable pressure conditions, lead to a sudden adiabatic release with vapourization (phreatic explosions). This phenomenon is well known and has been suggested as an explanation of the eruption of the Soufrière volcano, in 1976. It is also one of the possible explanations, with radiolysis, of the nuclear catastrophe reported by Medvedev (3), which is said to have occurred in the USSR in 1975, and where very large quantities of nuclear waste stored in the ground allegedly was spread out over an area or several thousands of km^2 following an explosion.

[4] one may easily show (6) that, due to the division of the porous medium, this time is indeed very short, of the order of one minute for a grain of 1 mm, of 2 hours for a fissured medium with 10 cm blocks.

[5] in fact, cold water was injected into a hot aquifer, which is thermally equivalent.

[6] the building produces a surplus of heat over the year, but uses energy for heating in winter.

[7] CEA-ETEA patent.

REFERENCES

(1) Dumon, R.: 1977, Energie solaire et stockage d'énergie, Masson, 134 p.

(2) Combarnous, M.: 1970, Convection naturelle et convection mixte en milieu poreux, thèse de l'université de Paris, 114 p.

(3) Medvedev, Z.: 1976, Two decades of dissidence, New Scientist, 4 p. Also: June 1977, Facts behind the Soviet nuclear disaster, New Scientist, 4 p.

(4) Marsily, G. de: April 1978; Peut-on stocker de l'énergie dans le sol ? Annales des Mines

(5) Carslaw, H., Jaeger, J.: 1959, Conduction of heat in solids, Oxford at the Clarendon Press, 510 p.

(6) Houpeurt, A., Delouvrier, J., Ifly, R.: 1965, Fonctionnement d'un doublet hydraulique de refroidissement, Mémoires et Travaux de la Société Hydrotechnique de France, n°1, 8 p.

(7) Laffitte, P.: 1958, Propagation de la chaleur dans les roches autour d'une source chaude sphérique, Bull. Soc. Franç. de Minéralogie et Cristallographie, n° LXXXI, pp. 147-150

(8) Brun, G.: 1965, La régularisation de l'énergie solaire par le stockage thermique dans le sol, Revue Générale de Thermique, n°44. Also: Le stockage thermique dans le sol en vue de la régularisation de l'énergie solaire, Conférence prononcée le 20 Avril 1967 au Centre de Perfectionnement Technique, 6 p.

(9) Harris, W., Davison, R., Morse, W.: Demande de brevet n° 7428393: Méthode de stockage d'énergie en aquifère liquide.

(10) Meyer, C., Todd, D.: June 1973, Conserving energy with heat storage wells, Environmental Science and Technology, vol.7, n°6, 5 p.

(11) Latil, M.: Mai 1976, Possibilités de stockage d'eau chaude en aquifère et près de la surface, rapport IFP pour l'Agence pour les Economies d'Energie, 31 p.

(12) Mise en service aux USA d'un système d'utilisation de la chaleur des roches. AGEFI 13 Juin 1977, p. 8 and also: Communications personnelles de MM. Sarda (IFP) et Batchelor (Camborne, School of Mines).

(13) Cazal, A.: 1963, Injection d'eau chaude dans une nappe à l'aide de forages, Assoc. Intern. des Hydrogéologues, publication n° 63, 11 p.

(14) BURGEAP: Nov 1973, Etude des échanges thermiques dans un aquifère au voisinage d'un puits, rapport pour le Secrétariat Permanent pour l'Etude des Problèmes de l'Eau, établi avec la participation de l'Ecole des Mines-CIG, 38 p.

(15) Clouet d'Orval, M., Ledoux, E.: 1975, Détermination in situ des paramètres du transfert de la chaleur dans les aquifères en écoulement monophasique, Bull. du BRGM, 2ème série, section III, n°1, 17 p.

(16) Mathey, B.: 1977, Development and resorption of a thermal disturbance in a phreatic aquifer with natural convection,

Journ. Hydrol., n° 34 (3/4), 19 p.

(17) Werner, D., Kley, W.: 1977, Problems of heat storage in aquifers, Journ. Hydrol., n° 34 (1/2), 9 p.

(18) Dieulin, A., Goblet, P., Iris, P., Ledoux, E., Marsily, G. de, Clouet d'Orval, M., Fabris, H., Gringarten, A.C.: Mai 1977, Détermination des paramètres thermiques d'un aquifère captif et de ses épontes, Journées J. Goguel, Orléans, 10 p.

(19) Lebrun, M.: 1975, Récupération d'énergie par stockage de calories dans une couche aquifère captive, rapport option Ecole des Mines-SERETE

(20) Despois, J., Nougarede, F.: 1977, Stockages souterrains de chaleur, Revue Générale de Thermique, n° 184, 8 p.

(21) Iris, P.: 1979, Expérimentation de stockage de chaleur inter-saisonnier en nappe phréatique, Campuget, Gard, 1977-78, rapport Plan Construction, Ecole des Mines-EDF. Also: Iris, P., Marsily, G. de, Cormary, Y.: Aug 1980, Heat storage in a phreatic aquifer - the Campuget experiment, 15th Intersociety energy conversion engineering conf., Seattle, Wash.

(22) Molz, F.J. and al.: 1978, Thermal energy storage in confined aquifers, Water Resources Research Institute, Auburn University, Alabama, USA.

(23) Andrews, C.B.: 1978, The impact of the use of heat pumps on groundwater temperature, Groundwater, vol. 16, n°6.

(24) Iris, P.: 1980, Expérience de stockage thermique intersaison-nier en happe phréatique. Remarque sur la valorisation énergétique par pompe à chaleur des aquifères à faible et moyenne profondeur, Revue Générale de Thermique, n° 224-225, Aug-Sept 1980, p. 687-697

(25) Iris, P.: Patent registered in February 1980 under n° 8002937: "Procédé d'exploitation saisonnière de la chaleur sensible naturelle d'une nappe aquifère.

(26) Antonelli, R., Iris, P., Cordier, E.: 1980, Héliogéothermie: utilisation énergétique des nappes peu profondes, XIème Journées d'Etude, Comité Français d'Electrothermie, Versailles, 4-5/12/80.

(27) Iris, P.: October 1981, Two concrete projects of space heating with solar energy and aquifer thermal use. Energy storage conf., Seattle, Wash.USA

(28) Conseil, B., Prieur, A., Ronc, M., Rousseau, G., Boissonnet, F.: 1981, Maison solaire expérimentale, extraction des calo-ries du sol et recharge par énergie solaire. Colloquium on storage of solar energy for space heating, INSA, ALEDES, AGEDES, Lyon, 21-22/1/81.

(29) Ausseur, J.Y., Menjoz, A., Sauty, J.P.: Jan 1981, Présenta-tion de deux projets de stockage d'énergie sensible en aqui-fère. Colloquium on storage of solar energy for space heating, INSA, ALEDES, AGEDES, Lyon, 21-22/1/81.

SHORT AND LONG-TERM THERMAL STORAGE IN SOLAR PONDS

H. Tabor

(The Scientific Research Foundation, Jerusalem)

1. INTRODUCTION

Fundamental to a wider use of solar energy are advances in materials technology and energy storage.

Materials technology is important because all materials used in solar devices have to be durable under highly adverse conditions. Furthermore, because solar collectors have to be large - if reasonable amounts of energy are to be harnessed - the materials must be low in cost. Storage is important because, except in the most primitive societies, the consumer does not wish his energy-use pattern to be dictated by the vagaries of solar radiation. An alternative to storage is an auxiliary energy source but this limits the total fractional solar contribution, especially if there is a large energy demand during periods of low insolation. Thus storage, particularly long-term (seasonal) storage is a key requirement for viable exploitation of solar energy for all applications where the time pattern of demand is not identical to the time pattern of insolation. The heating of buildings during the winter season is a clear case of mis-match between the solar supply and the load - and this is probably one of the major factors inhibiting wide-scale use of active solar heating systems for buildings.

The reference to materials technology and the need for

durable and cheap materials refers not only to the collectors but also to the storage system because, like the collectors, the system has to be large if it is to be effective. As an example, phase-change materials have been proposed for heat storage -as a substitute for water or rockpiles: the material costs more per kg than water (including its container), but may store more heat per kg for a limited temperature swing. For diurnal storage, the quantities of such materials needed are reasonable and a number of companies are promoting such heat-storage means, especially where the space available is limited. (One obvious case is where the material is inserted into the walls of the building.) However, for seasonal storage, the quantity of material needed is very large since a large amount of heat is to be accumulated during the summer months. The economics of such materials for seasonal storage is highly questionable.

It follows, therefore, that attempts to find a solution for long-term storage are concentrated on using the earth or masses of water, the latter including underground aquifers (which do not have to be paid for nor do they take up precious building space!).

2. COMBINED COLLECTION AND STORAGE

In most solar thermal systems, the collection and storage of solar energy are carried out separately: indeed, it has been a teaching, in classical solar collector theory, that the thermal capacity of the collector should be kept to a minimum since most of the heat "stored" in the collector would be lost to the environment during periods of low or zero insolation: the storage should be in a separate system that can be adequately insulated against heat loss.

Today, we will discuss a special case that violates this teaching, i.e. a collector where the storage is built-in. One justification for this is that, in accordance with the philosophy of using low-cost and durable materials, the present collector plus storage involves, primarily, earth and water for both collector and storage components. Furthermore, this system is technically justified because the heat loss coefficient is much lower than in conventional solar collectors. This case is the non-convecting solar pond.

The theory and technical aspects of solar ponds - as large area solar collectors - have been described in a number of reports (Refs. 1-6). If we consider the usual solar collector - as seen on roof-tops in Israel and many other countires - this is a small device of a few square metres in area: to harness solar energy by the square kilometre requires a different approach.

The oceans have been recognised as large-area solar collectors since they offer excellent answers to the questions of materials technology and energy storage - and attempts have been made to exploit the small temperature differences that occur between the upper water - where the solar radiation is absorbed and which is therefore hotter - and the deep water that is colder.

But the temperature difference, at best $20^{\circ}C$, is very small so that conversion to power by a heat engine leads to very low Carnot efficiencies and the absolute levels, 5 to $25^{\circ}C$, are of little practical use for heating purposes.

The non-convecting solar pond has a similar philosophy of using a mass of water as the solar collector, but the depth is limited, usually to a metre or two deep in its basic form, so that the solar radiation is absorbed at the bottom. In a normal pond, the heated water would rise to the top and the heat would be dissipated to the atmosphere.

By imposing a density gradient - by the dissolution of salt in increasing concentration with depth - convection can be suppressed and water heated at the bottom of the pond stays there and is insulated by the non-convecting mass of water above.

Because of absorption of some solar radiation on the way down, the temperature gradient is not linear but curved (see Fig. 1) and may show dT/dz, where T represents temperature and z is depth measured positively downwards - of near zero at the bottom. (The infra-red component, which comprises about 50% of the solar spectrum, is absorbed in the first few cms, water being substantially opaque to IR. Part of the visible spectrum is absorbed, dependent upon water clarity, so that between 15 and 30% usually reaches the bottom of a 1 m deep pond.) We thus arrive at an interesting case of a material of finite thermal conductivity behaving as an almost perfect insulator. This is seen at once from the Fourier conduction law for one-dimensional heat flow:

$$\text{Heat flow } q = kA \frac{dT}{dz}$$

i.e. heat flux upwards from the bottom of the pond is:

$$\text{flux } F = \frac{q}{A} = k \frac{dT}{dz}$$

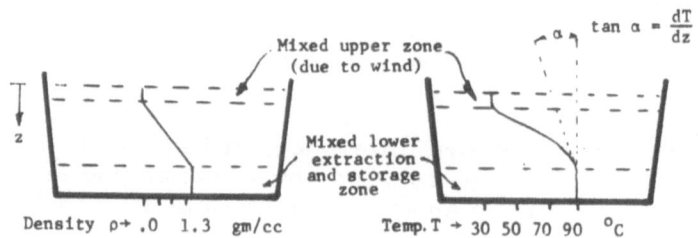

FIG.1 DENSITY AND TEMPERATURE GRADIENTS IN SOLAR POND

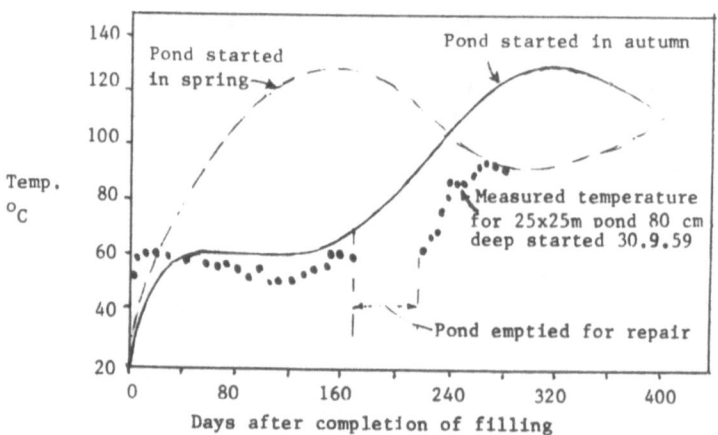

FIG.2 TEMPERATURE RISE ABOVE AMBIENT IN SMALL EXPERIMENTAL POND
COMPARED WITH COMPUTED TEMPERATURE RISE

If the gradient were linear,

$$F = k \frac{\Delta T}{Z}$$

where ΔT is the temperature difference between the bottom and top of the non-convecting zone, and Z is the depth. For $k \sim 0.6 \ W/m^{\circ}C$, $\Delta T = 60^{\circ}C$, $Z = 1 \ m$;

$$F = 36 \ W/m^2$$

With a (24 hour) mean insolation of 200 - 240 W/m^2 for sunny areas and, say, 25% transmission through the pond, some 50 - 60 W/m^2 would reach the bottom, over half of which would be lost upwards by conduction if the temperature gradient were linear. As indicated, in practice dT/dz is much reduced by the absorbed energy and indeed Weinberger (7) has indicated that optimum design conditions for a pond (depth, temperature of extraction, etc.) is when $dT/dz \sim 0$.

As a result of this strong insulating effect of the non-convecting layer of water, temperatures over $100^{\circ}C$ (boiling-point of the salt solution is around $110^{\circ}C$) have been recorded at the bottom of such ponds, though $90^{\circ}C$ is a practical design point temperature.

At the bottom of the pond, there must be a mixed layer (induced by the heat-extraction process) as otherwise only a very small fraction of the heat accumulating there could be exploited. Heat withdrawal can be carried out by laying a heat exchanger - in the form of an array of parallel pipes - in the bottom of the pond (Hipser + Boehm (8)), or by the method of decanting the hot layer which is possible because of the density gradient. (Hydrodynamics teaches that, where a vertical density gradient exists in a mass of fluid, it is possible to remove a horizontal slice without disturbing the fluid above or below. See the original paper by Elata and Levin (ref. 9) or the summary in ref. 4.) The decanted hot layer is passed through a heat exchanger external to the pond and then returned to the other end of the pond for reheating.

As a source of calories, the non-convecting pond is almost an order of magnitude cheaper than other solar collectors (ref. 3): indeed, the calories are so cheap that conversion to power, even at the low Carnot efficiency resulting from a $90^{\circ}C$ source temperature, is possible and viable. A 150 kW(e) solar pond power station (SPPS) was inaugurated December 1979 at Ein Bokek, on the shores of the Dead Sea.

Apart from low cost and some other advantages - such as no windows to clean and ease of collecting solar energy over large areas with little plumbing - of special interest to us here is the feature of built-in storage resulting from the mass of water involved and from the ground under the pond. As an example of the storage capacity of solar ponds, the turbo-generator at the Ein Bokek solar pond power plant has delivered 150 kW over prolonged periods whilst the pond itself is undersized by a factor of about 7, i.e. the pond delivers, on request, 7 times its rated capacity.

3. TEMPERATURE RISE IN A SOLAR POND

Weinberger (7) in what is now recognised as the classical paper on the physics of solar ponds, determined the theoretical rise of temperature at the bottom of a pond of given depth subject to a given annual insolation pattern. (In order to reduce the very large number of parameters, the thermal diffusivity of the pond and of the ground under the pond were assumed equal, a reasonable first assumption if the ground is satured with water.)

Figure 2 (taken from ref. 2) shows actual measured results from a small experimental pond built in 1959. On the same figure are shown Weinberger's computed temperature rises (a) for a 1 m deep pond, first exposed to solar radiation in the spring and (b) for the case of the pond being first exposed to insolation in the autumn. In crude terms, these two curves show a sine-wave of one-year period, resulting from the annual solar cycle, superimposed on an exponential-type rise of mean temperature as the mass of water, and the ground underneath, heats up. The general similarity between the computed curve (b) and the curve for the experimental pond, started in the autumn, is quite striking.

4. SHORT-TERM HEAT STORAGE

For short-term heat storage, the pond is very effective. For the parameter variables he used, Weinberger's Fig. 12, reproduced here as Fig. 3, shows an expected daily swing of temperature at the bottom of a 1 m deep pond of about $15^{\circ}C$ with no convective zone at the bottom. Since such a zone will invariably exist due to the heat-extraction process, Fig. 3 shows the effect of a 20 cm convecting zone: the 24 hour variation in temperature is reduced to just over $5^{\circ}C$, i.e. $\pm 2,5^{\circ}$ about the mean. In the various solar ponds built in the field to date in Israel (five in number), all somewhat deeper than 1 m, the output

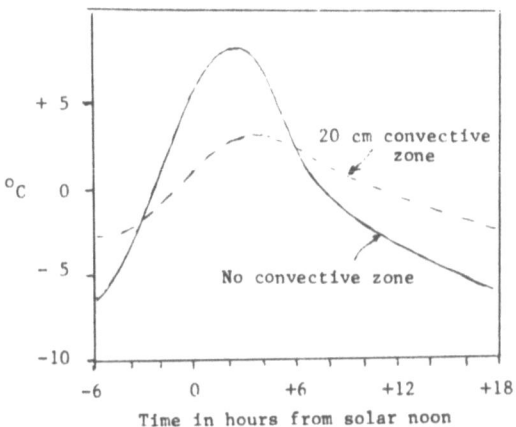

FIG.3 HOURLY VARIATION ABOUT MEAN TEMPERATURE AT BOTTOM OF A POND
1 M DEEP WITH AND WITHOUT A CONVECTIVE ZONE AT THE BOTTOM

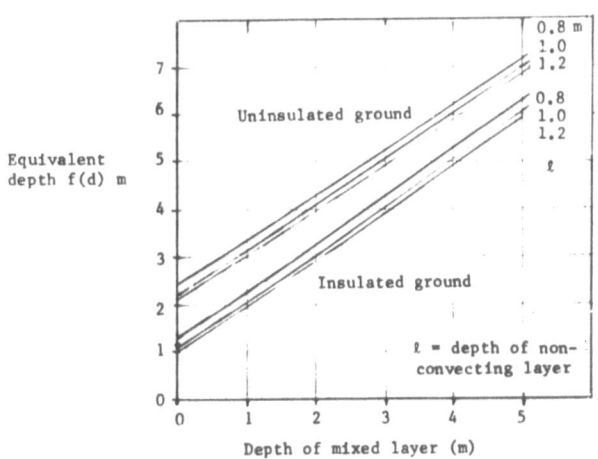

FIG.4 EQUIVALENT STORAGE DEPTH ALLOWING FOR STORAGE IN GROUND AND
IN NON-CONVECTING LAYER

temperature has shown little variation over a week or so whatever the insolation conditions. This short-term storage capability is also of great interest when a solar pond is used to produce electricity: the storage of electricity on an intermediate to large scale is technically very difficult, yet there is, in general, little or no correlation between the electrical load and the solar insolation. In the solar pond, we store the heat, so that, provided the turbogenerators and heat-exchangers are designed to handle peak power, the energy can be withdrawn as the load demands. The Israel Electric Corporation is considering solar ponds to provide peak power to replace expensive gas-turbine power now used (Bronicki et al. (10)).

5. LONG-TERM STORAGE

The possibilities of long-term storage have been indicated by Tabor + Weinberger (11) for large ponds, where only the heat-capacity of a thick mixed zone at the bottom of the pond was considered. In the treatment below, the effect of thermal diffusion into the static non-convecting zone above, and into the ground below, is considered. A paper by Rabl + Nielsen (12), where the interest was primarily in small ponds, offers a similar treatment except that they start with an assumed sinucoidal incident solar radiation wave (of one-year period) and assumed optical properties of the pond solution: in the present treatment, we assume a sinucoidal energy wave reaching the bottom of the pond.

We first note that, for large ponds, the extra cost of making the pond deeper is small, except for the increased inventory of salt. This is because large ponds are not made by digging out the ground but by flattening the ground and building a wall or embankment round: the length and hence cost of the wall per m^2 of pond area decreases as $1/A^{1/2}$ where A is the area. However, the cost per metre run of wall increases approximately as the square of the height because its average thickness also increases with height. The cost of the lining (used to prevent leakage) on the inner wall face will also increase linearly with height. As a rough guide, a pond of about 1 hectare (10,000 m^2) area and 1,5 m total depth and assumed to cost $ 12/m^2 with free brine, would increase in cost by $ 2-3/m^2 for each additional metre of depth added for storage. If salt has to be paid for at $S per ton, it is necessary to add $ 0.35S per m^2 for each additional metre of depth.

The simple first-order approach for large and deep ponds is to take the mass of water in the convecting storage zone, assume this to be well mixed and to be perfectly insulated at the bottom, top and walls; the annual insolation reaching and absorbed at the bottom is approximated as a sine-wave of one year period. (We can ignore daily variations that are, for all practical purposes, completely damped out by the large heat capacity.)

Then the rate of change of temperature θ of a volume V of liquid for a uniform extraction rate $A\bar{I}$ is:

$$V\rho c \frac{d\theta}{dt} = A(\bar{I} + \tilde{I} \sin\omega t) - A\bar{I} = A\rho cd \frac{d\theta}{dt} \qquad (1)$$

where ρ = density of liquid, kg/m^3, c = specific heat, $J/kg, {}^\circ K$, A = area of pond, m^2, t = time in seconds; t = 0 at spring equinox, ω = angular velocity = 0.1988 x 10^{-6} rads/sec for an annual cycle, $\tilde{I}$ = annual mean insolation reaching the bottom of the pond, W/m^2, $\bar{I}$ = amplitude of sinucoidal component of insolation reaching the pond bottom, W/m^2, d = depth of the mixed volume V, metres.

This at once yields:

$$\theta = \frac{\tilde{I}}{\omega\rho \, cd} \sin(\omega t - \pi/2) + \theta \, \text{mean} \qquad (2)$$

and the annual swing of θ about the mean is:

$$\pm \theta = \frac{\tilde{I}}{.1988 \times 10 \, (\rho c)} = 1.62I/d \text{ for } (\rho c) =$$

$$= 3.1 \times 10^6 J.m^{-3}K^{-1} \qquad (3)$$

The minimum is at the spring equinox (t = 0).

As an example, under Israel conditions (latitude 32°N) for a pond with a non-convecting zone 1 m deep, and a saline solution having transmissivity similar to sea-water No. 3 (ref. 7), the computed energy reaching the bottom is approximately $35W/m^2$ in mid-winter and 107 W/m^2 in mid-summer, i.e. $\bar{I} = 71$, $\tilde{I} = 36$. This gives $\pm \theta = 58/d°K$. With d, say, 5 m the temperature swing is $\pm 11.6°$ about the annual mean level.

In the above treatment, we have ignored the flow of heat into and out of the ground. We now consider the effect of this heat flow, i.e. the diffusion of heat into a semi-infinite body subject to a sinucoidal temperature at its surface, and will subsequently consider heat flow upwards into the non-convecting zone above the mixed storage zone. We assume the mixed zone to have an oscillating temperature with an annual cycle

$$\theta = \theta_m + \tilde{\theta} \sin(\omega - \epsilon) \tag{4}$$

where ϵ is an arbitrary phase angle that allows for the temperature cycle not being in phase with the incoming energy: its value is determined later.

We wish to determine the oscillating flux that enters and leaves the ground. From Carslaw and Jaeger (13), p. 65, the temperature at any depth x from the surface of a semi-infinite body subject to a harmonic surface temperature $\theta = \tilde{\theta} \sin(\omega t - \epsilon)$ is given by:

$$\theta = \tilde{\theta} e^{-kx} \sin(\omega t - \epsilon - kx) \tag{5}$$

where $k = (\omega/2D)^{1/2}$ d = diffusivity $K/\rho c$

$$= (\omega \rho c/2K)^{1/2} \tag{6}$$

The heat flux F_g at the surface is:

$$F_g = -K(\partial\theta/\partial x)_{x=0}$$

$$= \sqrt{2kK\tilde{\theta}} \sin(\omega t - \epsilon + \pi/4) \tag{7}$$

$$= (\omega \rho c K)^{1/2} \tilde{\theta} \sin(\omega t - \epsilon + \pi/4) \tag{8}$$

$$= \tilde{F}_g \tilde{\theta} \sin(\omega t - \epsilon + \pi/4) \tag{9}$$

Consider the ground to have thermal properties similar to water, i.e. K (conductivity) = 0.62 W/m°C, ρ (density) = 1000 kg/m^3, c (specific heat) = 4.18 x 10^{3}J/kg°C, ρcK = 2.592 x 10^{6}J^{2}sec^{-1} m^{-4} °C^{-2}, $\omega \rho cK$ = 0.515 for ω = 0.1988 x 10^{-6}rads/sec,

$$\text{Amplitude } \tilde{F}_g \tilde{\theta} = (\omega \rho cK)^{1/2} \tilde{\theta} = 0.718\tilde{\theta} \text{ W/m}^2 \text{ °C} \tag{10}$$

We now have the flux F_g into the ground.

There is also a heat flux F_u upwards into the non-convecting layer above the mixed storage zone. If this were infinitely thick, the sinucoidal flux F_u would be similar to F_g if we assume similar thermal properties. As, however, the non-convecting layer is of finite thickness, we have to use the theory of a slab of thickness   subject to a sinucoidal temperature wave at the surface.

For a sinucoidal temperature $\theta = \tilde{\theta} \sin(\omega t - \epsilon)$ at $x = \ell$ and zero fluctuation at the other face $x = 0$ (which is in contact with a well-mixed layer of high surface conductance) Carslaw + Jaeger (p. 105-6) give the solution for at any depth

$$\theta = A'\tilde{\theta} \sin(\omega t - \epsilon + \phi) \tag{11}$$

where $A' = [\dfrac{\cosh 2kx - \cos 2\,kx}{\cosh 2k - \cos 2\,k\ell}]^{\frac{1}{2}}$

$$\phi = \text{arg.}\ \frac{\sinh kx\ (1+i)}{\sinh k\ \ (1+i)} \tag{12}$$

The flux F_u at $x = \ell$ is:

$$F_u = K\ (\frac{\partial\theta}{\partial x})_{x=\ell}$$

$$= \frac{Kk[\tilde{\theta}\ \sin\omega t(\sin 2k\ell + \sinh 2k\ell) + \cos\omega t(\sin 2k\ell - \sinh 2k\ell)]}{\cosh 2\,k\ell - \cos 2\,k\ell}$$

$$\tag{13}$$

$$= [\frac{\omega pcK}{2}]^{\frac{1}{2}} \cdot \tilde{\theta}\,r\,\sin(\omega t - \epsilon + \psi) \tag{14}$$

where $r^2 = 2(\sinh^2 2k\ell + \sin^2 2k\ell)/(\cosh 2k\ell - \cos 2k\ell)^2 \tag{15}$

$$\tan\psi = \frac{\sinh 2k\ell - \sin 2k\ell}{\sinh 2k\ell + \sin 2k\ell}\ \ (\to 1 \text{ for } k\ell \text{ large}) \tag{16}$$

$$\widetilde{F_u\theta} = (\frac{\omega cpK}{2})^{\frac{1}{2}} r\tilde{\theta} \tag{17}$$

We can calculate r and ψ for a range of values of $k\ell$. For water, $k = (\omega pc/2K)^{1/2} = 0.819$. For the saline solution used for the non-convecting layer, the density will be up about 15%, the specific heat down by about the same amount and K will be down about 5%. We will thus assume k for the solution to be substantially the same as for water. The quantity $(\omega pck)^{1/2}$ for water is 0.718 and will not be very different for the saline solution.

We obtain Table I.

TABLE I

		=	0.8	1.0	1.2	m
ℓ						
r		=	1.6108	1.3791	1.2610	
$F_u = (\dfrac{\omega pcK}{2})^{\frac{1}{2}} r$		=	0.818	0.700	0.640	
ψ		=	0.2733	0.4022	0.5242	radians

Thus, summarising the diffusion effects for a temperature variation - at the surface of the ground - given by the variable part of equation (4)

$$\theta = \theta \sin(\omega t - \epsilon)$$

we have: a flux downward $F_g = \tilde{F}_g \tilde{\theta} \sin(\omega t - \epsilon + \pi/4)$ - equation (9) and a flux upwards $F_u = \tilde{F}_u \tilde{\theta} \sin(\omega t - \epsilon + \pi/4)$ - equation (14) where $\tilde{F}_g$ is given by equation (10) and $\tilde{F}_u$ is given by equation (17).

The phase angle ϵ will depend upon when and how useful energy is withdrawn from the pond related to the incoming energy.

For example, we assume that the energy withdrawn is not constant during the year but has a maximum (for example, in the winter for a house-heating application) and a minimum half a year away. By assuming a sinucoidal variation, we can obtain results analytically. We write the variable component F_L as:

$$F_L = \tilde{L}\sin(\omega t - \phi_L) \tag{18}$$

where $2\tilde{L}$ is the total range of energy extraction about a mean load L - which mean must be equal to $\bar{I}$ on a long-term basis - and ϕ_L is the phase lag of the load.

Thus the net or effective energy $\tilde{I}_e$ going into the mass of storage fluid is:

$$\tilde{I}_e \sin(\omega t - \delta) = \tilde{I}\sin\omega t - \tilde{F}_g\tilde{\theta}\sin(\omega t - \epsilon + \pi/4) -$$
$$\tilde{F}_u\tilde{\theta}\sin(\omega t - \epsilon + \psi) - \tilde{L}\sin(\omega t - \phi_L) \tag{19}$$

where δ is the phase angle (presently unknown) for the net energy wave.

From equation (2) the temperature wave $\tilde{\theta}$ lags the net energy input by $\pi/2$, i.e. the phase angle ϵ in (4) is $(\delta + \pi/2)$, since the phase of the input is $-\delta$. Equation (19) becomes:

$$\tilde{I}_e \sin(\omega t - \delta) = (\omega\rho cd)\tilde{\theta}\sin(\omega t - \delta - \pi/2) \quad \text{using (2)}$$
$$= \tilde{I}\sin\omega t - \tilde{F}_g\tilde{\theta}\sin(\omega t - \delta - \pi/4) -$$
$$F_u\tilde{\theta}\sin(\omega t - \delta - \pi/2 + \psi) - \tilde{L}\sin(\omega t - \phi_L)$$

i.e. $\theta[\omega\rho cd \sin(\omega t - \delta - \pi/2) + \tilde{F}_g \sin(\omega t - \delta - \pi/4) +$
$$F_u \sin(\omega t - \delta - \pi/2 + \psi)] = \tilde{I}\sin\omega t - \tilde{L}\sin(\omega t - \phi_L) \tag{20}$$

Summing the sine-waves on each side gives:

$$\tilde{\theta} R \sin(\omega t - \delta + \beta) = (\tilde{I}^2 - 2\tilde{I}\tilde{L}\cos\phi_L + \tilde{L}^2)^{\frac{1}{2}} \sin(\omega t + \Omega) \tag{21}$$

where

$$\tan\Omega = \frac{\tilde{L}\sin\theta_L}{\tilde{I} - \tilde{L}\cos\phi_L} \tag{22}$$

$$\tan\beta = \frac{\omega\rho cd + .707\tilde{F}_g + \tilde{F}_u\cos\psi}{.707\tilde{F}_g + \tilde{F}_u\cos\psi} \tag{23}$$

and $R^2 = (\omega\rho cd)^2 + 2\omega\rho cd(.707 F_g + \tilde{F}_u\cos\psi) + \tilde{F}_g^2 + F_u^2 + \sqrt{2}\tilde{F}_g\tilde{F}_u(\sin\psi + \cos\psi) \tag{24}$

For equation (12) to be an identity, i.e. true for all values of t, the phases must be equal, i.e.

$$\delta = (\beta - \Omega) \tag{25}$$

and (12) becomes: $\tilde{\theta} = \dfrac{(\tilde{I}^2 - 2\tilde{I}\tilde{L}\cos\phi_L + \tilde{L}^2)^{\frac{1}{2}}}{R} \tag{26}$

Writing $p = 1/\omega\rho c$ (= 1.62 in the present example) and $R = 1/p\ f(d)$, we have:

$$f(d) = [d^2 + 2p(.707\tilde{F}_g + \tilde{F}_u\cos\psi) + p^2\{\tilde{F}_g^2 + F_u^2 + \sqrt{2}\tilde{F}_g\tilde{F}_u(\sin\psi + \cos\psi)\}]^{\frac{1}{2}} \tag{27}$$

and (26) becomes $\tilde{\theta} = \dfrac{p(\tilde{I}^2 - 2\tilde{I}\tilde{L}\cos\phi_L + \tilde{L}^2)^{\frac{1}{2}}}{f(d)} \tag{26a}$

We note that the bracketted term of the numerator is a function of the incoming radiation and the load: the denominator is a function of the depth of the storage zone and the properties of the ground and the non-convecting layer: it may be regarded as the "equivalent" storage layer depth, i.e. we can write $f(d) = d + d_o$ where d_o is an additional contribution to the storage due to the mass of the earth below the mixed storage zone and the mass of the non-convecting layer above the mixed storage zone.

For $p = 1.62$ and using a value of $\tilde{F}_g = 0.718$ (for ground with thermal properties similar to water) and the values of $\tilde{F}_u$ and ψ in Table I for three depths of the non-convecting layer, and applying these to equation (27) gives $f(d)$.

For $\ell = 0.8$ m $\quad f(d) = (d^2 + 5.20\ d + 5.80)^{1/2}$

$\quad\quad = 1.0$ m $\quad f(d) = (d^2 + 3.73d + 5.09)^{1/2}$

$$= 1.2 \text{ m} \qquad f(d) = (d^2 + 3.44d + 5.76)^{1/2}$$

These yield Table II for f(d) in terms of d.

TABLE II

Depth of mixed layer, d = 0		1	2	3	4	5m
$\ell = 0.8$m f(d)	= 2.41	3.32	4.27	5.23	6.21	7.20
d_0 = f(d)–d	= 2.41	2.32	2.27	2.23	2.21	2.20
$\ell = 1.0$m f(d)	= 2.26	3.13	4.07	5.03	6.00	6.98
d_0 = f(d)–d	= 2.26	2.13	2.07	2.03	2.00	1.98
$\ell = 1.2$m f(d)	= 2.18	3.03	3.95	4.91	5.88	6.86
d_0 = f(d)–d	= 2.18	2.03	1.95	1.91	1.88	1.86

Thus when we express f(d) as $d + d_0$, d_0 varies from 1.9 to 2.4, i.e. is of the order of 2 m for ℓ in the expected range of 0.8 - 1.2 m.

(26a) becomes $\tilde{\theta} = \dfrac{1.62\,(\tilde{I}^2 - 2\tilde{I}\,\tilde{L}\cos\phi_L + \tilde{L}^2)^{\frac{1}{2}}}{d + d_0}$ \hfill (26b)

For constant load (or a load with small random variations during the year)

$$\tilde{\theta} = \frac{1.62\,\tilde{I}}{d + d_0} \tag{28}$$

For the extreme case where the load is a maximum in winter and zero in summer, i.e. $\phi_L = \pi$ and $\tilde{L} = \bar{I}$ (the load varies from zero to 2I during the year), (26b) gives:

$$\tilde{\theta} = \frac{1.62\,(\tilde{I} + I)}{d + d_0} \tag{29}$$

Thus for the example quoted of $\tilde{I} = 36$, $I = 71$

$$\tilde{\theta} = \frac{173}{d + d_0} \sim \frac{173}{d + 2}$$

d = 1	2	3	3.76	m
$\tilde{\theta}$ = 58	43	34.6	30	°C

If we assume that a safe maximum temperature is 100°C (somewhat below the boiling point of the solution) and 40°C is the lowest useful temperature, the maximum value of $\tilde{\theta}$ is 30°, i.e. the pond should have a storage zone of at least 3.8 m, for the extreme case considered.

If the bottom of the pond is well insulated so that $\widetilde{F}_g \to 0$, equation (27) leads to $f(d)$ as follows:

For $\ell = 0.8$ m $f(d) = (d^2 + 2.55d + 1.76)^{1/2}$

1.0 m $f(d) = (d^2 + 2.09d + 1.29)^{1/2}$

1.2 m $f(d) = (d^2 + 1.795d + 1.075)^{1/2}$

which yields Table III.

TABLE III

Depth of mixed layer $d = 0$		1	2	3	4	5m
For $\ell=0.8$m $f(d)$	$= 1.325$	2.30	3.30	4.29	5.29	6.29
d_0	$= 1.325$	1.30	1.30	1.29	1.29	1.29
$=1.0$m $f(d)$	$= 1.13$	2.09	3.08	4.07	5.06	6.06
d_0	$= 1.13$	1.09	1.08	1.07	1.06	1.06
$=1.2$m $f(d)$	$= 1.04$	1.97	2.94	3.93	4.92	5.92
d_0	$= 1.04$	0.97	0.94	0.93	0.92	0.92

i. e. $d_0 \sim 1$m.

Thus for practical cases $d_0 \sim 2$ m for the case of wet ground, ~ 1m for an insulated pond and an intermediate value for dry ground without insulation. The results of Tables II and III are plotted in Fig. 4.

Phase of the temperature wave

Note that, from (25) and (21) the phase of the flux wave into storage is Ω, as given by equation (22), i. e. for constant load $\Omega = 0$ and the temperature wave is in phase with the incident radiation. In the extreme case of heat extraction in winter ($\theta_L = \pi$), Ω is again zero. The temperature wave lags the flux wave by $\pi/2$, i. e. the temperature wave will lag the incident wave by $\pi/2$ for the two cases considered, i. e. will show a maximum in the autumn and a minimum in spring.

CONCLUSIONS

Treating the radiation I reaching the bottom of the pond as a sine wave of one year period and the extracted energy L as a sine wave lagging by a phase cycle ϕ_L, the temperature of the mixed layer at the bottom of the pond will show an annual (sinucoidal) variation, about a mean, of:

$$\pm\,\theta = \frac{p[\tilde{I}^2 - 2\tilde{I}\tilde{L}\cos\phi_L + \tilde{L}^2]^{\frac{1}{2}}}{f(d)}$$

where: $\tilde{I}$ and $\tilde{L}$ are the amplitudes of the oscillating components of I and L respectively.

$p = 1/\omega\rho c$ (= 1.62 for a typical pond)
$f(d) = d + d_0$

where d = depth of the storage zone.

$d_0 \sim 1m$ for a completely insulated pond, i.e. represents an additional storage effect due to the mass of the non-convective layer in the pond, assumed to be of the order of 1m deep.

$d_0 \sim 2$ m for a non-insulated pond, assuming the thermal properties of the ground to be fairly close to those of water.

For a substantially constant load ($\tilde{L} \sim 0$)

$$\pm\,\theta = \frac{p\tilde{I}}{d + d_0} \simeq \frac{1.62\,\tilde{I}}{d + d_0}$$

and for the extreme case where the load is a maximum in winter and zero in summer

$$\pm\,\theta = \frac{p\,(\tilde{I} + I)}{d + d_0}$$

where $\bar{I}$ is the annual mean value of I.

REFERENCES

1.   Tabor H., "Large-area solar collectors (Solar ponds for Power production", UN Conf. New Sources of Energy, Rome (1961), reprinted Solar Energy VII, No. 4, pp. 189-194, Oct. 1963.

2. Tabor H., "Solar ponds", Science Journal, pp. 66-71 (1966).
3. Tabor H., "Solar Ponds (Non-convecting)", UNITAR Conf. on Long-Term Energy Sources, Montreal (1979).
4. Tabor H., "Non-convecting solar ponds", Phil. Trans. R. Soc. Lon. A295, pp. 523-433. Reprinted in book "Solar Energy", published by Royal Soc. of London (1980).
5. Nielsen C. E., Chapter on "Non-convective salt gradient solar ponds" in Solar Energy Handbook, Eds. Dickenson + Cheremishoff, Marcel Decker (1979).
6. Savage B.S., "Solar ponds - a review", McGill Univ. Montreal, Techn. Rep. No. 75-J (FML) (Aug. 1975).
7. Weinberger H., "The physics of the solar pond", Solar Energy VIII, No. 2, pp. 45-56 (April 1964).
8. Hipsher M.S. + Boehm R.F., "Heat transfer considerations of a non-convecting solar pond exchanger", Am. Soc. Mech. Engrs. 76-WA/Sol 4 (1976).
9. Elata C. + Levin O., "Selective flow in a pond with density gradient", Hydraulic Rep., Technion, Haifa, Israel (1962).
10. Bronicki L., Lev-Er J., Porat Y., "Large solar electric power plant based on solar ponds", World Power Conf., Munich (1980).
11. Tabor H. + Weinberger H. Z., Chapter on "Non-convecting solar ponds", Solar Energy Handbook, Ed. Kreider, Mc-Graw-Hill, N.Y. (1980).
12. Rabl A. + Nielsen C. E., "Solar ponds for space heating", Solar Energy, 17, No. 1, April 1975, pp. 1-12 (1975).
13. Carslaw H.S., Jaeger J.C., "Conduction of heat in solids", Second Edition, Oxford Univ. Press (1959).

THERMAL ENERGY STORAGE IN COMMUNITY SYSTEMS

Gunnar Wettermark

Division of Physical Chemistry
THE ROYAL INSTITUTE OF TECHNOLOGY
S-100 44 STOCKHOLM 70   Sweden

From ensured supply to energy conservation

When problems of energy came to the foreground of attention at
the beginning of the seventies, it started as a question of ensuring sufficient supply. First, after some years of price
climbing, the use of energy as such was questioned. Saving energy became a goal and soon the center of campaigns and even mass
movements. It was the first law of thermodynamics that was being
paid attention to: the sum of all supplied energy must balance all
energy carried away or lost if no storage takes place. In many
regions, space heating is a considerable part of the total energy budget and several countries have concentrated their efforts
on energy conservation in the building sector. Fig. 1 shows
some of the important means available to lessen the demand for
energy supply and illustrates the type of approach given to energy problems today.

Another major energy consumer today is the transportation sector.
Here, the possibilities of reaching great savings may not be as
high as in the space heating area without radically changing our
way of living. To substitute travelling by car by collective
means of transportation must be accompanied by different a way
of town planning than prevails today and this implies that considerable time is needed for such an adjustment. Fig. 2 may
illustrate the short term approach given to these problems.

Oil, the ideal energy store

What has happened during recent years is that we have started
applying the first law of thermodynamics in a more conscious way.
However, we still maintain an oil economy, *i.e.* an economy where
oil is the primary source of energy. Oil is an excellent medium

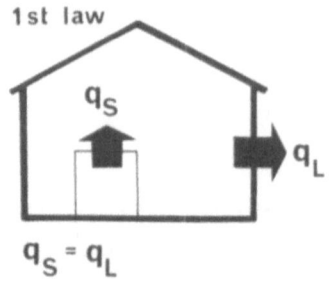

$q_S = q_L$

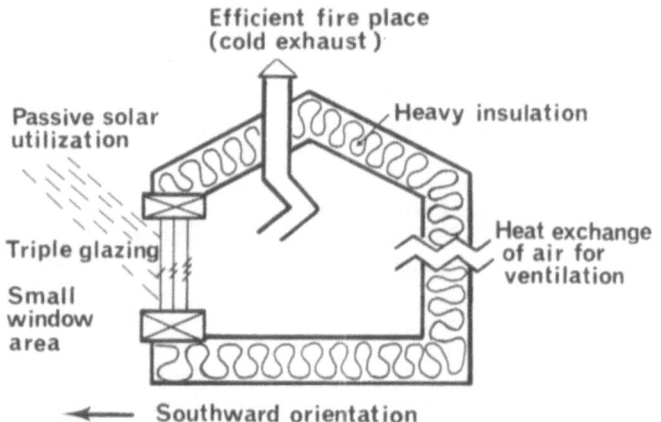

Fig. 1   In energy conservation the first approach is to pay
attention to the first law of thermodynamics, *i.e.* that
supply and demand must match. Space heating is a considerable part of many nations' energy budget and the
needs in that sector can be considerably reduced through
insulation measures and the like.

for energy storage: it is easy to transport, long term losses
are small, there is no necessity for applying high technology or
having an advanced infrastructure. It may also easily be distributed and transformed to the desired form of energy using small
or large units for transformation as desired. All this means
that energy storage using other methods than storing energy as a
fuel has received little attention for many years except for very
particular applications. An example of such an application is
the storing of electricity in lead batteries for starting engines
or as reserve power etc. Another example is the construction of
great dams for hydroelectric power systems. Special storage
units for heat have seldom been in demand except for the storing
of hot water over short periods of time.

Fig. 2   Major energy conservation in the transportation sector
can only be obtained by means of drastic structural
changes in today's society. Therefore, in the short-
term, energy savings will have to take the form illustrated above, *i.e.* to reduce the comfort of travel.

Oil today

Today, energy storage has become of considerable interest, not
least the storage of heat. Sometimes, the newly awakened activities in the field give the impression of resulting from an in-
tuitive feeling that storage will be of great importance in the
future. Any clearly motivated role for the storage in specific
systems is in many cases lacking. I believe, however, that it
is important to outline how future energy systems may sensibly
look and from there judge what kind of storage they will demand.
When doing so, it may be suitable to start from the present situation. Fig. 3 summarizes some important characteristics of contemporary society. In our oil economy, transportation is relatively inexpensive. That allows people and goods to be moved
quite freely over long distances in daily activities. The whole
of modern society is based on this fact. It lets us have a dispersed society with long distances between various functions, *i.e.*
living quarters, nurseries, commercial centra, places of work,
hospitals etc. This kind of society prevails in the heavy industrialized nations and it is on this structure that we apply
the first law of thermodynamics when trying to lower the various
demands for energy. This structure is, however, such that possible energy conservation measures are limited. In other words,
community planning and steering of energy resources go hand in
hand.

Coal, nuclear power, and the second law of thermodynamics

More coal, nuclear power, and application of the second law of
thermodynamics means a substantial change from today's society,
Fig. 4. For reasons of processing and environmental concerns,
it is necessary to centralize the energy production to a very high
extent. In the case of coal, it is in particular the environmental concerns that induce this centralization. Smoke and ashes
cannot properly be taken care of in small units. Transporting coal
from the mines also has to be minimized. In the case of nuclear
power, it is primarily the security aspects which result in large

OIL

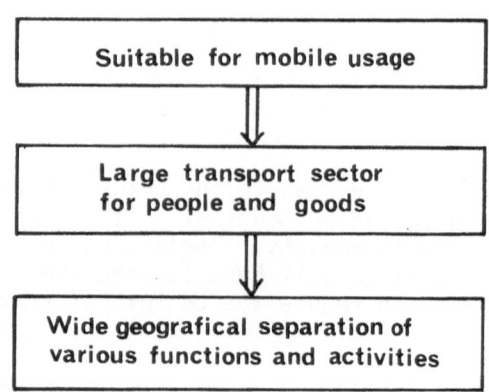

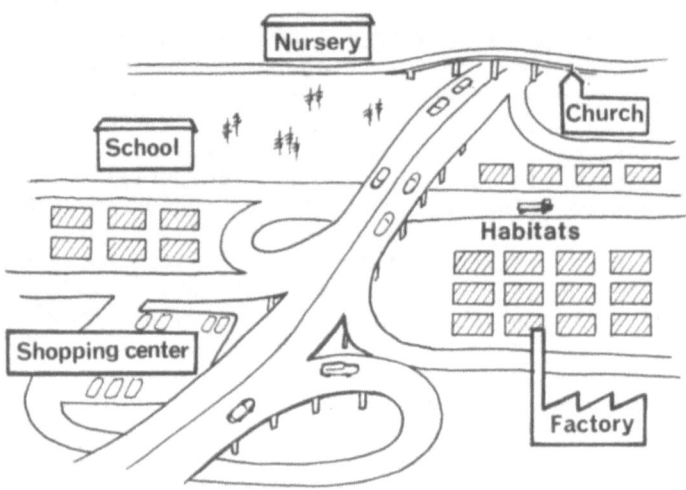

Fig. 3   Important characteristics of a society based on oil (in
words and in appearance). "Separation of functions".

plants. In the future, we will probably have to pay relatively
more for the energy going to the transportation sector when fuel
must be synthesized from coal or with the help of nuclear power.
This implies that there will be forces trying to minimize this
sector in relation to other kinds of energy consumption. This
also leads to a highly concentrated society, $i.e.$ towards a
Megapolis structure. Another factor of maybe greater importance

COAL, NUCLEAR ENERGY & 2nd LAW

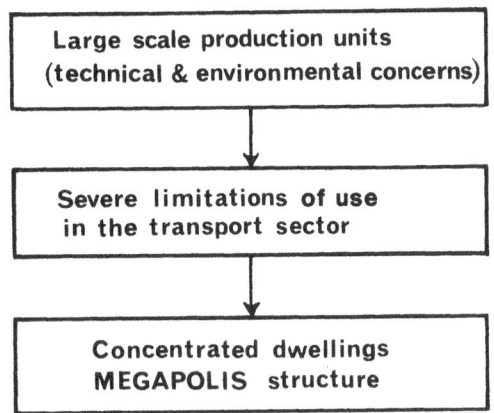

Large scale production units
(technical & environmental concerns)

↓

Severe limitations of use
in the transport sector

↓

Concentrated dwellings
MEGAPOLIS structure

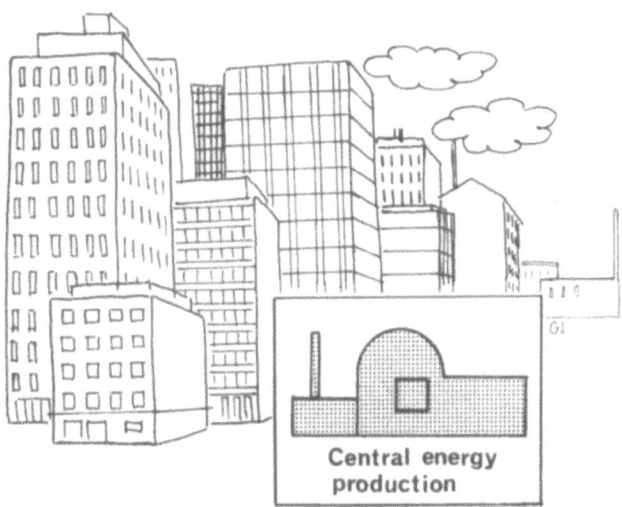

Central energy
production

Fig. 4   A future society based on coal, nuclear power and the
second law of thermodynamics is likely to result in a
"Megapolis" structure. Megapolis provides for an efficient use of exergy by linking industrial and domestic uses of energy.

is the integration of different activities of the society. This
provides for a better application of the second law of thermodynamics thereby improving the energy household of the society.
The second law of thermodynamics introduces energy quality. It

should be observed that there is an enormous potential in the world for a better use of energy quality: western society uses only 5% of the exergy. Merely by better use it is theoretically possible to satisfy all of our energy demands and correspondingly drastically reduce our energy consumption. A concentrated structure of society makes this possible as we get better means of using the energy several times over, right quality at right place. Heat pumping is also a key factor here. A suitable way of illustrating this is the energy triangle, Fig. 5, which relates energy to temperature, the energy thus being assigned a quality. The triangle in Fig. 5 shows some examples of different qualities of energy required in a household starting with the high quality demanded by the vacuum cleaner, the electric mixer and lighting, followed by the lesser quality needed for hot tap water and space heating, and ending with the heat required to preheat air for ventilation or to serve as a heat sink for the heat pump.

The ultimate society

Fusion could be the joker leading society in another direction, but it is more probable that the sun and its derivatives, wind, waves etc. will become the primary energy sources in a long-term perspective. The derivatives are here of particular importance as they often represent high quality forms of energy. They also amount to a considerable portion of the total solar energy influx. About 23% is converted in the hydrological cycle, 0.2% is trans-

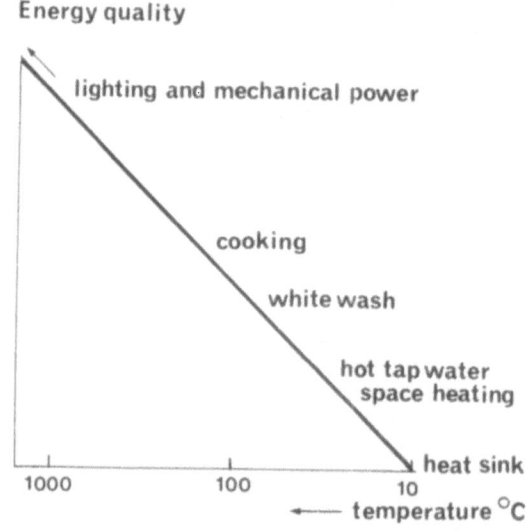

Fig. 5   Right energy quality for the demand.

formed into wind and waves, 0.02% is fixed in the plants through photosynthesis. Solar energy is, however, basically diffuse in its nature. A centralized structure of the society therefore appears to be incompatible with the wide-spread use of solar energy even if, from a purely technological point of view, solar energy may be concentrated. Solar power transforming satellites and solar power stations based on mile wide mirror parks may be a result of today's search for a substitute for oil, replacing oil by sun as a fuel. It is more probable that the solar receivers and the windmills will have to be built together with other structures, *i.e.* houses, roads, parking lots etc. A high degree of adjustment between the living environment and local autonomous solar energy systems will naturally be fundamental, Fig. 6. We may expect solar collectors to be part of structures in the community covering not only houses but also roads and parking lots. It is simply impossible to reserve enough ground for mirror parks.

In the future, intelligence communication and information will not, like today be based on transportation of people and goods. Instead, it is likely that information (intelligence) itself has to be transmitted more widely. Extensive computorization implies that many of our present distribution systems will no longer be needed, *e.g.* post-office service. Work places such as banks and administrative offices are also likely to disappear. All functions will be managed from the home and thus there will be no reason to keep nurseries or schools as separate social functions. They will become a matter of local interest. To transmit information with computers through our telephone circuit and to work from terminals in the media room of our homes is cheaper with regard to both energy and cost.

Demands for heat storage today and tomorrow
We have outlined three types of societies: that of today, based on oil; another based on coal, nuclear power and better use of energy quality; and a third, entirely solar. A more likely future will be societies blending all this in various proportions but the needs for storage may preferably be analyzed in relation to the extreme cases. Several situations are clearly seen where thermal storage will be in demand or even a necessity. A list of applications may be long but I shall limit myself to those below:
1. Solar heating
2. Heat sinks, heat pumps
3. Electrical off-peak
4. Levelling of thermal power plants
5. Time matching of different consumption patterns in cascade systems
6. Distribution of heat
7. Total energy systems based on sun and its derivatives

SUN and WIND

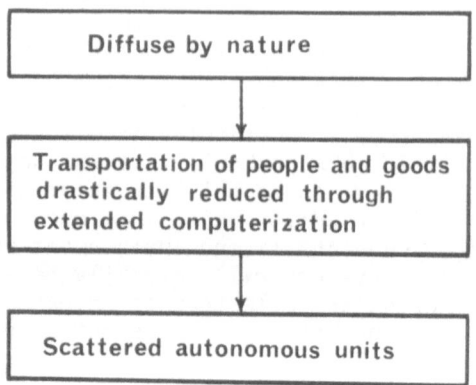

Fig. 6   The ultimate society is likely to be based on sun and
         its derivatives.

The first three applications are of great concern already today.
The fourth, storage in power stations, used to be an important
application for thermal storage when the stations were coal fired.
Nowadays, thermal storage is used particularly in connection with
district heating. In a society based on coal and nuclear power,
thermal storage is more necessary in order to control and integra-
te the complicated energy flows of the Megapolis structure. Application 4 but especially 5 and 6 will then be very important.
When entering a society based on the sun it is necessary to try
to form units which are autonomous from the point of view of

energy supply. Since demand and supply are often entirely out of phase with each other when the supply is sun or wind, long term storage, even seasonal storage, may be needed. This means that the kind of solar houses we are trying out today may not be very usable in a solar society. Contemporary solar houses are partial systems built on the assumption that a stand-by system capable of taking over the entire load when it is at maximum is accessible.

Solar heating

The newly awakened interest in heat storage has, to a great extent, its roots in the solar energy technology. The link to solar energy may be so strong that most nations' thermal energy storage program form a part of the solar energy program. As mentioned above, there is reason to distinguish between different forms of solar technology in relation to the society for which it is created. One goal may be to lessen the strain on other energy sources in a present day type society, another to develop a technology for a future society entirely based on solar energy. In this paragraph we will only deal with the first case and come back to the storage problems associated with a solar society later.

It is for hot tap water and for heating of swimming pools that we today see the great commercial applications of solar energy. In swimming pools, Fig. 7, the store is innate. In countries such as Australia and Israel solar heated storage units for hot water dominate the town scape. Such units, which allow for a ra-

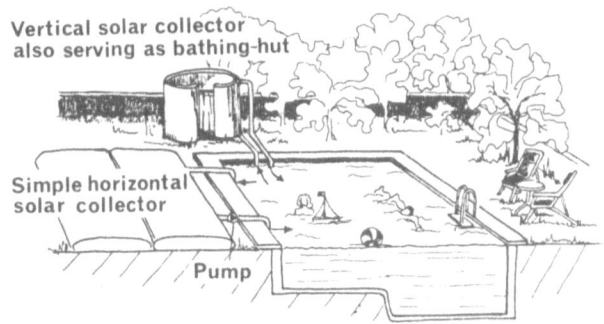

Fig. 7   In solar heating of swimming pools the store is innate (drawing from F Peterson and G Wettermark: "Solenergiboken", distributed by Ingenjörsförlaget AB, Box 27315, 102 54 STOCKHOLM, Sweden).

pid market penetration can be easily installed on existing buildings. For house heating the situation is quite different. It is usually difficult to retrofit buildings with solar heating.

Solar heated homes are usually described as passive or active depending on whether or not they are equipped with active components such as fans, pumps etc. In passive systems, it becomes a question of creating a large thermal mass with a high heat capacity at room temperature. As a rule, heavy construction materials such as concrete, bricks, or adobe clay are employed but

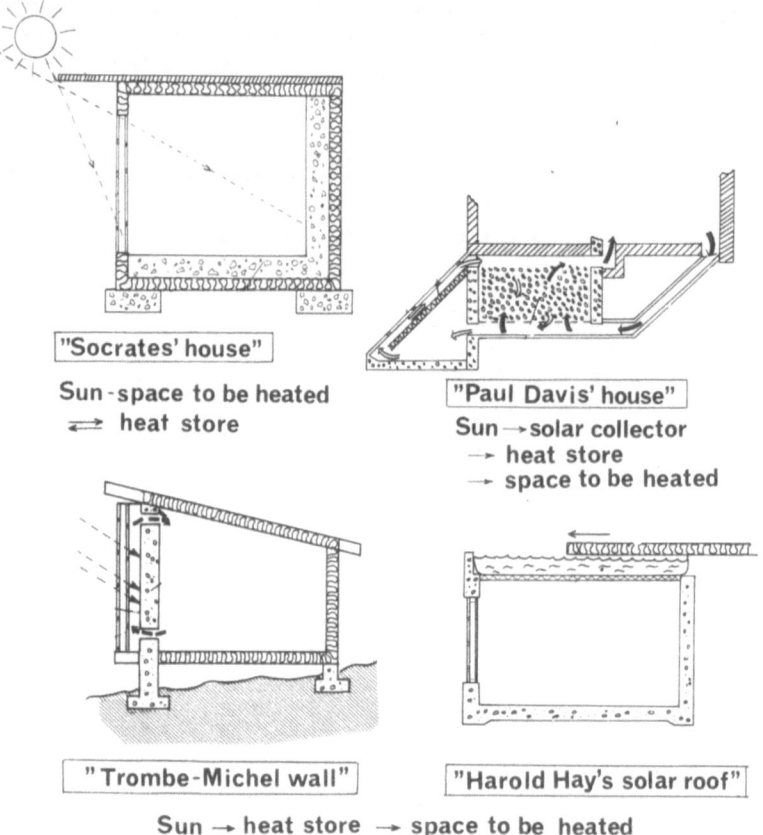

"Socrates' house"

Sun-space to be heated
⇌ heat store

"Paul Davis' house"

Sun → solar collector
→ heat store
→ space to be heated

"Trombe-Michel wall"   "Harold Hay's solar roof"

Sun → heat store → space to be heated

Fig. 8   Solar heating using passive systems employs a large
         thermal mass (in the construction materials) (from
         G Wettermark, B Carlsson and H Stymne: "Storage of heat.
         A survey of efforts and possibilities", Swedish Council
         for building Research, Document D2:1979, distributed by
         Svensk Byggtjänst, Box 7853, S 103 99 STOCKHOLM, Sweden).

even water stores may be used in the passive mode, Fig. 8. Latent heat storage should be capable of yielding better characteristics than sensible heat storage particularly for passive solar
applications and better maintain a desired temperature in the
building. Different attempts to incorporate heat-of-fusion
stores in construction materials have, to a great extent, failed
but no doubt this is an exciting area for experiments.

Actively heated solar buildings have a water tank as the storage
unit in most cases. A common system also contains water as the
transfer medium from solar collector to tank. That is why a separate heat exchanger is not needed. This may be all right in a
climate where there is no risk of frost but in colder regions
measures have to be taken to prevent freezing. This can be
accomplished by emptying the solar collectors during cold spells
or by adding antifreezing agents, but this again makes heat exchanging between the store and a separate transfer unit advantageous since only a small amount of antifreezing agents will be
needed.

There is seldom any reason to increase the storage capacity above
a few hundred kWh for a partially solar heated one-family house.
The extent of coverage rapidly reaches a plateau as the storage
size increases and a radically higher degree of solar self-suffi-
ciency can only be obtained by using seasonal storage, Fig. 9.

The solar collector is, as a rule, the most costly part of a solar heating system. However, solar collectors and stores cannot
be chosen independently of each other. Instead, it is the case
that, certain solar collectors are easier to combine with certain
types of storage. This means that the development of solar collectors also has a great influence on the types of stores being
developed. Today, the market is dominated by very simple solar
collectors where storage efficiency is drastically reduced when
the collector temperature increases, Fig. 10. In a future with
mass production of evacuated solar collectors and inexpensive
electric solar cells, there may be good reason to look for other
forms of storage. On the whole, solar energy development today
is that of handicraft, whereas a future solar energy society is
likely to be based on mass produced items. This is a good reason for being careful when extrapolating experiences from contemporary solar energy technology for application with future
systems.

In addition to the common system where water is the transfer medium and a water tank the store, there is almost only one other
system which is wide-spread. This is the system using an air-
cooled solar collector and storage in a pebble bed. The heat exchange to the store is automatically solved as the storage medium itself acts as a heat exchanger and gives the desired high

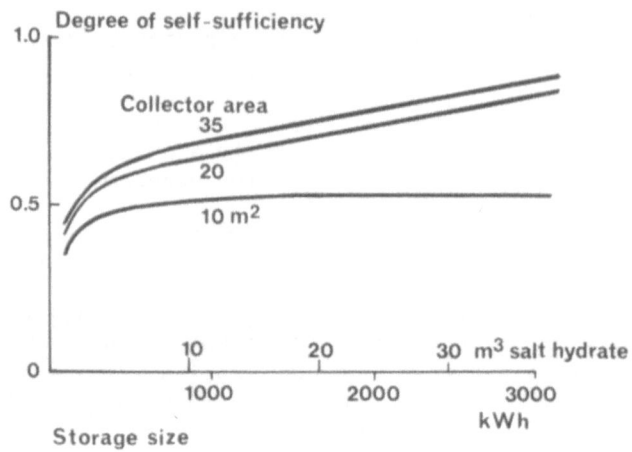

Fig. 9   It is seldom worthwhile to increase the storage size
        above a few hundred kWh for a solar heated one-family
        house. The example given above shows the situation for
        a house in Stockholm, latitude 60°N having the standard
        insulation prescribed in the sixties and requiring 21.4
        MWh annually for space heating - retrofitted with solar
        heating and equipped with a non-leaking heat-store. The
        collector is operated at 45°C. It is double glazed and
        absorbs 0.75 of the direct and 0.67 of the diffuse in-
        solation   Its emission factor for long wave-length ra-
        diation is 0.1. Storage is a heat-of-fusion in a salt-
        hydrate $CaCl_2 \cdot 6H_2O$, transition temperature 30°C (1 m^3 =
        83 kWh). (From V Girdo, Swedish Council for Building
        Research, Report R108:1978, distributed by Svensk Bygg-
        tjänst).

surface area. The pebble bed has, however, a lower storage density mainly because of the dead space which, of course, is also
needed for the heat exchanger medium.

Vigorous attempts are made in seeking other forms of storage,
especially heat-of-fusion and absorption storage. Theoretically,
considerably less storage volume and better characteristics should
be obtainable using such stores but the technology is new and has
not yet reached commercial maturity.

It should be added that availability of inexpensive photovoltaic
cells may completely change the present day picture of solar
space heating technology. It will be most difficult for thermal
collectors to gain a marked in such a situation. Even if hydrogen storage with cheap solar cells is likely to be attractive,

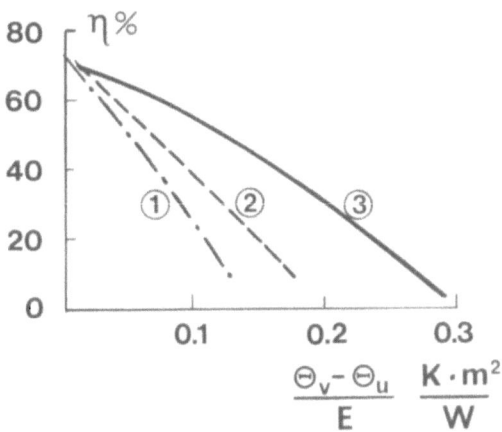

Θ_v = Temperature of collector

Θ_u = Temperature of surroundings

E = Irradiance W/m^2

η = Collector efficiency

Fig. 10   Simple solar collectors must be used at low operating
temperatures in order to avoid excessive losses but
with selective coatings and evacuation it is possible
to obtain a good performance even at higher operating
temperatures. 1 ordinary double glazed flat-plate
collectors; 2 the same provided with additional selective absorber coating; 3 and evacuated (from P Isakson,
Swedish Council for Building Research, Report R35:1978,
distributed by Svensk Byggtjänst).

high temperature thermal storage may also be an interesting option.

Heat sinks
The heat pump, as mentioned above, is essential for obtaining a
better exergy use. The heat pump, however, performs best when it
is allowed to work continuously under a constant load. This demands storage preferably both on the cold and the hot side, $i.e.$
bottom and top end storage. Furthermore, it is better that such
stores work at a constant temperature thus making chemical storage particularly attractive. Compressor-driven heat pump systems
today are constructed where water-ice transformation is used at
the heat sink, $i.e.$ in principle a heat-of-fusion store being
used as heat sink. The ice is melted with the help of waste heat
or solar energy. We should also remember that when using heat

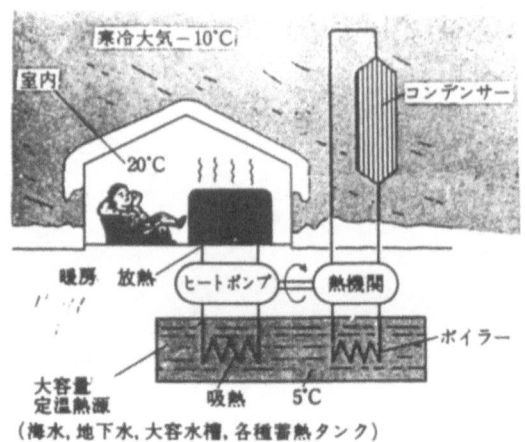

Inside the figure:
東冷大気 – 10°Ci
室内
20°C
コンデンサー
暖房　放熱　ヒートポンプ　熱機関
ボイラー
大容量
定温熱源　吸熱　5°C
（海水，地下水，大容水槽，各種蓄熱タンク）

Fig. 11   Heat sinks are believed to be of great value in the
          future as this picture illustrates (obtained from N
          Isshiki).

pumps, low temperature sources become valuable. I have a Japanese
picture here exaggerating this, but undoubtedly it illustrates
that the value of ground heat may increase with a colder climate,
Fig. 11.

Electrical off-peak

Utility companies have become aware of the often great differences in the cost of producing base power and meeting peak demands. For the power companies, it is not enough just to produce
energy according to load. Rather, they have to take active measures for load levelling. Here, small storage units placed with
the consumer, have proved efficient. In many countries different
insulated sensible heat systems for diurnal storage of resistance
heat are marketed. One instance is England where units built
from iron oxide loaded bricks are being sold. In advanced systems,
the charging of the storage units is controlled directly from the
power companies. The consumer is guaranteed a certain time of
inexpensive power during the day, let us say eight hours, but the
power company may freely decide when during the day these hours
should be given. Off-peak storage of electrical power is a field
of application which can be expected to have a rapid growth. It
is also likely that more advanced forms of thermal energy storage
should have a good market in this field of application. Heat-
of-fusion storage and managing the discharge through steering of
thermal conductivity of the encapsulation are interesting advanced storage concepts.

Levelling of thermal power plants

Today's power plants are often powered with oil or gas. Thermal energy storage at the plant is of only minor interest here because investments for boilers, fuel handling and storage of the fuel represent, relatively speaking, a small cost. The economically optimal solution from the point of view of the operations at the plant is therefore to continuously match supply and demand by increasing or decreasing the supply of fuel. Load control by fuel control (or neutronic control) is much more difficult to achieve in a coal fired or nuclear powered plant. There is also a stronger incentive for thermal storage in these plants as well as in solar power plants because of the low cost of the "fuel". This incentive to apply thermal energy storage in order to reach a better load balance increases when various power systems are brought together. Systems for cogeneration of power and heat are often equipped with substantial storage facilities but even pure district heating systems usually contain considerable storage facilities, Fig. 12.

Cascading (time matching of different consumption patterns in cascade systems)

Several industrial processes end up with smoke gases or waste water having temperature levels of interest for domestic use. In the Megapolis structure particularly, there is a great potential to form industrial-domestic combined energy systems. These have to be designed in such a way that the noble forms of energy are stepwise converted to various utilities. In order to stimulate such a development, it is likely that the price of energy in the future will be set in relation to quality (exergy). This is not the rule today. An example of a cascade system is shown in Fig. 13. An industrial process such as the melting or iron or enamelling is the point where the primary energy enters. Thereafter, the energy can be used to produce high pressure steam and to generate electricity in a steam turbine or to be used as process heat for the industry. The low pressure side of the turbine can also be designed so as to provide valuable process heat. In the next step, the waste heat from the industry is distributed for the domestic use of hot water and space heating. In order for such a cascade to work, there is need for storage at several levels. It is unrealistic to design a system where the different uses are entirely matched in time. In particular, it is difficult to match industrial and domestic demands. It may be necessary to shut down the industrial activities for a week or more because of vacations, repair etc. In order to avoid a stand-by system often requiring heavy investments, one must turn to better energy storage. But, note that the heat has to be stored at different temperature levels preserving its quality. Thermochemical storage opens very interesting perspectives in this connection.

Plant	Dimension of tank (m)		Volume (m³)	
	diameter	height	total	active*
Västerås	42	18	25 000	21 000
Fyen	23	33	13 700	12 000
Uppsala	42	21	29 000	25 000

---

*   1 m³= 58 kWh (50-95°C)

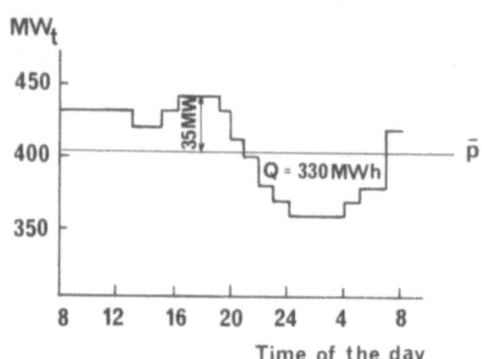

Total energy load = 9720 MWh /24 h

$\bar{p}$ = 405 MW$_t$

Fig. 12   Stratified hot water storage is commonly used with
district heating systems to even out the load. The
table gives some examples of this from Scandinavia together with the record from the Västerås plant Jan.
5-6, 1976, mean outside temperature -7°C (T J Hedbäck,
Swedish Engineering Academy, Rapport 124, 1978).

Transportation and distribution of heat

The storage of heat by reversible chemical reactions, *i.e.* thermochemical storage, offers interesting alternatives for a simultaneous storage and distribution function. The possibility of
entirely eliminating heat leakage makes the transport of heat
over long distances not unreasonable. This may be one way of
joining systems for industrial and domestic use. In a project
I know, industrial waste heat is used to charge containers filled with a thermochemical storage medium, Fig. 14. The whole
container is transported to the place of consumption and there
connected as a chemical heat pump. In this case, hydration
reactions are employed and storage densities up to 1 MWh thermal

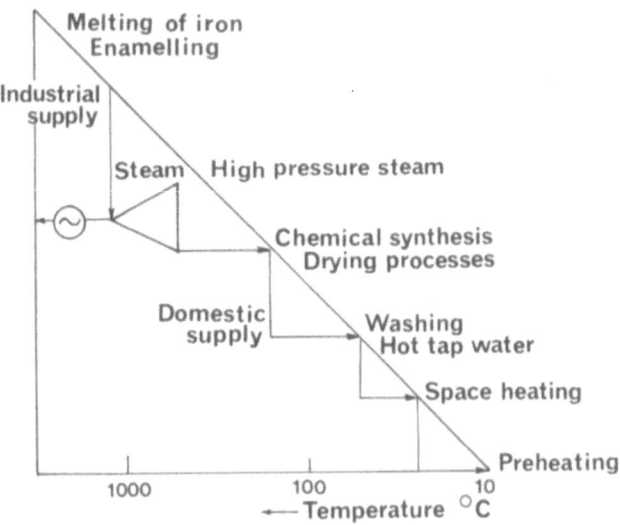

Fig. 13   An example of a combined industrial-domestic energy
           system

energy per cubic meter are attainable in the heat pumping mode.
Note that when water is used as the transfer medium, there is no
need for storing the distilled liquid. It is only the dehydra-
ted salt which is transported after charging and returned hydrated after discharging.

Continuous transport of chemically stored thermal energy is a
technology which has already reached a certain degree of maturity. In fact, a good portion of the central investments in the
field of thermochemical storage has been directed towards this
application. The development with has primarily concentrated on
the steam reforming reaction $CH_4+H_2O \rightleftarrows CO+3H_2$. The system is
called Eva-Adam with Eva as the energy absorbing part. (Ein-
zelspaltrohrVersuchsAnlage). The storage takes place through a
pressure increase in the distribution network or in special high
pressure containers. Both above earth steel containers and underground caverns are envisaged. Eva-Adam is a high temperature
system and its primary field of application may not be domestic
use. Normally, it demands temperature between 800-1200°C for
charging, and delivers heat in the 350-700°C range. Distribution of nuclear energy to various industrial uses is its primary field of application. It is, however, possible to think
of an integrated distribution in a Megapolis structure where the
pipe network has the different functions of fuel distribution,
distribution of stored heat, and where domestic uses also come

Fig. 14   Industrial waste heat being utilized to charge the che-
           mical heat pump for space heating and employing con-
           tainer transport for charged and discharged working
           material, 20,000 kWh on one truck/trailer (for refe-
           rence see Fig. 18).

in to play, Fig. 15.

Eva-Adam is thus the most explored chemical heat pipe system.
However, there is also a great interest in developing systems
operating at lower temperatures to be powered by light water
reactors or focusing solar collectors. The development of such
systems has only started and it is uncertain that such systems
with good characteristics and economy can be developed.

Total energy systems based on the sun and its derivatives
A future society based on the sun and wind will place great demands on seasonal storage of thermal energy. The heat leakage

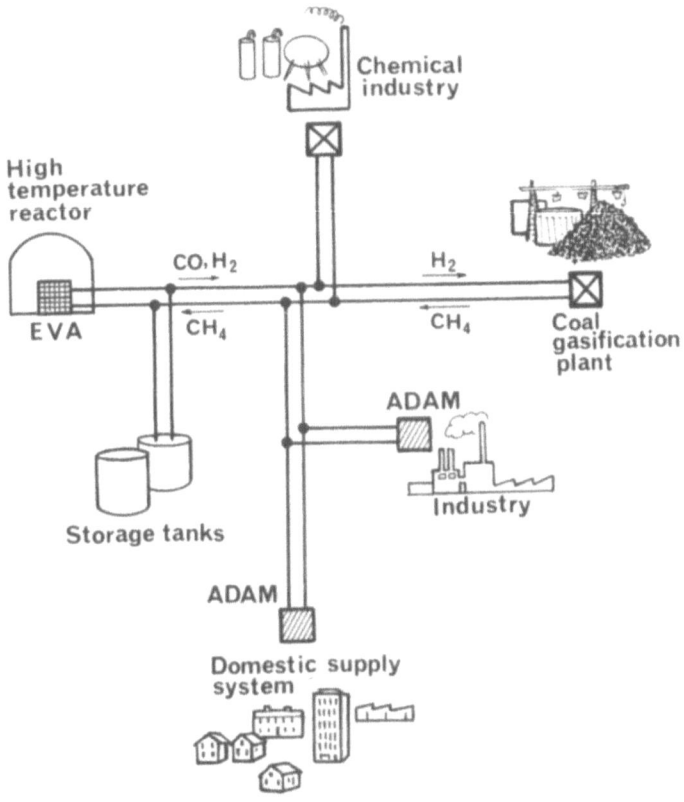

Fig. 15   The EVA-ADAM chemical heat pipe system extended to include other transport functions.

from the store will be the primary problem. It is possible to
insulate simple water stores so well that seasonal storage of
energy can be achieved for a small store also but it is as a
rule necessary to have a higher ratio between volume and surface
for a heat leaking store. This implies that the stores must be
built as central stores for use with groups of buildings rather
than for individual one family houses, Fig. 16. It becomes also
a matter of using natural materials such as water and stone in
order to maintain reasonable costs. Particularly in the United
States, seasonal storage using aquifers is believed to be an
appealing concept. Also, ponds, natural as well as artificial
lakes, are judged suitable for seasonal storage of low temperature heat. Great water stores, ground storage, aquifers, will
however set strong constraints on town planning. They are also

MODEL STUDIES OF DUCT STORAGE SYSTEMS

SWEDISH FIELD TESTS

Site	Year of construction	Size (m³)	Number of pipes	Temperature range (°C)	Ground
Sigtuna	1978	6000	40	25-30	Rock
Kungsbacka	1980	50000	600	8-20	Clay
Luleå	1981	800	20	20-70	Rock
Alnarp	1979	1300	50	10-45	Clay
Utby	1979	800	40	0-20	Clay

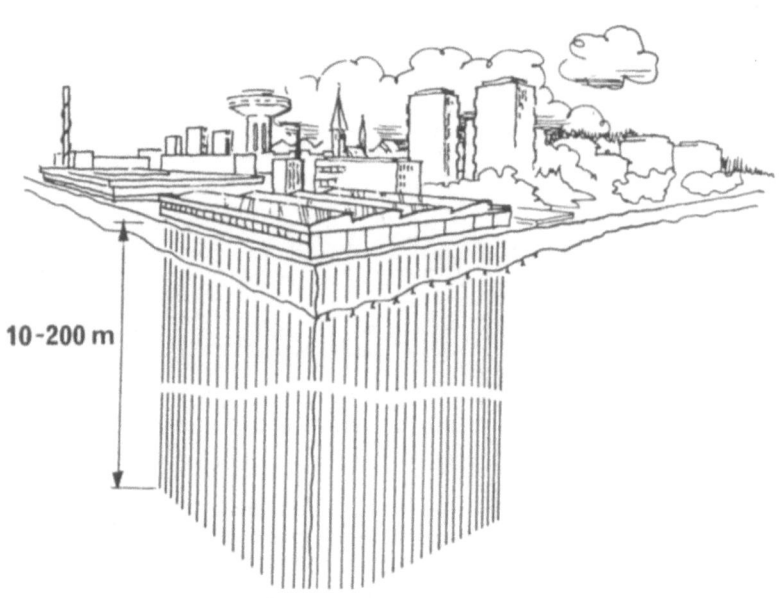

10-200 m

Fig. 16   To give sufficient volume-surface ratio for seasonal
storage, the stores must be large and combined with
district heating or the like (J Claesson and G Hell-
ström: "Model studies on duct storage systems" IEA
Congress Berlin, April 6-10, 1981).

difficult to apply to existing buildings. They place very defi-
nite demands on the heat distribution system and consumption pattern and must be planned to yield maximal solar energy usage.
Older settlements usually have a heat distribution system deman-
ding much higher temperatures than those optimal for solar heat
distribution. Furthermore, the uncertainty regarding the long
term effects of using large thermal reservoirs are notable.
Aquifer storage could, for instance, have an undesirable effect
on biological activity and pollute the groundwater.

For the reasons given above, it is quite likely that a solar and
wind society must depend on other forms of energy storage than
large sensible heat stores. Thermochemical storage and heat
pumping will probably be important. In case photovoltaic cells
become cheap, thermochemical storage may compete with other types of storage such as hydrogen. Thermochemical stores can be
shut off through a "simple valve" and thus the whole problem of
leakage is eliminated. They can also be designed to return high
quality energy and be used to yield mechanical energy and electricity. A total energy system may be thought of in this way.
Fig. 17 describes such a wind based system now being tested.

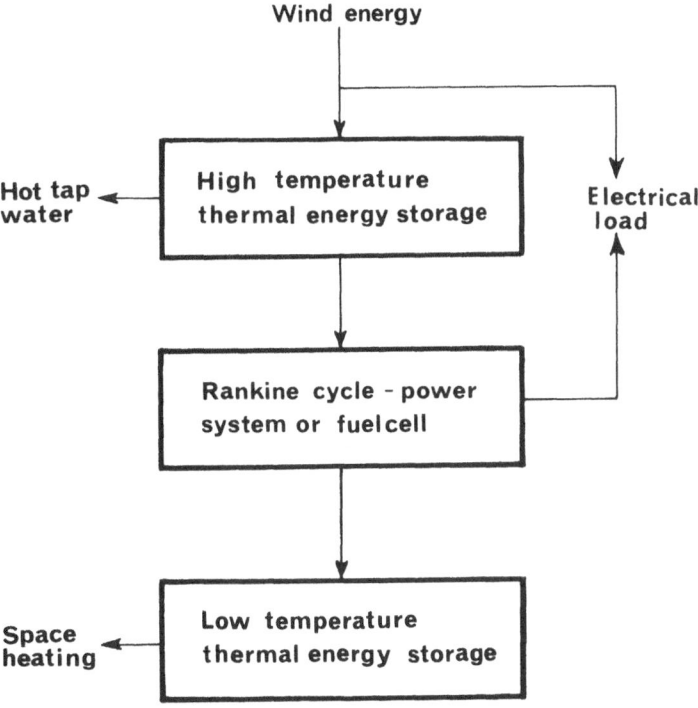

Fig. 17   A domestic total energy system based on wind.

Let me illustrate further possibilities by adding that thermally regenerative fuel cells and heat engines are likely options in a future society. In order to arrive at a proper match between building structure and solar energy exploitation, advanced storage is necessary. The last figure describes one possible line of direction for development of thermochemical storage showing a recently developed system for space heating where solar energy is seasonally stored thermochemically, Fig. 18. This is the so called distillation storage of low temperature heat. In a cold Nordic climate, there is a need for a salt store of about 20 cubic meters for a small one-family house. This may be thought of as a large volume but remember that many houses nowadays reserve much greater volumes for transportation (the garage).

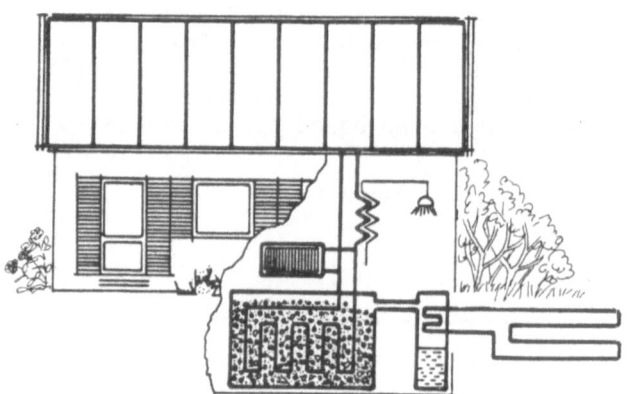

Fig. 18   Distillation storage for seasonal storage of solar energy. Such a system has been developed by the Tepidus Company based on dehydration-hydration of sodium sulphide (E-A Brunberg: "The Tepidus system för seasonal heat storage and for cooling", International Seminar on Thermochemical Energy Storage, Stockholm, January 7-9 1980. Proceedings edited by G Wettermark, Swedish Council for Building Research, Document D25:1980).

THERMAL ENERGY STORAGE FOR THE RECOVERY OF INDUSTRIAL WASTE HEAT

R.J.Wood

Applied Energy Group,
School of Mechanical Engineering,
Cranfield Institute of Technology,
Cranfield, Bedfordshire, United Kingdom

Waste heat recovery systems are of prime importance to the efficient use of energy in industrial processes. Although most waste heat recovery systems have been of the "non-storage" type, future trends may well see a significant increase in the use of Thermal Energy Storage (TES). This will be necessary in order to expand the rational use of large waste heat sources, to match temporally the cyclic variations of hitherto independent thermal processes and to permit the possibility for energy cascading and storage to overcome imbalances in plant energy systems.

This paper seeks to describe some of the characteristics of industrial energy use, the currently available information on temperature levels, working fluids, process periodicity and technological options available for storing industrial waste-heat.

1. INTRODUCTION

There are few other investments which would appear to offer such substantial and well-guaranteed financial returns as the application of energy conserving equipment. Furthermore, as fuel supplies inevitably become restricted the amount of "pay-back" on installed capital equipment with escalate.

The introduction of technically more complex energy conservation techniques should only be considered when all possible low cost and technically simple measures have been applied.

Waste heat recovery systems are often of primary consideration in industrial process plant. These include the reduction in primary energy demand by recycling rejected waste heat using, for example, recuperative or regenerative heat exchangers. Although most industrial waste heat recovery systems have been of the "non-storage" type, future trends may see an increased use of Thermal Energy Storage (TES). This will be necessary in order to expand the rational use of large waste heat sources, to match temporally the cyclic variations of previously independent thermal processes and to permit the possibility for energy cascading and storage to overcome imbalances in plant energy systems.

2. INDUSTRIAL ENERGY REQUIREMENT

Today, oil and natural gas provide over 60 percent of the world's energy consumption. The world primary energy demand is expected to rise to about $40,000 \times 10^6$ tonnes of coal equivalent per year by the year 2050, Ref.(1). Even for relatively low energy growth projections it is probably that energy consumption will double during the period 2000 to 2050 A.D.

The major energy storage requirements of the developed world are presently fulfilled by the stockpiling of fossil fuels by both suppliers and consumers of energy. These types of commodity stores are critically stocked in order to ensure that supplies are available to meet sudden and seasonal variations in demand. However, shortages still occur and very often only small deviations from the normal use patterns can create rapid shortages of primary energy supply. Although current levels of industrial production are relatively low and current fuel availability high, future uncertainties with respect to both supply and price make the consideration of energy conservation of primary importance to industry.

The components of industrial product cost that can be directly attributed to the energy consumed in its production has increased significantly in the last 10 years. However, the wide diversity of both products and the processes by which they are made, creates a barrier to the clarification of available statistical data. Most European nations now publish three- to six-monthly updates on primary energy production and consumption (Ref.2). These data allow the major trends in total industrial consumption and production to be mapped as in Table 1. They do not, however, contain the information most relevant to the application of storage technologies with recycled waste heat.

As can be seen from Table 1, the total industrial energy consumption for the UK has not changed dramatically from year to

year with the exception of two major factors:
1) fuel switching historically from coal to oil and then to natural gas,
2) economic factors such as the level of industrial production which has been reducing significantly over the last two years.

The need to safeguard production against costly stoppages either occasioned by fuel interruptions, shortages, or the addition of energy recovery plant means that the tight investment criteria now being applied in industry may prevent a considerable number of attractive schemes from coming to fruition. However, a typical plant built in 1970 on the basis of energy conservation would be at least five times more cost effective today than when it was built. If this trend continues, the message must be build sooner rather than later, Ref.(3).

3. THE CHARACTERISTICS OF ENERGY USE

One of the most relevant forms of statistical data available is that based on the end use of energy. An end use of UK energy utilisation shows that over 60 percent of all fuel consumption is in the form of thermal energy (heat), see Figure (1). Of this, some two-thirds is used at temperatures below 120°C, mainly for space heating, hot water and low temperature process heat. Of the remaining one-third, utilised at temperatures above 120°C, almost half is used for iron and steel production, and the remainder is predominantly used for lower temperature steam processes. It is clear, therefore, that there is a large concentration of energy use in the form of thermal energy.

Knowledge of the temperature, transport fluid quantity and periodicity of waste heat sources and their relationship with potential sinks is vital when assessing the value of storage. The most comprehensive study of industrial waste heat profiles has been carried out by Drexel University for the US Energy Research and Development Agency, Ref.(4). This source of data has been widely used for considering global statistics on energy usage including the US Department of Energy programme on Industrial Energy Storage, Ref.(5). An indication of the relative energy usage is given in Table 2, Ref.(6), based on the SIC (Standard Industrial Classification) number. The Drexel work quantifies waste heat in five categories; these are:

Liquid streams   - cooling water at $\simeq 45^{\circ}$C
                      - contaminated process water $\simeq 50^{\circ}$C
                      - condensate $\simeq 85^{\circ}$C
Gaseous streams - boiler exhaust gases $\simeq 250^{\circ}$C
                      - furnace exhaust gases $\simeq 350^{\circ}$C

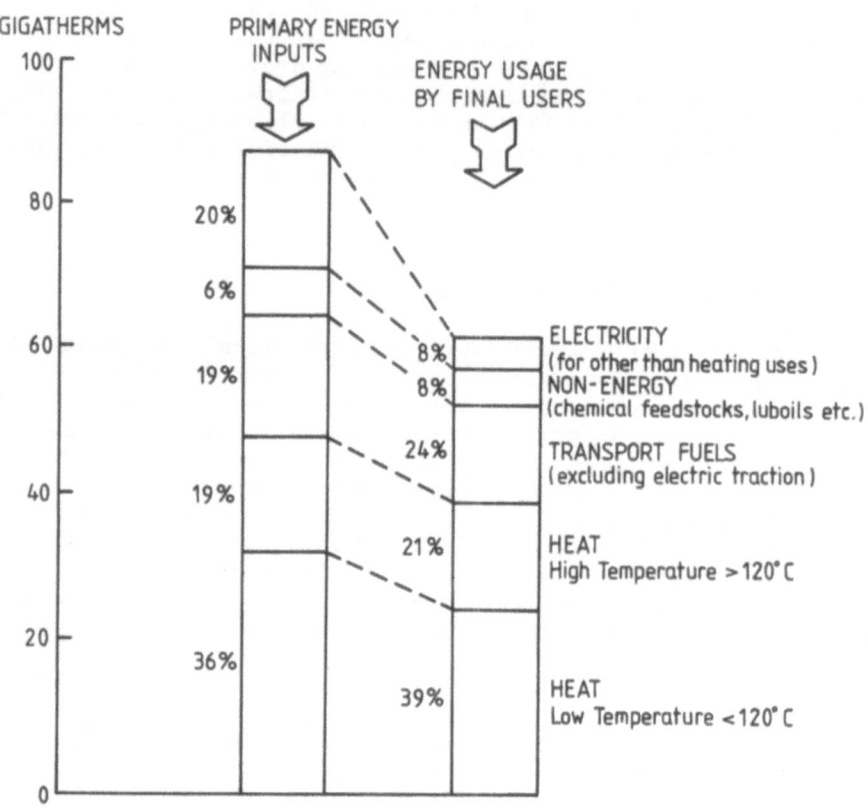

FIG. 1   End use Analysis of U.K. Fuel Consumption 1976 (Ref. 1)

	1971	1972	1973	1974	1975	1976	1977	1978	1979
Coal (1)	15272	11325	11716	10717	9423	8794	8794	8369	8954
Coke and Breeze (2)	944	807	690	634	556	425	530	458	492
Other Solid Fuel (3)	82	70	67	109	81	72	81	100	89
Coke Oven Gas (4)	22	36	46	50	66	62	64	61	68
Town Gas (5)	300	382	300	189	84	25	11	6	7
Natural Gas (6)	1831	2869	3901	4503	4617	5216	5447	5568	5736
Electricity (7)	63170	62974	68998	64965	64510	68521	69830	70875	74076
Petroleum (8)	21555	22142	22182	19820	17884	17879	18056	17769	17719
Creosote/Pitch Mixture(9)	81	49	47	36	34	38	28	23	21

(1), (2), (3), (8), (9) = Thousands of Tonnes

(4), (5), (6) = Millions of Therms (1 Therm ≈ 0.1 GJ)

(7) = GWh

Table 1   ENERGY CONSUMPTION BY U.K. INDUSTRY   (Ref.(2)

SIC	INDUSTRY	INPUT ENERGY $(GJ \times 10^6)$	BOILER FLUE GAS LOSSES $(GJ \times 10^6)$	FURNACE FLUE GAS LOSSES $(GJ \times 10^6)$
20	Food and Food Products	861	115	140
21	Tobacco	16	2.5	1.5
22	Textile Mill Products	308	52	18
23	Textile Apparel	41	4	2
24	Lumber and Wood	179	31	34
25	Furniture and Fixtures	22	3	–
26	Paper and Pulp	1254	129	210
27	Publishing	84	6	15
28	Chemicals	2500	135	125
29	Petroleum	3240	370	715
30	Rubber	197	24	33
31	Leather	26	5	4
32	Stone, Clay and Glass	1606	120	300
33	Primary Metals	2465	114	635
34	Fabricated Metals	273	3	75
35	Non-Electrical Machines	302	22	60
36	Electrical Equipment	246	15	50
37	Transport Equipment	361	11	100
38	Instruments	60	10	7
39	Miscellaneous	59	5.5	13
	Totals	14135	1177	2537

Table 2   1973 U.S. INDUSTRIAL SECTOR ENERGY USE
AND GAS LOSS SOURCES   Ref. (6)

Recent work initiated at Cranfield for the EEC has explored the same basic avenues for UK industry. Our pilot survey covered 502 industrial organisations and highlighted 147 significant thermally activated processes. The extention of this work to a second survey is underway and will include 3000 industrial organisations with a maximum potential of 15,000 different processes. The generally poor response rates to surveys will probably dictate that the final data (available by the middle of 1982) will cover some 1500 to 2000 processes.

Figure (2) shows the distribution of input-output temperature ranges for our pilot survey. Significant peaks occur at temperatures greater than 500°C, in the range 85 to 250°C, and less than 50°C. The groups shown are fundamentally dependent upon the working fluids used within industry and can be typically grouped as follows:

$T < 50^{\circ}$C	Contaminated process water
51°C $< T < 85^{\circ}$C	Condensate and hot process water
86°C $< T < 150^{\circ}$C	Low pressure steam - condensate and flash steam
151°C $< T < 250^{\circ}$C	High pressure steam - high pressure hot water and boiler exhaust gases
251°C $< T < 500^{\circ}$C	Boiler exhaust gases and furnace exhaust gases
501°C $< T$	High temperature exhaust gases and direct combustion processes

The distribution of heating mediums from the pilot survey is shown in Figure (3). It is clear that steam and air/combustion products are the most significant categories with process hot water being the smallest proportion. Oil and electricity represent significant proportions of primary energy input to processes rather than going through a secondary energy transport phase. Natural gas usage appears in both the air/combustion product and steam categories.

The third most significant category relevant to energy storage with waste heat recovery is concerned with the cyclic operational characteristics of process plant. Figure (4) shows the distribution of operating hours for the pilot survey. It can be seen that only a few processes appear to be operated with durations which fall outside the classic one, two and three "shift" systems of industrial working. Further information on this cyclic variation of operation within the various shift splits is currently being obtained.

4. TES COMBINED WITH WASTE HEAT RECOVERY

Thermal energy storage (TES) is not an end in itself, it must be viewed as an aid to energy management. Its application, therefore, is only appropriate in special circumstances where in particular the "process" performance can be improved or at least maintained, and the energy saving potential is significant.

Recycling industrial waste heat is clearly important. However, industrialists are particularly unwilling to invest in "non-productive" equipment unless payback periods are very short.

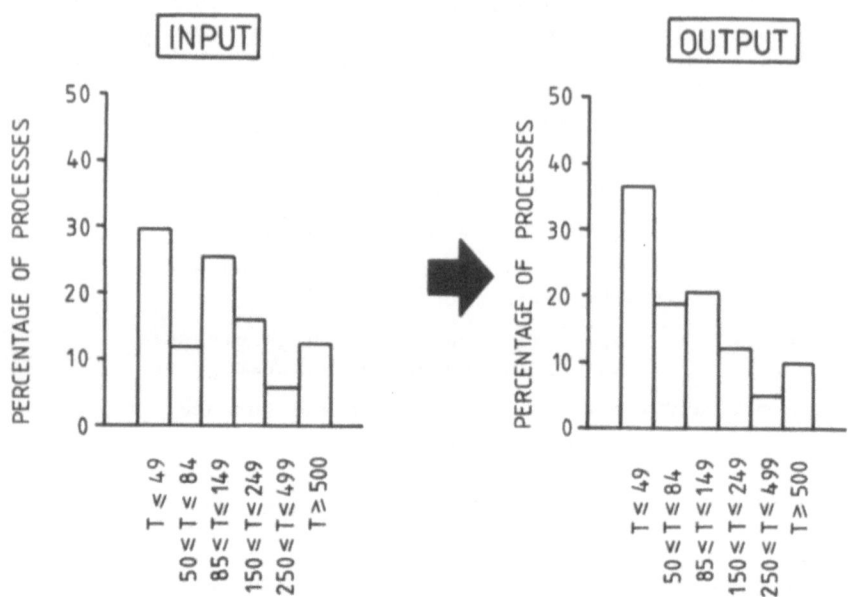

FIG. 2 Distribution of Input-Output Temperatures for the Pilot Survey

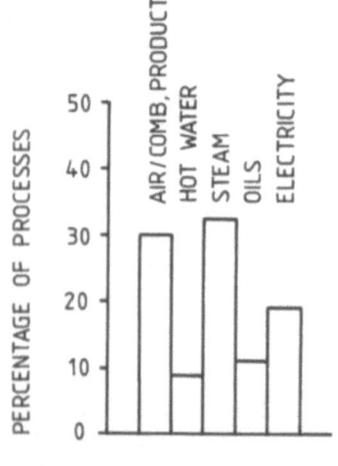

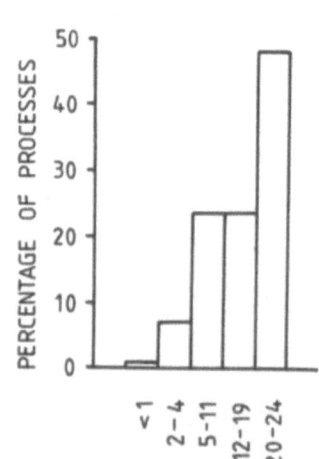

FIG. 3 Distribution of Heating Mediums for the Pilot Survey

FIG. 4 Distribution of Process Operating Hours Per Day

The waste-heat produced within an industrial plant can either be stored for use within the plant or sold to external users who may well possess the storage systems. Most high temperature heat (i.e. > 120°C) can find applications within process plant either by direct heat exchange or through conventional regenerators. Low temperature heat (i.e. < 120°C) is usually very difficult to recover and use in large quantities within a plant complex. This latter category may find increasing applications in district heating especially in the case of chemical process plant.

4.1 Application Identification

Application identification is the single most important aspect of the use of storage systems with industrial process heat.

Application identification consists of two primary factors:
1) the isolation of useful heat sources and sinks either within existing processes or in order to cascade energy to adjacent processes,
2) the identification of temporal mismatches between heat sources and sinks.

The most rational energy management approach is to remove, if possible, the need for storage by modifying the temporal mismatch between the heat source and sink. However, because of the large diversity of existing process stock this is not always possible.

Load charts should be constructed over representative periods of time, (e.g. steam supply and consumption charts). These can then be used as an aid to sizing heat-exchange equipment and any storage plant required. Very often load variations can be severe as shown in Fig.(5), but this has often been overcome by oversized and inefficient boiler installations in preference to the application of some form of thermal energy storage.

4.2 Typical Application Areas

4.2.1 **Cement Industry**. Approximately 80% of the energy used by this industry is consumed in kiln operation. It has been estimated that 75% of a plant's electricity requirements could be met with an on-site waste heat boiler for power generation. The rejected waste heat from the kilns is obviously not available during routine maintenance work. At such times the necessary electrical power of 5-10 MW for periods of 2 to 24 hours must be obtained by other means, for raw and finish milling. The problem could be overcome using thermal energy storage, Ref.(7).

The storage system that has been proposed for use with the dry process kilns involves a solid sensible heat storage material such as magnesia brick or cement clinker. A two store system is

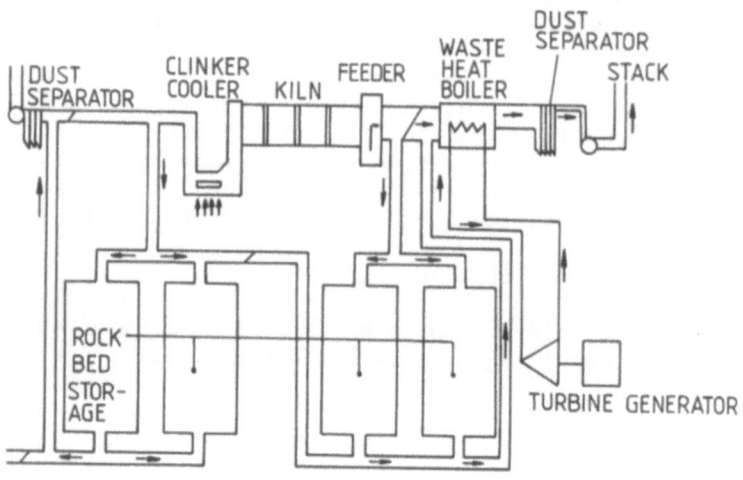

FIG. 6   Waste Heat Recovery in the Cement Industry   Ref.(7)

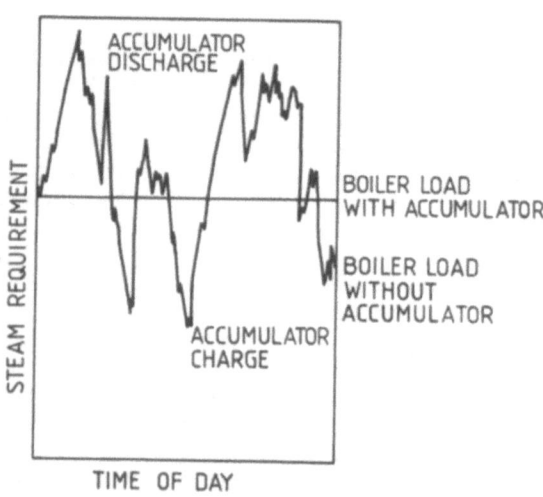

FIG. 5   Typical Steam Load Fluctuating with Time

envisaged, similar to the one shown in figure (6), one accepting high temperature reject heat from the kiln, exit gas at approximately 870°C, and the other storing heat from the clinker at approximately 230°C. These systems would be charged independently but discharged in series.

A further important area is in the use of kiln off-gas to preheat incoming wet cement prior to conversion to clinker.

The preheat gas is continually re-cycled. The alkali content in the gas builds up to unacceptable levels at which time portions of the gas are bled off in a by-pass. When the alkali content is reduced to acceptable levels the by-pass is terminated. This periodic heat rejection represents about 5% of the kiln's total energy demand and could be stored in regenerative units for future drying operations or for use in other inplant processes.

4.2.2 Iron and Steel Production. Approximately 70% of the total energy input to the steel making process is dissipated to the environment as heat in a variety of forms, as shown in Table (3).

The difficulty of recovering heat in a form that can be effectively utilised without creating imbalances within the existing plant energy supply and demand network has traditionally forced the iron and steel industries to the two main areas of combustion air preheating and steam generation with high temperature flue gases.

Form of heat rejection	Percentage rejected
Waste gases and cooling water	29%
Radiation from hot solids	19%
Radiation from surfaces of plant	17%
Iron and Steel plant slag	6%
TOTAL *	71%

(* specific energy consumption per tonne of steel produced $\simeq$ 25 GJ/tonne)

TABLE 3   DISTRIBUTION OF REJECTED HEAT FROM STEEL MAKING

The traditional application of switched regenerators has tended to evolve into the hot blast stove of the modern blast furnace, fired by low calorific value "off-gases". Switched regenerators are still used in open-hearth furnaces of older works for recycling heat to the melting process.

On a typical installation incoming combustion air is heated from ambient up to 1050 - 1100°C at a rate of 20-30 tonnes/hour by outgoing combustion products at a temperature of 1650°C for a silica or 1700°C for a basic furnace. Under normal operating conditions the combustion products leave the regenerator at a temperature of 500 to 550°C with gas firing, and 550°C to 600°C with oil firing, allowing for further heat recovery in waste heat boilers.

The bulk, cost and operational disadvantages of large switched regenerators in the iron and steel industry means that most effort is now being placed on ceramic and metallic recuperators of the convective and radiative type. The gas exit temperatures from recuperators are higher, but the capital cost is lower. Waste heat boilers are more expensive because of water treatment, and because of the variable supply nature of the steam produced. However, a waste heat boiler will give rise to a much higher level of heat recovery than a recuperator where exit temperatures of 250°C instead of 500°C occur.

One typical application of steam storage with a waste heat boiler for a UK steel works was estimated as costing £50,000 with an annual saving in excess of £1,000,000. A simple payback calculation shows this type of installation to pay for itself in under one month of full production. However, the steelworks in question was itself closed last year!

Other applications include the use of the intermittent electric arc furnace fume stream. This heat would be charged into a solid sensible heat storage medium capable of withstanding temperatures up to 1650°C (i.e. refractory brick, slag), and discharged on demand through a heat exchanger to produce steam. This steam would be used to drive a turbogenerator to produce electricity. At the moment in the steel industry, the dust-laden fumes stream from the electric arc furnace, is water quenched, and then ducted to the dust collection system before discharge to the atmosphere. The unquenched fume stream should flow from the furnace through the energy storage medium prior to being discharged through the existing dust collecting system. During peak demand periods, the combined streams from the furnace and the store should flow through the heat exchanger to create the steam for the turbogenerator.

4.2.3 Paper and Pulp Industries. Paper mills have large steam demands. Pulping processes in batch digesters are very cyclic and create sudden surges in steam supply. However only a small proportion of European industry (with the exception of Sweden) requires pulping.

Fuel substitution (i.e. wood waste) can be used to operate current waste fuel boilers at high base loads whilst discharging the steam to accumulators during periods of low demand. The economics of steam accumulation in preference to thermal regenerators are superior for storage periods of less than one hour.

Paper finishing mills use vast quantities of steam which is cascaded from one roller bank to another. Steam is fed to the inside of the polishing rollers where it accelerates the drying of the paper. Although the process of paper finishing is intended to be continuous, frequent breaks in the paper occur. These can be up to 20 per day, each taking up to 10 minutes to repair. Because the main boiler must be kept on load, the steam is currently vented to atmosphere. Steam recovery could take place and it would be stored in conventional steam accumulators for either integration into the low pressure steam network or used with a feed water storage unit.

4.2.4 Primary Aluminium Smelting. Most of the waste energy available from this industry is of a low grade, but in large quantities. At the present time the only possible use for this low grade energy is for space heating in dwellings. Since space heating is a cyclic energy demand, energy storage is necessary in using the constant waste energy supply from aluminium processes.

The electrolytic alumina reduction cells used in the process reject air at about 150°C. It has been proposed, Ref.(8), that this rejected heat be used to heat water that is stored in conventional storage tanks for use in district heating systems. It has been estimated that one plant can supply enough space heating for approximately 12,000 homes.

4.2.5 Glass Industry. The glass industry furnaces currently use high temperature regenerators. The checkerwork in a regenerator is normally about 8m high with a flow rate of about $1.0 m^3/s$ and with a velocity of $0.6 - 1.5$ m/s. The design of the regenerator is governed largely by the choice of fuel for the furnace, since if oil is to be used, the regenerator will have larger passages and less height than one used with a furnace burning natural gas. This is because the sulphur in the oil (less than 0.2% sulphur content in low sulphur oils) combines with the sodium salts in the furnace to produce sodium sulphate (Na_2SO_4) which blocks the checkerwork passages. The natural gas regenerators need to be taller, since the lower luminosity of the flame makes it more difficult to force the heat into the glass and makes it necessary to recover more heat from the exhaust gas.

The regenerators are not regarded as fuel economy devices but as essential parts of the furnace, to enable the intake air which will already have been taken from a warm area, probably from under the furnace with a temperature of about 50°C and to preheat it to 1100°C and allow the furnace to reach the operating temperature of 1600°C. The furnace inlet air is passed through one regenerator for about 20 minutes while the hot exhaust gas is passed through the other regenerators; the flow is then reversed by the moving of dampers and the burners on the opposite side are then fired. While the flows are being reversed there will probably be a dead time of about 20 seconds.

Another potential area where TES could become important is in glass production runs in day tanks (batch furnaces). These are required for high quality crystal glass or special orders requiring stringent quality control. Furnace exhaust heat can be stored for use in the subsequent annealing and tempering phases of manufacture.

4.2.6 Food Processing. This type of industry is a major user of low temperature ($< 120^{o}$C) process heat. Approximately 85% of the fuel utilised by this industry is used to produce process steam and hot water at low temperature. This is used for cooking, sterilization and equipment and work area washing and clean-up. Much of the heat is then rejected to the environment as steam or hot water. It is feasible that the majority of this waste can be recovered and used through storage.

The low temperature process water storage that many industries require dictates that storage will only be economically practicable when there is a large immediate-need application. Food processes of interest for TES include repeated pressure cooking cycles for quick processing and then canning, blanching for vegetables, periodic hygenic operations using hot water, grain drying and tobacco curing. These are batch processes where TES could be used to cascade energy from one operation to the next at a lower temperature with storage to provide the necessary temporal and instantaneous demand matching.

4.2.7 Textile Industry. The major thermal energy demand within this industry sector is in fabric colouring and finishing. The process fluid used is steam. Historically dye becks are used as batch fabric colouring vats using hot water, heated by steam, to condition the material for accepting dye addition to specification. The batch nature of the process means that sudden surges in steam load occur. There is clearly a requirement for steam storage in order to minimise the size of the boilers and maximise the efficiency. In addition "dirty" water is dumped, wasting this contained energy. TES would allow the re-use of this heat for later batches.

Current trends towards jet dying and continuous colouring processes result in much smaller load variations and so the need for steam accumulation is less pronounced.

4.2.8 Clay and Ceramic Industries. The characteristics of this type of industry are fundamentally different from the cement industry in that periodic operation of tunnel kilns often occurs.

Approximately 40% of the heat produced by curing bricks and ceramic products in periodic kilns is at the moment recovered and used for curing and drying processes. The remaining 60% is rejected as hot exhaust gases containing impurities from the processes. Conventional gas to gas recovery equipment could be used with storage to recycle a large proportion of this high temperature heat for combustion air heating.

4.2.9 Plastics and Rubber Industry. The wide diversity of plastics and rubber curing processes used throughout industry necessarily dictates that major applications within this industry are difficult to isolate.

Many thermosetting plastic processes are batch moulded at temperatures in the range 120°C to 215°C. Tyre moulding also involves step inputs of steam to soften rubber "layups" to form and cure tyres. Steam accumulation could find applications in this area.

One classic example which exposes the problems of cyclic steam loads is the curing of polystyrene products. These products require relatively low pressure steam at about 2 to 4 bar for short periods of time (of the order of 10 minutes). Many early installations of steam accumulators were made, however, as production generally increased during the late 1960's, multiple machines correctly phased with respect to steam demand was found to be a more economical solution.

4.2.10 Secondary Metals Industries. Metal products, i.e. castings, require heating in soaking pits or tunnel furnaces prior to normalisation, tempering and other forms of thermal conditioning. Very often large masses of metal products are quenched in oil vats. Heat recovery from the quenching oil could provide useful quantities of low grade heat for subsequent processes. Storage, however, is unlikely to provide much scope for improvement in the utilisation of this waste heat. Table (4) gives a brief summary of industrial processes suitable for thermal energy storage applications.

TABLE (4)   SUMMARY OF INDUSTRIAL PROCESSES SUITABLE FOR TES

	OVERALL MERIT	PROCESS KNOWLEDGE	PROCESS DIVERSION	ENERGY SAVINGS	SPECIAL HEAT EXCHANGE	TEMP °C	TES FUNCTION	PROCESS TES TECHNOLOGY
CEMENT								
- WHR from kilns	F	G	Sm	L	No	870	S	Regen
- WHR from clinker	F	G	Sm	L	No	230	S	Regen.
- WHR drying	G	G	Sm	Sm	No	230	D	Regen.
IRON AND STEEL								
- Combustion air pre-heat	Sm	G	Sm	L	No	500 - 1100	S	Regen.
- Steam generation from combustion products	G	G	L	L	No	250	S	Steam Acc.
- WHR from electric arc furnaces	F	G	Sm	L	No	500 - 1650	S	Regen.
PAPER AND PULP								
- Waste material hog boilers	G	G	L	L	No	250	D	Steam Acc.
- Batch pump digestors	P	P	Sm	Sm	No	145	S	Steam Acc.
- Paper breaks	G	G	Sm	L	No	120 - 150	S	Steam Acc.
ALUMINIUM INDUSTRY								
- Bauxite drying	F	F	M	M	No	120	D	UNK
- Aluminium cooling	P	F	L	M	UNK	160	S	UNK
- WHR District heating	F	F	L	L	Yes	90 - 60	D	Hot Water
GLASS INDUSTRY								
- WHR Furnaces	G	G	Sm	L	No	50 - 1100	S-D	Regen.
- WHR Batch furnaces	G	G	L	Sm	No	50 - 1100	S-D	Regen.

FOOD PROCESSING								
– Washing	G	G	L	UNK	No	60	D	Hot Water
– Scalding	G	G	L	UNK	No	50 – 60	D	Hot Water
– Pasteurisation	G	G	L	UNK	No	65	D	Hot Water
– Blanching	G	G	L	UNK	No	85	D	Hot Water
– Cooking	F	g	l	UNK	No	100 – 115	D	Steam Acc.
– Sterilisation	F	G	L	UNK	No	110	D	Steam Acc.
– WHR from wort kettles	G	G	Sm	G	Yes	120	S	Steam Acc.
TEXTILE INDUSTRY								
– Dye Becks	F	F	M	M	No	90 – 120	D	Steam Acc.
CLAY AND CERAMIC INDUSTRY								
– WHR from periodic kilns	P	G	L	Sm	No	250 – 350	S	UNK
PLASTICS AND RUBBER								
– Thermosetting processes	F	F	L	UNK	UNK	120 – 216	D	UNK
– Polystyrene curing	F	F	L	UNK	No	120 – 150	D	Steam Acc.
SECONDARY METAL INDUSTRY								
– Heat treatment	F	F	L	UNK	UNK	UNK	S	UNK
CHEMICAL INDUSTRY								
– Batch processes	U	P	L	UNK	UNK	100 – 220	S	UNK

KEY:

S	=	Variable Supply	Sm	= Small
D	=	Variable Demand	F	= Fair
UNK	=	Unknown	G	= Good
			P	= Poor
WHR	=	Waste heat recovery	L	= Large

Regen. = Regenerator
Steam Acc. = Variable pressure steam accumulator
Hot Water = Atmospheric hot water tanks

5. TES TECHNOLOGIES IN USE IN INDUSTRY

From the preceding section on typical TES applications it is clearly evident that three TES technologies dominate current and possible future applications; namely:
1) High temperature regenerators
2) Steam accumulators
3) Hot water storage

5.1 High Temperature Regenerators

Regenerators have two main elements, the material which is thermally cycled and the containment vessel with its associated insulation, cooling, foundations etc.   Both have a significant impact on the overall cost effectiveness of the TES as does the manner in which the storage material is used in the store. Direct contact TES is generally preferred, in which solid materials, such as bricks, checkers, pebbles or fine (fluidised) particles, are in direct contact with the pressurised air or combustion products, as shown in Figure (7).

Table (5) lists some candidate materials for high temperature regenerators, Ref.(9).

The general material requirements for regenerators are:
1) the ability to withstand oxidising atmospheres of up to 1500°C
2) high resistance to thermal shock
3) resistance to condensation and water evaporisation
4) high density, specific heat and thermal conductivity
5) low coefficient of expansion
6) low cost.

An approximate ranking of some refractory types on the basis of volumetric heat capacity (product of density and specific heat) and thermal conductivity are given in Table (6), together with an indication of their thermal shock resistance and cost.   On the basis of these rankings silica refractories have low thermal shock resistance and fireclay (low cost), high magnesite and high alumina (improved heat capacity and conductivity) are good, whilst materials such as Feolite (sintered iron oxide) and cast iron offer significant improvement in volumetric heat capacity.

Matrix configuration: The TES matrix is ideally arranged to offer the largest surface area to volume ratio with the smallest void fraction possible within cost constraints.   Standard square "checkers" are in common use in glass and steel industry regenerators, however pebbles with diameters less than 50mm show significant advantages in terms of surface area per unit volume and cost. Pebble beds have a further advantage in that surface heat transfer

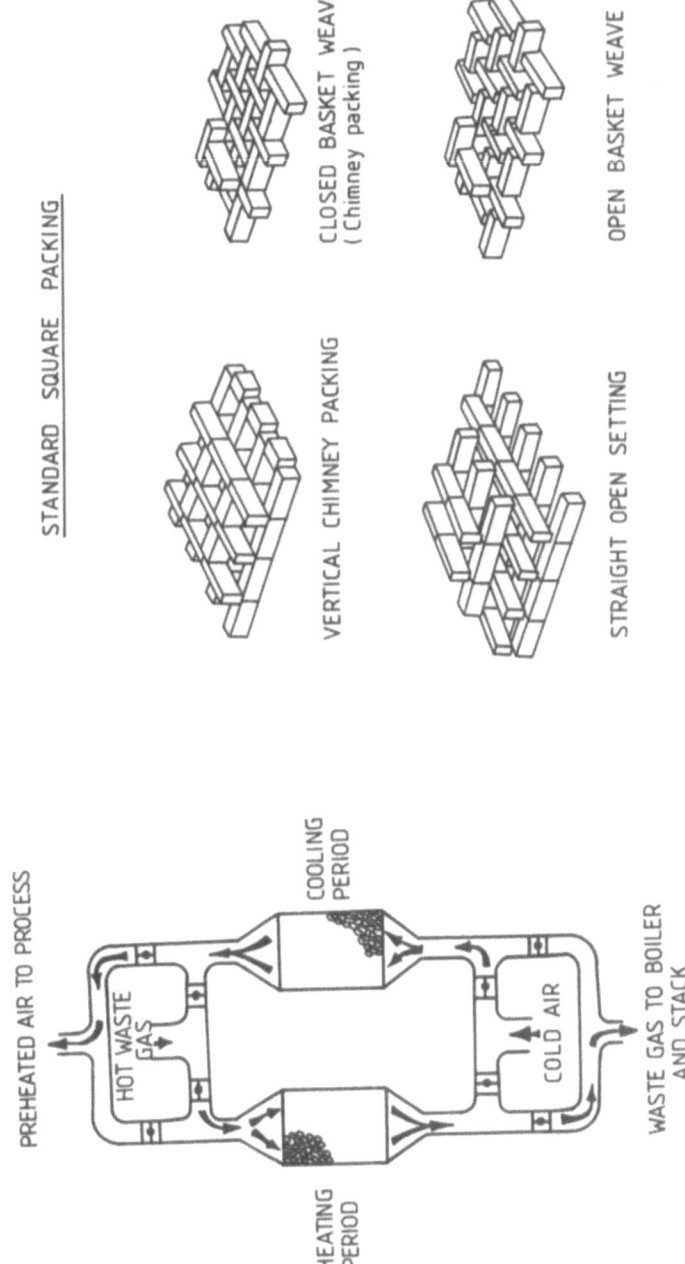

FIG(7) General arrangement of regenerators and typical checkers used.

Table (5) CANDIDATE HEAT STORAGE MATERIALS FOR SENSIBLE HEAT TES Ref (9)

Material	Operating Temperatures (°C)	Volumetric Heat Capacity $(\rho.c)$ (kJ/m³ °C)	Mass Heat Capacity (c) (kJ/kg °C)	$\frac{1}{\rho.c}$ Expressed as a Fraction of that for Dense Ceramic Pebbles	Cost (K) (£/kg)	$\frac{K}{c}$ Expressed as a Fraction of Dense Ceramic Pebbles
60% LiF/40% NaF (mol %)	Solid < 652	4130	1.63	0.56	2.2	6.63
46.5% LiF/11.5% NAF/ 42% KF. (mol %)	Solid < 454 / Liquid > 454	3480 / 3820	1.34 / 1.88	0.66 / 0.61	1.8	6.60 / 4.70
Na	Liquid 98-881	1100	1.3	2.10	0.512	1.94
NaK 56% Na/44% K (mass %)	Liquid 19-826	848	1.05	2.72		
NaOH	Liquid 186-1000	3987	2.27	0.58	0.122	0.264
Thermal Fluid (Shell Thermia 23)	< 320	1870	2.40	1.24	0.256	0.524
Dense Ceramic Pebbles	< 1100	2311	0.963	1.00	0.196	1.0

Table (6) RANKING OF SOME REFRACTORY MATERIALS ON VOLUMETRIC HEAT CAPACITY AND THERMAL CONDUCTIVITY (TOGETHER WITH COST AND THERMAL SHOCK GUIDE) Ref (9)

Material in Ranking Order (worst first)	Volumetric Heat Capacity (kJ/m³ °C)	Thermal Conductivity (W/m °C)	Cost (in standard squares) (£/m³)	Thermal Shock Resistance
Silica	1500-1900	1.0-1.5	130-220	low especially up to 500°C
Fireclay	2000		80-140	good to excellent
High alumina	2400-3300	1.6-4.0	420(55% Al$_2$O$_3$) -1640(90% Al$_2$O$_3$)	good to excellent
High magnesite	3700	3.0-6.0	600(90% MgO)	Moderately good to excellent
Feolite	5829	5.0	255	capable of continuous cycling between ambient and 1000°C
Chrome Cast Iron		20.9	-	excellent
Berrylia	5400	70-120	very expensive	-

coefficients are generally larger than for checkers, due to the disturbed flow conditions created by the pebbles.

Table (7) shows typical surface area to volume ratios and void fractions for various matrix geometries in common use.

Pebble beds offer the greatest simplicity of design for a wide variety of applications which differ from the traditional applications.

Despite the relative novelty of the large pebble bed regenerators there are some examples which give confidence to the concept. NASA Ames, Ref.(10) have operated a pebble bed 6 cycles per day for around 15 years to provide preheat to a hypersonic, re-entry vehicle testing tunnel. A pebble bed regenerator has been used as a high temperature supersonic combustion chamber test facility at Cranfield since 1967, Ref.(10).

The layout of this type of regenerative heat exchanger is shown in Figure (8). As can be seen, the facility consists of a pressure vessel lined with several grades of refractory materials and with a bed of randomly packed refractory pebbles at its centre, as illustrated by Figures (9) and (10).

The ceramic elements are heated by means of the kerosene burner shown mounted in the roof of the heater. The combustion products from this burner are blown down through the pebble-bed and are then exhausted to atmosphere. The heating cycle usually lasts three to four hours until the required temperature distribution is obtained. At this point the kerosene burner is closed off, the exhaust valve closed and test-air, supplied from the compressor house at the required pressure, is passed upward through the bed. This test-air, heated by the pebbles during its passage through the bed, is then discharged through the appropriate test-section. A pebble diameter of 0.5 inches was used which is small by comparison with many previous applications, however good heat transfer was obtained.

The limit in air mass-flow which can be heated in any particular facility is reached when the aerodynamic lift force acting on each pebble exceeds the pebble weight and the pebbles "lift-off".

The insulating lining is made up of individual units in three layers as shown in figure (8). The inner layer is constructed of Norton RA 5190 high density alumina in small interlocking shapes. The lower half of this layer is self-supporting from the floor and the upper half is supported from the steel ring welded inside the pressure vessel. Thus the lower elements do not have to support all of the weight of the roof lining. The RA 5190 is backed by a layer of Norton RA 4058 bubble insulating alumina, and behind the RA 4058 is a layer of low-temperature firebrick.

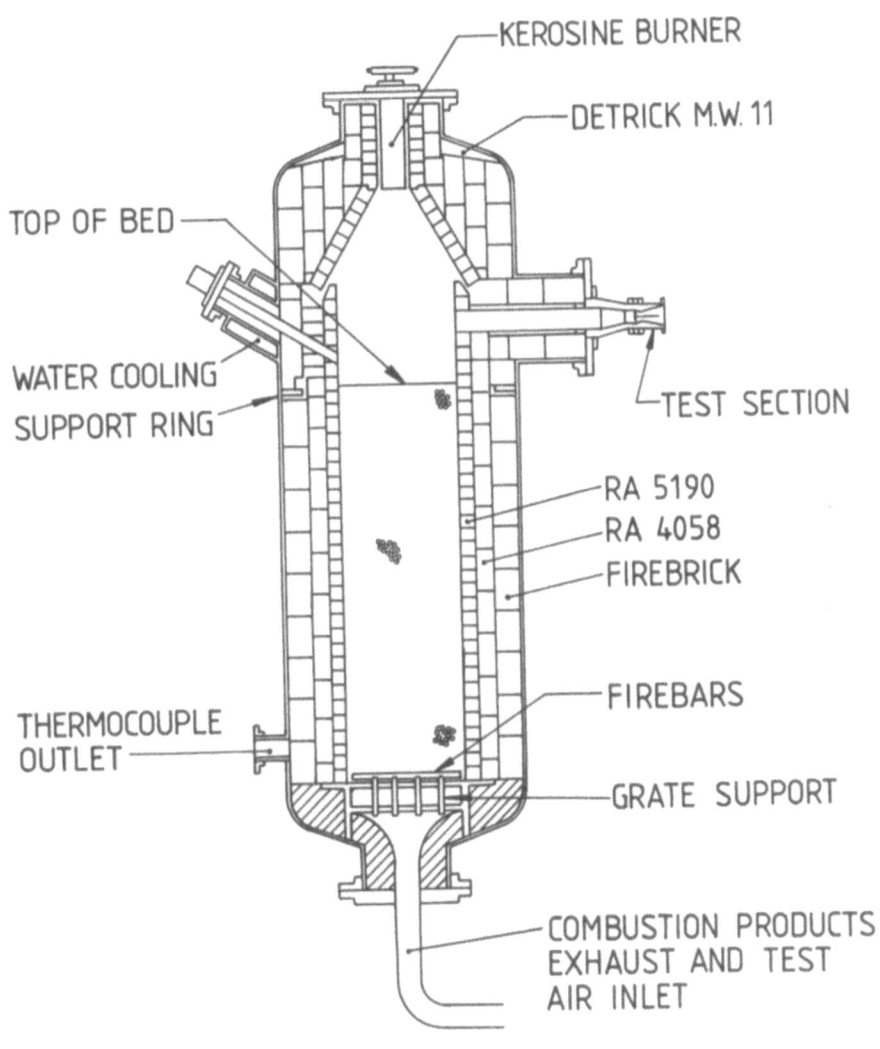

KEROSINE BURNER

DETRICK M.W. 11

TOP OF BED

WATER COOLING
SUPPORT RING

TEST SECTION

RA 5190
RA 4058
FIREBRICK

FIREBARS

THERMOCOUPLE
OUTLET

GRATE SUPPORT

COMBUSTION PRODUCTS
EXHAUST AND TEST
AIR INLET

FIG. 8   CIT Supersonic Combustions Test Facility Regenerator

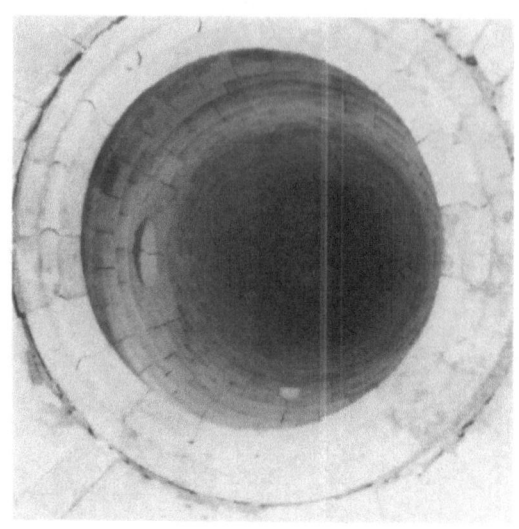

FIG. 9 Internal Refractory Insulation on regenerator

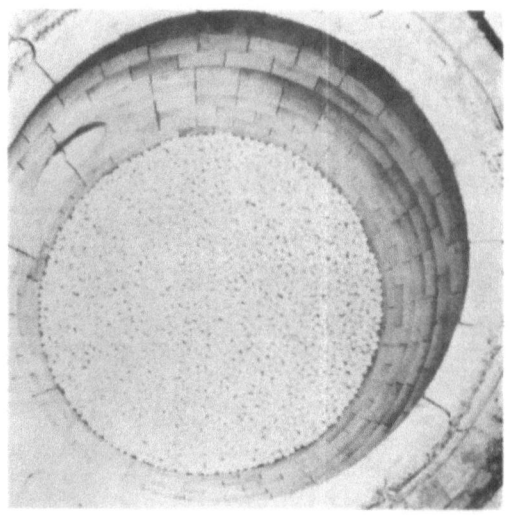

FIG. 10 Randomly packed regenerator pebbles

Matrix Geometry	Surface Area/Volume Ratio (m^2/m^3)	Void Fraction
Typical matrix constructed from standard squares (152 mm square gas passages on 229 mm pitch)	11.7	0.44
M&P checkers (40 mm dia. gas passages)	38.3	0.377
M&P checkers scaled to 25.4 mm passages	60.3	0.377
Pebbles 50.8 mm dia.	73.2	0.38
Pebbles 19.1 mm dia.	194.8	0.38
Small channel checkers: 6 mm channels on 26 mm triangular pitch	32.2	0.0483
6 mm channels on 16 mm triangular pitch	85.0	0.1275

Table (7)   SURFACE AREA/VOLUME RATIO AND VOID FRACTION OF VARIOUS MATRIX GEOMETRIES

The support for the pebbles is shown in Figure (11). This consists of 1" diameter Incaloy D.S. round bars supported by four Incaloy D.S. support bars bolted together in the configuration shown. This total assembly is mounted into a support ring which in turn is set in a cast base of Norton 33-I castable cement and through which the load is transmitted to the pressure vessel. Stress calculations made on the Incaloy support bars indicate that they will withstand the load at temperatures up to 1000°C. More importantly, creep calculations indicate that the bars have a life expectancy of up to 10,000 hours operation providing that their temperature does not exceed 900°C.

A major cause for concern arose when pebbles "lifted-off" at the beginning of one of the full-scale tests. This occurrence was as mystifying as it was drastic since the bed pressure was well below the operating pressure at the time. Upon investigation it was found that water, which had been used to cool the exhaust gases, had been trapped in the outlet pipe. Thus, when the cycle was reversed this water was carried into the bottom of the pebble-bed by the flow of test-air. The resulting sudden expansion of steam so produced was equivalent to a mass-flow far greater than the maximum allowable.

FIG. 11 Pebble Support Plate

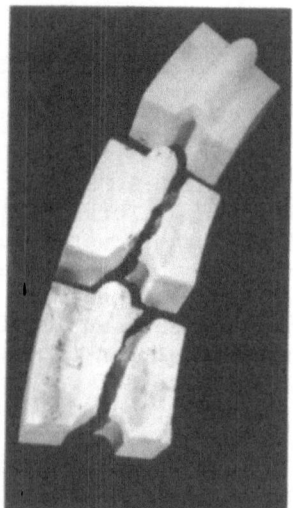

FIG. 12 Typical ceramic insulation failure

The most recent of the problems associated with the operation of the facility still lacks a solution. This takes the form of severe cracking of some of the lining bricks. This phenomena is in no way connected with over-heating of the bed, but has all the appearances of thermal shock. The manufacturers are still investigating this occurrence but have stated that it has not taken place in any of the USA facilities for which they have supplied the ceramics. The bricks shown cracked in Figure (12) are not load-bearing members and have been replaced as they are, for further test runs. If this problem is due to thermal shock it is surprising, since the bricks lower down the bed are subject to greater temperature gradients yet remain undamaged. This problem could well be associated with cyclic thermal expansion and pebbles being "locked" across the diameter of the bed, thus inducing high local expansion stresses and causing ceramic insulation failure.

5.2   Regenerator Thermal Design Methods

There are two classic approaches towards the analysis of the transient performance of regenerators. These are generally associated with the way in which the regenerator is used. Very often a pair of high temperature regenerators are switched in order to produce preheat to a continuous process (e.g. in the iron and steel industry). It is clear that in such an application switching between heating and cooling for a fixed period of time (of up to 20 minutes) will eventually establish a state of operation in which the temperature at any place in the gases or walls is the same as it was at a time one full cycle earlier. This state is called a "state of cyclic operation". In contrast to this state a second fundamental "state of starting operation" or "single blow" is often used where the charge and discharge periods may not be the same and there is a possibility of dwell periods occuring between successive cycles.

A number of analytical methods have been evolved for the design of regenerators under states of cyclic equilibrium operation. The principal method is due to Hausen, Ref.(11). The main differences in the various methods are due to the approximations made, the time dependent temperature of the fluid and the solid matrix.

Numerical methods are available and are split into two types of solution, those in which a separate study of the heat transfer material and those in which conduction through the matrix material is included.

Reference should be made to the wide body of literature available on the design of regenerators and only a critical synopsis will be given here based on the well-established

analytical method of Hausen as applied to TES sizing. The
Hausen theory expresses performance in terms of an overall heat
transfer coefficient together with a regenerator correction term
(F) presented as a function of the modified bed parameters (the
"reduced length" (Λ) and the "reduced period" (Π)).

Under cyclic equilibrium operation the thermal performance
of the regenerator can be determined by the effectiveness
expressed in terms of the end of period temperature difference
for a given period of time applicable to both the charge and
discharge cycles. The effectiveness (ε) is given by:

$$\varepsilon = 1 - \frac{\Delta T_e}{\Delta T_o} \qquad \qquad \ldots (1)$$

where ΔT_e is the end of period temperature difference and ΔT_o is
the temperature difference between the hot and cold fluid streams.
It is clear that when $\Delta T_e \to 0$ the effectiveness approaches 100%.
$\Delta T_e = 0$ is obtained when the charge period is long (for no heat
loss) or when the number of transfer units equals infinity.

Very often limits must be set to the end period temperature
difference in order not to heat critical parts of the regenerator
beyond safe temperature limits (i.e. such as condensation and
incipient corrosion limits when cooling hot gases and end point
maximum temperature levels on packed bed support structures).

The key variables of interest to a designer are the surface
area for a given regenerator duty and more importantly the volume
of the regenerator.

An expression for surface area (S) is given by Ref.(12):

$$S = \frac{1}{F} \left\{ \frac{1}{h_c \, P_c} + \frac{1}{h_d \, P_d} + \frac{\phi d_e}{3k} \left(\frac{1}{P_c} + \frac{1}{P_d} \right) \right\} \frac{\Delta T_o W_c \, P_c}{\Delta T_e} \ldots (2)$$

where

      h_c, h_d = air to regenerator material heat transfer
                coefficients for the charge and discharge cycle

      P_c, P_d = charge and discharge periods

      d_e = equivalent matrix semi-thickness

      k, ρ, c_p = matrix conductivity, density and specific heat

      ΔT_o = step change in temperature

ΔT_e = thermal steady state end temperature difference

W_c, W_d = $\dot{m}Cp$ for charge and discharge

ϕ = correction factor to allow for reversal of temperature profiles

The characteristics of regenerators are generally described in terms of the reduced length (Λ) and reduced period (Π) given by:

$$\Pi = \frac{2}{\rho C_p \, d_e \left\{ \dfrac{1}{h_c \, P_c} + \dfrac{1}{h_d \, P_d} + \dfrac{\phi d_e}{3k} \left(\dfrac{1}{P_c} + \dfrac{1}{P_d} \right) \right\}} \qquad \ldots (3)$$

and

$$\Lambda = \frac{4S}{(W_c \, P_c + W_d \, P_d) \left\{ \dfrac{1}{h_c \, P_c} + \dfrac{1}{h_d \, P_d} + \dfrac{\phi d_e}{3k} \left(\dfrac{1}{P_c} + \dfrac{1}{P_c} \right) \right\}} \qquad \ldots (4)$$

An approximation for the correction factor F for $\Pi > 15$ is

$$F = 1 - 0.7 \frac{\Pi}{\Lambda} \qquad \ldots (5)$$

Checker matrices are generally within the range of published solutions (typically $\Pi = 20$, $\Lambda = 40$) but pebble beds are typically outside current published data ($\Pi \approx 130$, $\Lambda \approx 70$).

By substituting equation (5) into (2) and using the fact that

$$\frac{1}{1 - \varepsilon} = \frac{\Delta T_o}{\Delta T_e} \qquad \ldots (6)$$

the total surface area is given by:

$$S = W_c \, P_c \left\{ \left[\frac{1}{h_c \, P_c} + \frac{1}{h_d \, P_d} + \frac{\phi d_e}{3k} \left(\frac{1}{P_c} + \frac{1}{P_d} \right) \right] \left(\frac{1}{1 - \varepsilon} \right) + \frac{0.7}{\rho \, C_p \, d_e} \right\} \qquad \ldots (7)$$

The factor ϕ is dependent on matrix geometry, thermal properties of the matrix and lies in the range $0.6 - 1.0$. A value of 0.8 is generally acceptable. Equation (2) can be further transformed in terms of matrix volume by using the following expressions:

$$V = \frac{S}{\sigma} \qquad \qquad \ldots(8)$$

$$d_e = \frac{1 - e}{\sigma} \qquad \qquad \ldots(9)$$

$$\text{and} \quad \sigma = \frac{(1 - e)6}{d} \qquad \qquad \ldots(10)$$

where V is the matrix volume,
    d is the pebble diameter, and
    e is the bed voidage.

The matrix volume is therefore given by

$$V = \frac{W_c\, P_c}{(1 - e)} \left\{ \left\{ \frac{d}{6} \left(\frac{1}{h_c\, P_c} + \frac{1}{h_d\, P_d} \right) + \frac{d^2\phi}{108k} \left(\frac{1}{P_c} + \frac{1}{P_d} \right) \right\} \frac{1}{1 - \varepsilon} + \frac{0.7}{\rho\, C_p} \right\} \qquad \ldots(11)$$

Using expressions (11) and (7) it is possible to establish initial design estimates of the regenerator and pebble size required for a given application. In addition the sensitivity of important design variables can be investigated. For instance the changes in matrix volume resulting from changes in pebble diameter can be seen to be dependent on the relative magnitude of the three dominant terms of equation (11). These are respectively the "convective" resistance term:

$$(1 - \varepsilon)\frac{d}{6} \left(\frac{1}{h_c\, P_c} + \frac{1}{h_d\, P_d} \right)$$

the wall resistance term:

$$\frac{d^2\phi}{108k(1 - \varepsilon)} \left(\frac{1}{P_c} + \frac{1}{P_d} \right)$$

and the heat capacity term:

$$\frac{0.7}{\rho C_p}$$

Typically the wall resistance term is small in all cases and for high thermal capacity materials larger reductions in total volumes can be obtained by reducing pebble sizes or the bed cross-sectional area to flow. Clearly it is important to establish the relation-

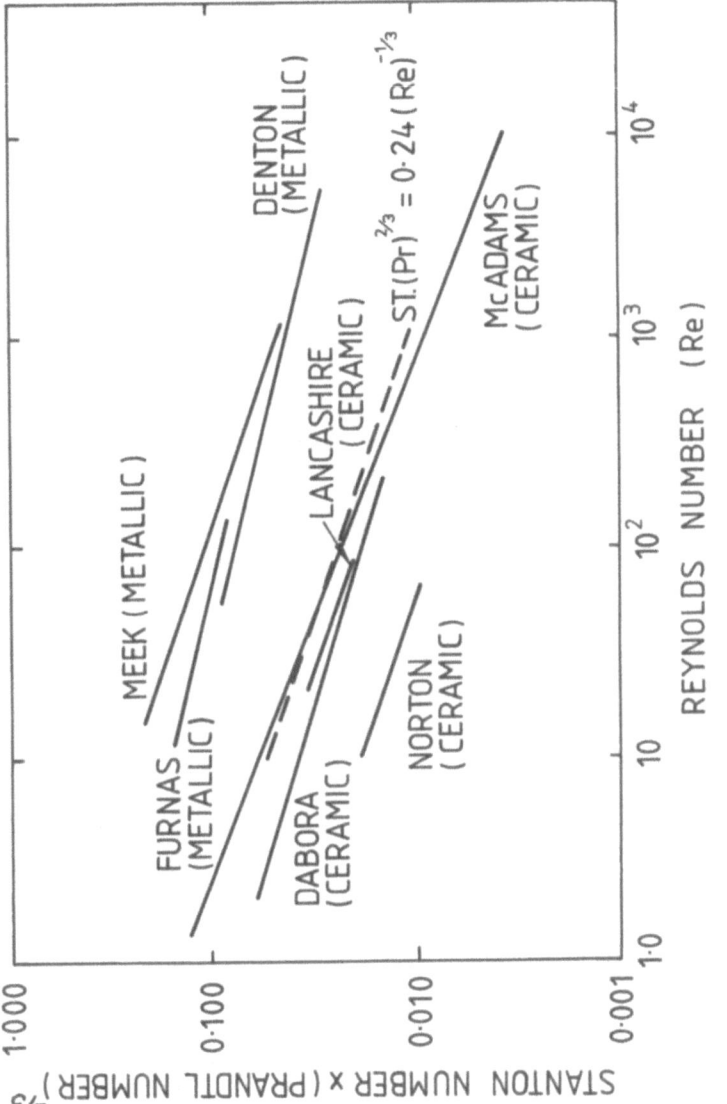

FIG. 13   Packed Bed Heat Transfer Correlation

ships available for determining the variation in heat transfer
coefficients with bed Reynolds number. Figure (13) illustrates
some of the correlations available by which h_c and h_d can be
determined.   The correlations are all of the type:

$$St \ (Pr)^{2/3} = M \ Re^{-(n)} \qquad \qquad ...(12)$$

where n is typically (1/3). The constant (M) depends on the
regenerator material. For Ceramic particles the constant lies in
the range $0.05 < M < 0.1$ and for metalic particles in the range
$0.2 < M < 0.6$.

Generally the heat transfer coefficient increases as the bed
Reynolds number increases. However, for a fixed size regenerator,
greater thermal effectiveness can only be achieved by reducing the
air mass flow rate. This is because the gradient of the Nusselt
number versus Reynolds number characteristic is always less than
unity and increasing values of flow rate reduce the fluid residence
time within the bed and increases the significance of the wall
resistance term in equation (11). Care should therefore be taken
not to seek to increase the heat transfer coefficients at the
expense of thermal effectiveness. In addition the increased
viscid pressure loss inevitably associated with increased air mass
flow rates means that for many applications the overall heat
transfer coefficient must be kept low.

Johnson Ref.(12) obtained numerical solutions for the
"single blow" or "step input" conditions which is of general
validity. His results are shown in Figure (14) where the gas outlet temperature is plotted in non-dimensional form ($\bar{T}$) against
the utilisation factor (UF); where

$$\bar{T} = \frac{T - T_o}{\Delta T_o} \qquad \qquad ...(13)$$

and

$$UF = \frac{W_c}{V \ \rho_p} \ t \qquad \qquad ...(14)$$

ρ_p is the density of the pebbles and t is the time

This is one of the most convenient ways of representing the
fluid temperature changes through the regenerator matrix and
shows the characteristic "S"-shaped curves of the transient
response, with the temperature ultimately approaching the final
equilibrium condition of the gas inlet temperature. By comparing
observed temperature-time variations with those predicted, the

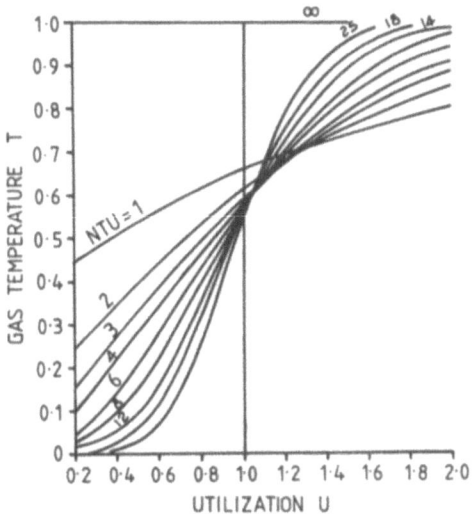

FIG(14)   Single-blow outlet response of a regenerator Ref.(12)

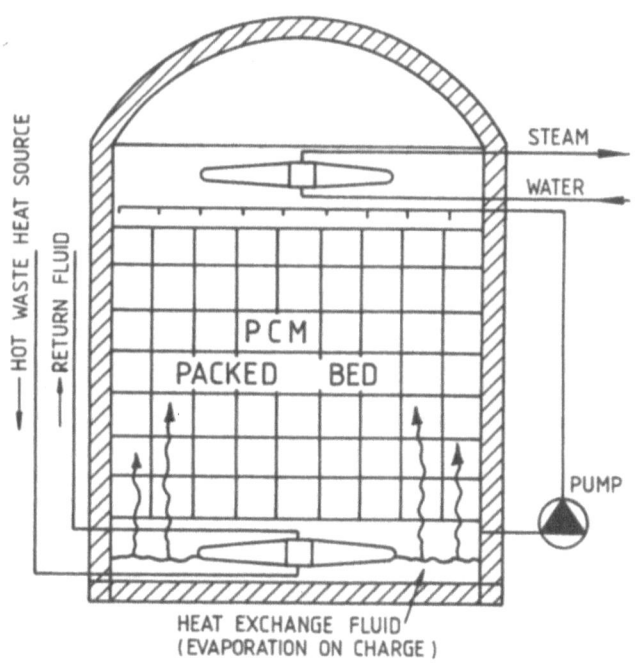

FIG(15)   High temperature Phase-Change Reboiler Ref.(15)

value of the Number of Transfer Units (NTU) can be determined; that is

$$NTU = \frac{US}{W_{c,d}} \qquad \qquad \ldots(15)$$

The case for NTU $\neq \infty$ is shown in Figure (14) and the corresponding charge effectiveness is 100%. Thus the single blow effectiveness (for the no-heat loss case) is given by:

$$\varepsilon = 1 - \int_0^1 \bar{T} \, dUF \qquad \qquad \ldots(16)$$

The second term of equation (16) is the mean value of transient temperature distribution over the interval $0 < UF < 1$. Since the mean value of the transient temperature distribution is always greater than the end period temperature difference, the "single blow" effectiveness is always less than that resulting from cyclic equilibrium operation evaluated at a utilisation factor of one.

High temperature continuous waste heat recovery has been the only widespread use of regenerator technology to date. It is clear that two regenerators but only one recuperator are needed for continuous operation. Continuous waste-heat recovery using ceramic recuperators tends, therefore, to be a more economic method. In the lower temperature range of 250°C to 500°C the regenerator volume becomes excessively large if only the sensible heat is used. A large potential exists for "phase change regenerators" operating in this temperature range.

5.3 Future Trends in Medium Temperature Regenerators

A storage system under development in the USA, Ref.(13) comprises a regenerator matrix where each of the matrix elements consists of a container in which is housed a salt or salt eutectics as shown in Figure (15). Although this particular design includes an indirect heat exchange process in a form similar to a conventional heat pipe, direct systems using steam condensation and evaporation are also possible. In this latter form, such a storage system could be used as a waste heat steam boiler with integral storage capacity. The clear advantage is that such a system would produce constant pressure steam at a pressure corresponding to the saturation temperature at which this phase change occurs.

The main problems associated with this technology are those of containment material corrosion Ref.(14). Table (8) shows some typical candidate phase change materials along with the "break-even" operational temperature range within which the volume of such a store would be less than that of an equivalent sensible heat regenerator.

Material /mol %	Melting Point (°C)	Latent Heat kJ/Lt	Break-even Sensible Temperature Swing (°C)*	Operational Temp. Range (°C)
51% KCl - 49% ZnCl$_2$	235	491	133	168 - 302
Na NO$_3$	307	381	105	255 - 360
38% MgCl - 62%NaCl	435	708	191	327 - 530
M$_g$Cl$_2$	714	967	261	583 - 844
N$_a$Cl	800	1062	287	657 - 944
KF	857	1071	289	713 - 1001

*   Reference sensible heat material Feolite $\simeq$ 3700 kJ/m^3 °C

TABLE 8   TYPICAL HIGH TEMPERATURE LATENT HEAT MATERIALS
            AND "BREAK-EVEN" OPERATIONAL TEMPERATURE RANGES

It is clear that in order to maximise the benefit of such phase change waste heat boilers, the temperature swing should be kept as low as possible.

Several high temperature sensible heat storage systems which are based on heat transfer in a fluidised form have been reported, Ref.(15) and Ref.(16). The storage dwell period generally occurs with material, such as sand, at rest in a bulk container. Heat transfer during charge and discharge is achieved by conveying the sand into a fluidised bed heat exchanger where direct or indirect contact heat exchange takes place. Direct contact heat exchange can also be achieved by a falling cloud heat exchanger, Ref.(17), where sand or other suitable particles are allowed to fall through an upward moving air stream.

In a fluidised bed system, the heat exchanger can be built into the containment vessel or the fluidised medium can be transported to bulk containers. The material (sand) can be moved via a conveyor or by blowers where the hot sand is transported in ducts. The ability to transport the storage material adds greatly to the value of waste heat recovery since very often a waste heat source may be remote from a potential application.

Table (9) illustrates some of the advantages, disadvantages and capacity cost estimates for schemes using sand particles as the storage medium.

5.4   Steam Storage Systems

With all steam accumulators high pressure water is the storage medium and steam is produced in the so-called variable pressure accumulator by flashing the water to saturated steam when the pressure is reduced. The present-day designs have not changed much since the   1920s.

Varying-pressure accumulators are in widespread use. They are used as integral components within steam service systems for factories and process plants. Steam accumulators are sited between the high and low pressure steam mains. Although constant pressure accumulators (sometimes called feed water heaters) have also been used, by far the largest number of installations are of the varying pressure type.

Steam accumulators are very nearly maintenance-free, requiring only occasional inspection. In spite of the advantages of steam accumulation, the industry (especially within the UK) has been in decline for the last 15 years after a rise in the number of installations that were necessary when large capacity Lancashire boilers were replaced by more efficient low capacity boilers.

Steam storage within the framework of waste heat recovery falls into two main categories:
1)   Waste heat boilers, and
2)   Flash steam recovery
Because the steam produced by waste heat boilers are generally used by low pressure steam consumers, with fluctuating loads, the sizing of an accumulator must be made by investigating the relationship between steam production and demand. Steam charts are therefore the only practical way in which this process can be achieved. Correct sizing of an accumulator can, for instance, increase the cost effectiveness of a waste heat boiler by increasing the number of users that can be supported by the system as well as ensuring the availability of an instant steam reserve.

Flash-steam (or low pressure vented steam) is probably one of the largest untapped sources of industrial waste heat. Very often flash steam is vented since if it were condensed and the thermal energy recovered, low temperature condensate would have to be disposed of with often prior cooling to acceptable temperatures. Steam recompression with subsequent re-introduction into the low pressure main could become increasingly important. This would have the effect, however, of reducing the required steam storage capacity or at best increasing the scope of an existing installation to supply more consumers.

Storage System Concept	Advantages	Disadvantages	Estimated Capacity Cost £/kWh (1979)
Fixed Bed Sand	Simple with few moving parts. High reliability Low containment costs	High heat exchanger tube density for high charge and discharge power density	£3.9/kWh
Fluidised Bed sand	High Charge and discharge power density	Large containment cost	£1.95/kWh
Conveyer system sand	Low containment cost	High transport cost	£2.8/kWh
Rapid transport system using blowers	Good thermal response. Low containment cost.	High transport cost. High errosion on pipes	£1.8/kWh

Table (9)   RELATIVE PERFORMANCE AND COST OF TES SYSTEMS USING SAND AS THE STORAGE MEDIUM, Ref.(15)

The pressure vessel necessary for steam accumulation is the most expensive component of the installation. The object of vessel design should be to minimise the weight of the vessel consistant with simple manufacture and low floor space occupied. However, many industries will find room for accumulators even though their floor space is at a premium because installations can be remote from the end use and either inside or outside the buildings.

Typical peak storage capacity costs for three possible types of steam accumulator vessels are shown in Table (10).

Pressure (bar)	Steam Accumulator peak capacity ratio		
	Mild Steel (£/kWh)	Alloy Steel (£/kWh)	Concrete Pressure Vessel (£/kWh)
20	2.5	1.76	1.06
40	5.6	3.7	1.56
60	10.6	5.9	2.38

Table (10) PEAK STORAGE CAPACITY COST FOR STEAM

ACCUMULATORS

5.5 Hot Water Storage Systems

Hot water storage systems are the simplest form of industrial thermal storage. Surplus steam may be charged directly into the storage water itself, either by condensing it in the water space or mixing in the steam space, the heat of the steam of transferred to the hot water. Alternatively the charging process can be done indirectly through a heat exchanger. The most frequently used is indirect charging and direct discharging. Because the maintainance of discharge temperature with time is vital to the effective utilisation of water storage units, careful control of the discharge temperature is important.

The two most fundamental modes of operation are:
1) fully mixed
2) perfectly stratified.

For a given store volume a fully mixed store (i.e. one in which perfect mixing occurs between the cold make-up water and the remaining hot water) the discharge temperature will fall off exponentially with time. For a perfectly stratified store the discharge temperature will remain constant with time until all the hot liquid has been discharged and replaced by cold fluid. In terms of effective utilisation of the storage volume a fully mixed store will only discharge 63% of the energy that could be discharged by a perfectly stratified store in one "fill period". Fig.(16) shows the relationship between storage effectiveness and dynamic response where:

$$T_c^* = \frac{T - T_{IN}}{\Delta T} \qquad \qquad ...(17)$$

CHARGE CYCLE

DISCHARGE CYCLE

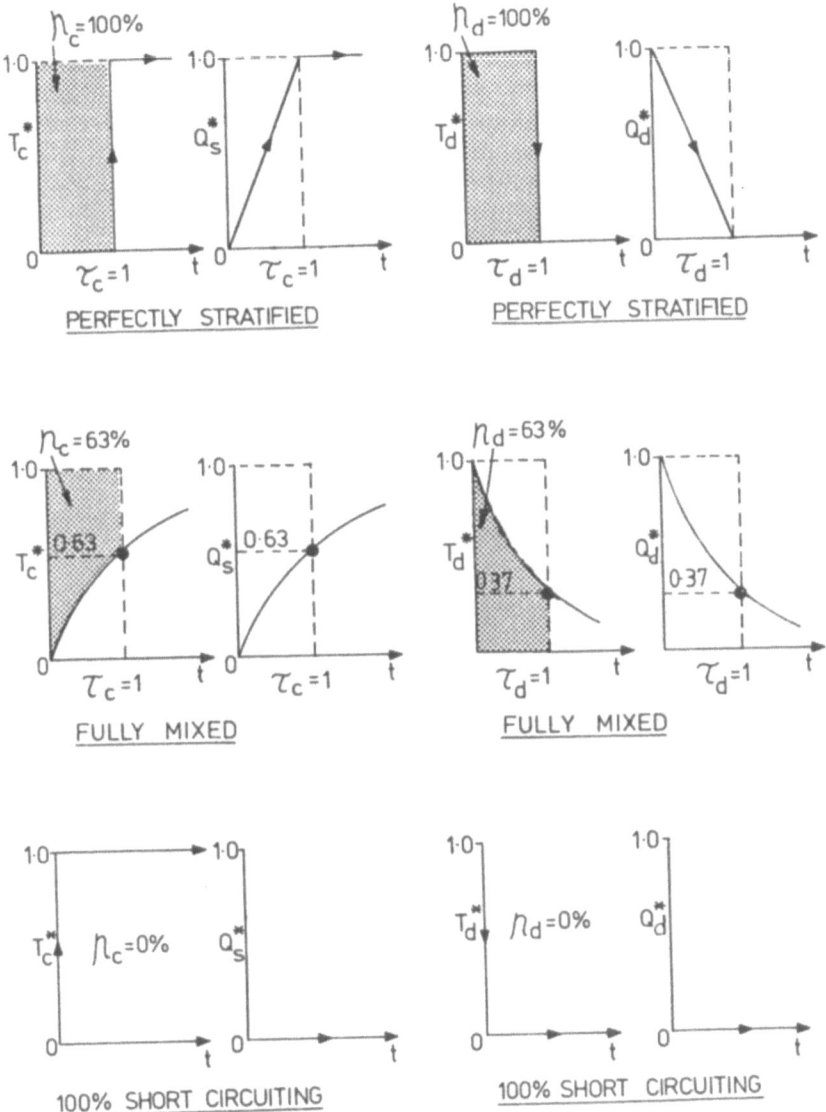

FIG(16)   Schematic Relationship between store dynamic
Response and thermal effectiveness.

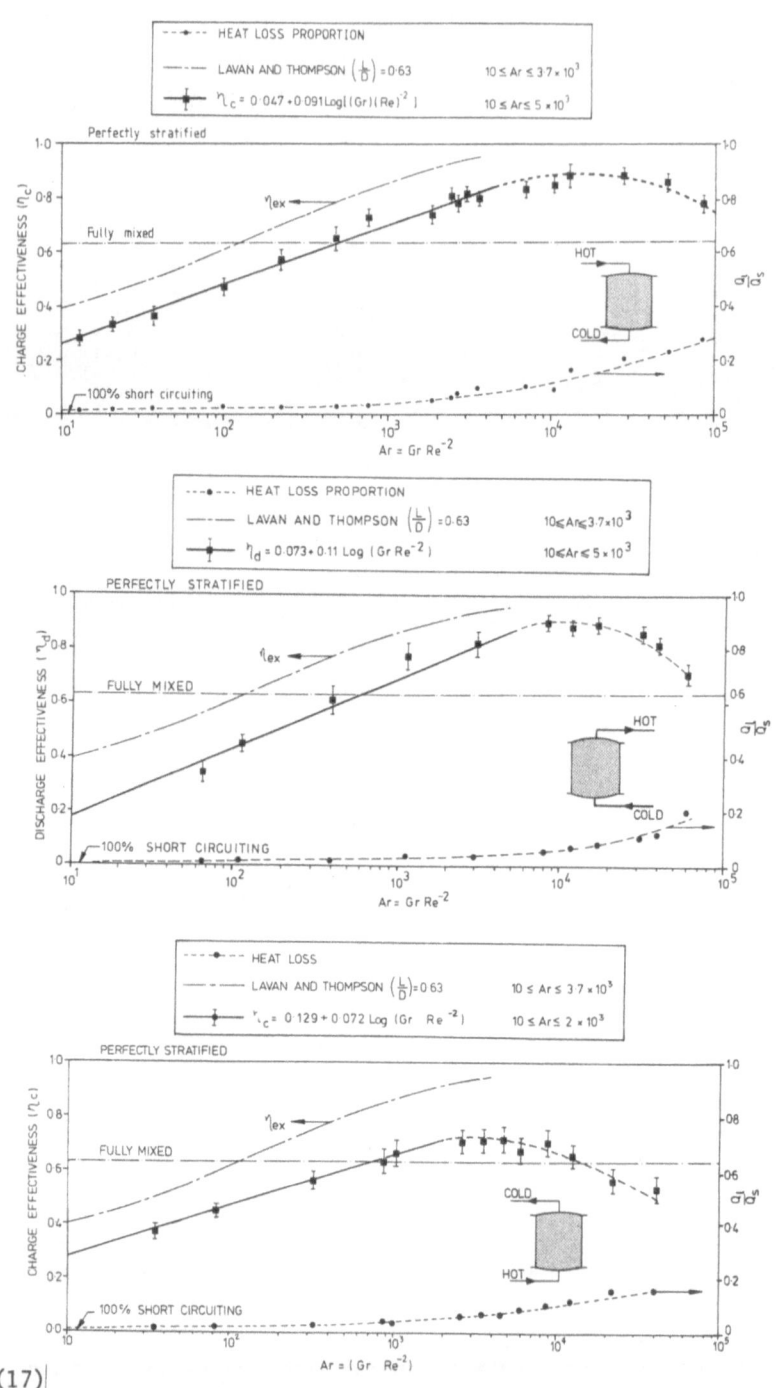

FIG(17)

Experimental Thermal Effectiveness of Stratified Water Stores.

$$T_d^* = \frac{T - T_0}{\Delta T} \qquad \qquad \ldots(18)$$

T_{IN} = Inlet fluid temperature

T_0 = Intial storage temperature.

One further category is of importance, namely 100% short circuiting, where the cold make-up fluid short circuits the storage tank by flashing to the exit port.

The storage volume reduction that can be made by operating the system in a stratified mode is very significant, however, when an indirect heating method is used (say with a waste heat recovery heat exchanger), it is very difficult to prevent mixing. The typical solution has been to accept mixing within the water storage tank with return fluid mixing on discharge to achieve the desired outlet temperature control.

Design correlation are now available by which the flow parameters can be selected in order to maximise the thermal effectiveness. These basic forms of correlation are shown in Fig.(17), Ref.(18).

The main problem with operating hot water storage tanks in a stratified mode is that very often industrial systems require very small volumes with rapid heating and this constrains the design to operate as a fully mixed store. Some new designs are now available for either connection to low pressure steam or waste heat recovery from low pressure hot water, which go part-way towards producing stratified stores.

The full potential for hot water storage can seldom be achieved. This is primarily because of the large quantities involved. As the temperature of the waste water is generally low, few direct applications can be found. The possibility of increasing the temperature of the waste water by using a heat pump which discharges the condenser heat to a process water storage tank significantly improves the value of hot waste water. Simultaneous production of hot water and chilled water for batch heating and cooling operations is also a significant advantage.

6. APPRAISAL OF TES ECONOMICS

The main question to be asked when a TES system is employed is whether the savings obtained through use of a TES system can cover the additional investment of the TES system and other incorporated costs. It is an advantage at the planning stage of a project to be able to establish the desired storage system effectiveness (efficiency) and the annual number of storage cycles through which the system will be taken since these are the main factors which determine the economics of the store. The simplest index of acceptability of a design is the straight-payback-period (SPBP). Although this index dose not take into account such factors as fuel inflation rate, the cost of capital and the intrinsic cost benefits resulting from improved operation of other system components such as boilers, any successful installation would have to show a small SPBP before a more detailed study were undertaken.

For most installations, if the cost of charging and discharging equipment is small in comparison to the capital cost of the store itself, and the cost of waste heat is assumed to be zero then

$$SPBP = \frac{C_s'}{C_f} \frac{1}{N} \qquad \qquad \ldots (19)$$

where C_s' is a modified storage system capacity cost given by

$$C_s' = \frac{C_s}{\eta} \qquad \qquad \ldots (20)$$

and C_s = storage capacity cost £/kWh stored
η = charge-dwell-discharge effectiveness
C_f = cost of fuel £/kWh
N = number storage cycles per year.

Table (11) illustrates the typical values of SPBP to be expected for the storage system alone. It is clear that the SPBP is very sensitive to the number of charge/discharge periods per year, however, for the number of charge cycles greater than about 500 per year, the SPBP will be acceptable to most industries. Care should, however, be taken since as the charge/discharge period becomes small the required charge/discharge power density becomes high and the relative cost of the heat exchange equipment rises.

7. CONCLUDING REMARKS

Although the application of energy storage systems in industry has been small in comparison to other forms of conservation, the future will probably see a significant increase in the number of installations. The process of increasing the number of installations will be aided by well documented demonstration projects and the development of technologies which "fill-the-gap" between the current high temperature regenerators and steam accumulators. In particular, the development of phase-change waste-heat boilers operating with waste heat sources in the temperature range 250 to 500°C which have a significant degree of demand sensitivity, will do much to improve the application of other forms of waste heat utilisation such as turbogenerators for on-site power generation.

Charge/Discharge Period		1 week	1 Day	12 hr.	8 hr.	4 hr.	2 hr.	1 hr.
Number of charge cycles per yr.		26	183	274	248	1095	2190	4380
Storage to fuel cost ratio $\dfrac{C_s}{C_f}$	56 (Low)	2.15	0.31	0.2	0.1	0.05	0.03	0.01
	833 (High)	32	4.6	3.0	1.5	0.76	0.38	0.1

PAY BACK PERIOD IN YEARS

Table (11) TYPICAL RANGE OF SPBP FOR CURRENT INDUSTRIAL TES/WHR APPLICATIONS

It must be remembered that TES is not an end in itself; it must be viewed as an aid to energy management, and a plant designed with forethought now, will become increasingly cost-effective throughout its operational lifetime.

Acknowledgement

The author wishes to express his thanks to the fellow members of the thermal energy storage research team at Cranfield in particular Dr. P.W. O'Callaghan, Mr. D.T. Baldwin, and Mr. R.E. Manning.

REFERENCES

1. "Energy Technologies for the United Kingdom", Energy Paper 39, Vol.1, HMSO, London.

2. Digest of U.K. Energy Statistics, HMSO, London.

3. Flower, S. "Industrial Steam Raising and Power Generation: economic considerations. I.Mech.E. Conference "The Generation and Utilisation of Steam", May 1981.

4. "Industrial Applications Studies", University of Drexel, Pennsylvania, U.S.A.

5. Olszewski, U. "The U.S. Industrial Thermal Energy Storage Programme", International Conference on Energy Storage, Brighton, U.K., April 29th - May 1st, 1981.

6. Musgrove, G. et al. "Energy Recovery: A Multidisciplinary Approach and its Ramification related to Industrial Strategies." IEE Conference 192, "Future Energy Concepts", pp.208 - 213, 1981.

7. Hoffman, H.W. et al. "Thermal Energy Storage for Industrial Waste Heat Recovery". 14th IECEC Conference 789071, 1979.

8. Hoffman, H.W. et al. "Thermal Energy Storage for Industrial Waste Heat Recovery". 14th IECEC Conference 789071, 1979.

9. Glendenning, I. et al. "Technical and Economic Assessment of Advanced Compressed Air Storage (ACAS) concepts" Electric Power Research Institute Report EM-1289, December, 1979.

10. Cookson, R.A. "Design of a Pebble-bed Heater as a High Temperature Gas Turbine Combustion Chamber Test-Facility", Cranfield Institute of Technology, Memo No. 18, October, 1970.

11. Hausen, H., Tech. Mech. Thermodynam. 1, 219, 1930.

12. Johnson, J.E. Regenerator Heat Exchanger for Gas Turbines, A.R.C.R. and M. No.2630. 1952.

13. Nemecek, J.J. et al. Demand Sensitive Energy Storage in Molten Salts. Solar Energy, Vol 20, pp.213 - 217, Pergamon Press, 1978.

14. Heine, D. et al. "Physical and Chemical Properties of Latent Heat Storage Materials", New Ways to Save Energy, October 23rd - 25th, 1979. pp.536 - 545.

15. Turner, H. et al. "High Temperature Thermal Storage in Moving Sand". Jet Propulsion Laboratory, Pasadena, California, 1978.

16. Babaz, G. et al. "High Temperature Heat Storage in a Fluidisable Media: Paper F1, International Conference Energy Storage", Brighton, U.K. 1981.

17. Sagoo, M.S. "The Development of a Falling Cloud Heat Exchanger", "New Ways to Save Energy", Brussels 23 - 25th October, 1979, pp.436 - 444.

18. Wood, R.J. et al. "Thermally Stratified Hot-water Storage Systems", Applied Energy 6, October, 1981.

THERMAL ENERGY STORAGE FOR PEAKING POWER GENERATION

P.V. GILLI

Head, Institute of Thermal Engineering,
Graz University of Technology
A-8010 Graz, Austria

G. BECKMANN

Industrial Plant Dept., Waagner Biro AG,
A-1221 Vienna, Austria

Abstract

Energy storage in electric grids leads to better utilization
of base load plants and thus to the substitution of premium fuel,
to operational advantages and to lower system cost, in particular
when the proportion of nuclear or solar power plants increases.
Thermal energy storage (TES) plants, integrated in the power
plant itself, are mainly based on pressurized hot water in conventional or advanced pressure vessels, above or below ground,
and on the steam storage, the feed water storage or the cascading
arrangement. Advantages of TES include: no additional energy conversions; practically no limitation of charging power; high turn-
around efficiency.

1. INTRODUCTION

In an electric grid a balance must be found at any time
between demand and generation. This problem of load balancing
can be solved in two different ways (Fig. 1): by adapting generation to instantaneous demand, or by controlling demand according

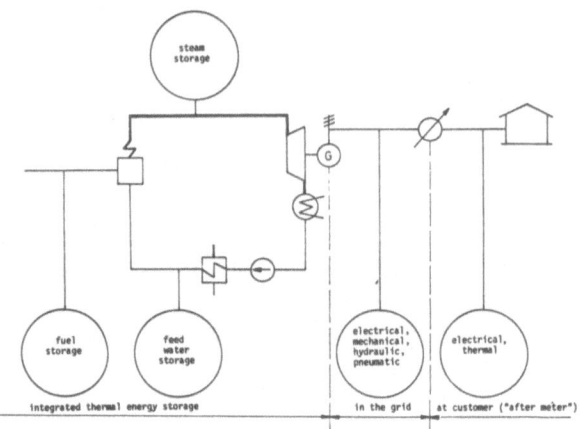

Fig. 1   Possibilities of load balancing by means of engery
         storage /23/

to possible generation. The latter means customer energy storage
("after meter"), usually heat storage, e.g. high temperature heat
of refractories for space heating, or low temperature water
storage for hot tap water supply. Adapting generation to demand,
on the other hand, either entails integrated energy storage in
the power plant itself - i.e. storage of energy before electric
surplus power is generated /1...24/ - or storage somewhere else
in the grid ("pumped storage"), i.e. producing electric surplus
energy, transporting it to the location of the pumped storage
plant, transforming it into storable energy and transforming it
back to electric energy during times of peak demand /22, 25/.

    Storage in the grid ("pumped storage") may be achieved by
means of
- hydraulic pumped storage plants (above ground or underground),
- pneumatic pumped (compressed air) storage plants,
- steam pumped storage plants,
- electro-chemical storage (conventional batteries or advanced
  batteries such as the sodium-sulphur cell or the lithium-
  sulphur cell),
- magnetic storage (with superconducting magnets),
- mechanical storage (fly-wheels),
- or even by the hydrogen cycle, i.e. by producing hydrogen with
  off-peak power, storing it as metal hydride or as pressurized
  gas, and generating peak power, e.g. by a fuel cell /26/.

Integrated energy storage in the power plant itself may, in principle, be fuel storage, thermal energy storage (TES), mechanical storage, or electrical energy storage. Fuel storage and electrical storage are borderline cases: with fuel storage, the plant becomes a peaking plant rather than a storage plant in the true sense; electrical or magnetic storage, on the other hand, should already be considered "pumped storage" in the grid as in this case the product of the power plant is stored (which is possible anywhere in the grid). Mechanical storage will not be feasible in the power plant since the sychronous speed of the turbo-generator does not permit any appreciable storage of kinetic energy even if its rotating mass was increased: Mechanical energy storage by fly-wheels, therefore, will only be possible with asynchronous or direct-current units fed from the grid, i.e. by "pumped storage" plants. The only feasible case of integrated energy storage, therefore, is thermal energy storage. In a steam plant (Fig. 1), thermal energy storage may be located between steam generator and turbine ("steam storage") or between turbine and steam generator ("feed water storage").

An advantage of integrated thermal energy storage is that in the chain of energy transformations from the fuel via heat (of flame and flue gas), internal energy (of steam or gas), flow energy (of steam or gas) and mechanical energy (blade, shaft) to electrical energy there is no additional transformation, although there will be exergy (available energy) losses and heat losses. In pumped storage systems, on the other hand, there are many additional energy transformations and losses, for instance, in a hydraulic pumped storage plant: motor, pump, friction in the penstock, evaporation and trickle losses, friction again, turbine, generator. Another advantage of integrated TES is that usually the charging power may be rather high (several times the discharging power) without appreciable first cost since the charging equipment consists only of piping, valves, nozzles, and - only for some types - pumps, whereas the charging equipment for, say, pumped hydro or compressed air permits only a charging rate of the order of the discharge rate. Integrated TES therefore, is well suited to accept high surplus power bursts that may occur in special conditions such as load shedding.

An advantages of pumped storage plants may, on the other hand, in certain cases be that they can be located in or near the center of demand.

This paper deals with thermal storage only. Therefore, integrated thermal energy storage in the thermal power plant is discussed, but also two cases of pumped storage: compressed air storage which is - at least partly - to be considered thermal storage, and steam pumped storage. Sensible heat, latent heat, and heat of sorption and of thermo-chemical reactions are considered.

For the purpose of providing peaking capability, thermal storage may be applied to steam power plants with fossil fuel, solar energy, or nuclear energy (fission and fusion) as the energy source. It may also be applied to gas turbine plants, either for enabling a very quick start up of conventional peaking gas turbines by means of compressed air storage, or for making — within the limits of the storage capacity — solar gas turbines independent of the stochastic and/or diurnal changes of sunlight ("bottoming" application).

2. ECONOMIC INCENTIVES

Energy storage for electric power generation — integrated thermal energy storage or pumped storage — may serve one or more of the following purpose:

(1) Peak load coverage with lower installation cost than a peaking plant without storage;

(2) Transformation of surplus energy availalbe at times of low demand from base load thermal power plants with low fuel cost (such as nuclear or coal-fired plant) into valuable peak load or medium energy that would otherwise have to be genera- ted from plants with high fuel cost (due to the use of scarce fuel and to low efficiency) such as gas turbines, with the consequences of savings of fuel cost, substituting scarce fuels such as oil, increasing the load factor of the base load plant, and avoiding or minimizing thermal cycling (e.g. of the fuel elements of a nuclear reactor);

(3) Transformation of surplus energy from thermal power stations with low or zero fuel cost but periodic and/or stochastic availability of the energy source (e.g. solar power plants or waste heat from industrial batch processes) into electrical energy available at times of low or zero energy input, there- by enabling not only a better coverage of the demand profile but also a less costly plant (as all parts after the storage system are smaller) with higher load factor;

(4) Short term storage for buffering and for facilitating control (integrated storage only);

(5) Stand-by duty in order to increase the availability of the plant (integrated storage only);

(6) Quick start-up of turbomachinery (gas turbine plants).

Whether or not the storage system will be able to fulfill one or more of these purposes economically will, therefore, depend on:

- the (additional) investment cost of the storage system – generally consisting of a power-dependent part (e.g. storage turbine set or larger main turbine set, charging and discharging equipment) and an energy-dependent part (storage vessels), relative to the investment cost of a peaking plant without storage;

- the specific cost of the energy used for charging the storage systems, in relation to the fuel cost of a peaking plant without storage;

- the turnaround-efficiency, defined as the ratio of net electrical energy produced from the storage system during discharge, to the electrical energy required for charging (in the case of pumped storages) or to the electrical energy produced less due to the charging operation (in the case of integrated storage);

- the gains due to operational advantages (e.g. continuous full-load operation of a nuclear reactor even at times of very low demand, avoiding thermal cycling of fuel elements);

- the shape of the daily (and weekly) load diagram throughout the year; and

- the actual proportion of base load and peaking power stations in the grid considered.

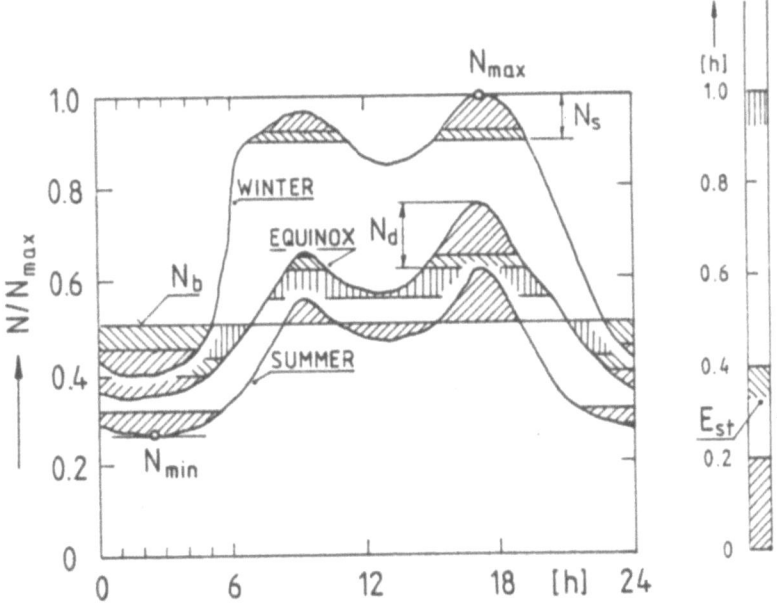

Fig. 2   Load diagram

A daily load diagram typical for Central Europe is shown in Fig. 2 for the two extreme cases (mid winter, mid summer) and the medium case (spring and fall equinox). In mid winter, load reaches the highest value (N_{max}) of all the year; in mid summer it reaches the lowest value (N_{min}). Spring and fall equionx are in between. Peaks occur in the morning and in the late afternoon.

The available power of base load stations N_d is indicated in the diagram; this may be construed as being the installed capacity or - more correctly - a somewhat reduced capacity due to availability limitations. For annual load profiles like the ones in the diagram it will be different for winter and summer conditions: Maintenance of fossil-fired stations as well as maintenance and refueling of nuclear stations will be done preferably during the summer (low demand) season, with the consequence of a considerably lower available power during the summer season. It may also be preferable to shut some base load power stations deliberately down during the low load season and to run the remaining ones with a higher load factor employing the storage system; in this way low part load efficiency and/or fatique due to thermal cycling is avoided. However, for the sake of simplicity only one line of available power is shown in Fig. 2.

What are the savings due to the introduction of storage systems into a grid ?

- During the winter peak, thermal storage saves installation cost as well as fuel cost due to a reduced requirement and reduced operation of peaking plants. Utilization of the storage plants will be limited by the available charging energy at low demand hours during the night.

- In the mid summer case, relatively little peak energy (above the line of available base load power) will be required. Thermal storage saves fuel cost. Utilization of the storage plant will be limited by the peak energy required. If the reduced availability of base load plant during the low demand season due to maintenance and refueling is taken into account, the saving of fuel cost is increased.

- In the two equinox cases, the utilization of storage plants may or may not be limited by either the available charging energy (as for the mid winter condition) or by the required peaking energy (as for the mid summer condition).

In Fig. 2, the possible storage utilization (indicated by different shading in the three cases) turns out to be about 0.2, 0.4, and 1.0 hours of the highest load N_{max}. The maximum possible storage utilization occurs at the equinoxes. The following characteristic values for different installed storage capacities

in the particular case are apparent:
- possible substituted power of peaking plant N_s during the highest load at mid winter;
- maximum discharging power N_d (e.g. storage turbine power) at or near equinox;
- maximum charging power N_c (also at or near equinox);

Actually, the storage utilization will even be better than shown due to the weekends that introduce an added periodicity of the demand.

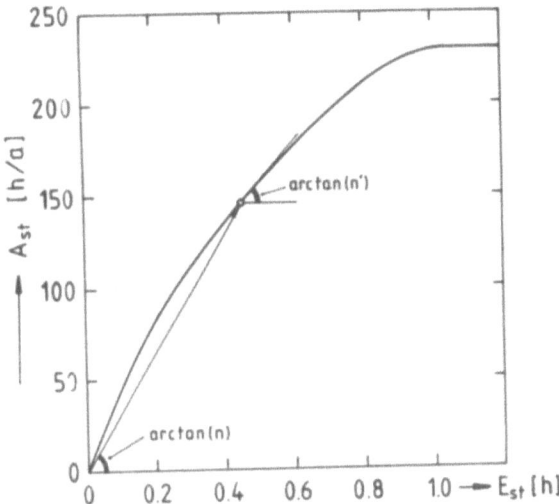

Fig. 3 Annual stored energy A_{st} vs. installed storage capacity E_{st}

From an analyse of the daily load curves throughout the year - represented by the special cases in Fig. 2 - the maximum annual stored energy A_{st}(h/a), again measured in hours of max. load, may be found. It is plotted in **Fig. 3** against E_{st}(h), where E_{st} stands for the total installed storage capacity in the grid. The average numbers of full discharge per year $n = A_{st}/E_{st}$ (1/a) of all these storage plants are represented by the slope of the radius vector, and the number of discharge n' of an incremental storage capacity corresponds to the slope of the tangent. For the case considered, A_{st} increases with E_{st} steeply at the beginning, more slowly at higher values of E_{st} up to a limit of 235 h/a. The values of n and n' decrease continuously with increasing E_{st}; thea are plotted in **Fig. 4** against E_{st}.

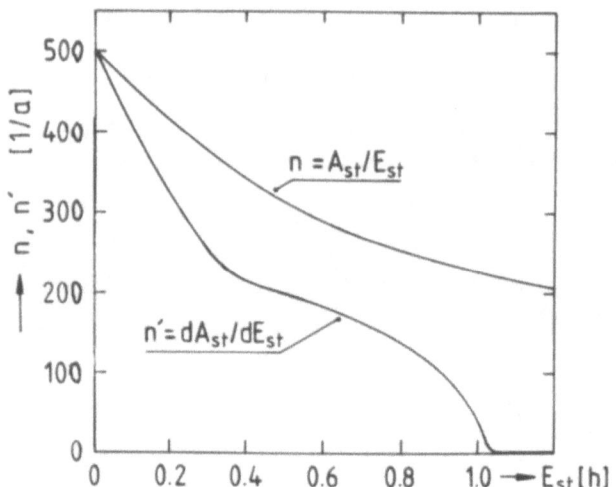

Fig. 4   Average number of discharges n and incremental number of discharges n' vs. E_{st}

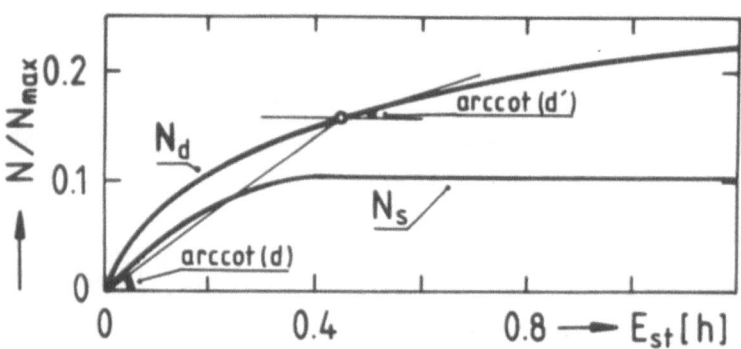

Fig. 5   Discharge power N_d and substituted power of peaking plant N_s vs. E_{st}

Fig. 5 shows against E_{st} the maximum relative discharge power N_d/N_{max} and also the relative substituted power of peaking plant N_s/N_{max}; they both can be found from Fig. 2, using the maximum discharge power N_d during the year and the discharge power N_s during the day with the highest load, respectively.

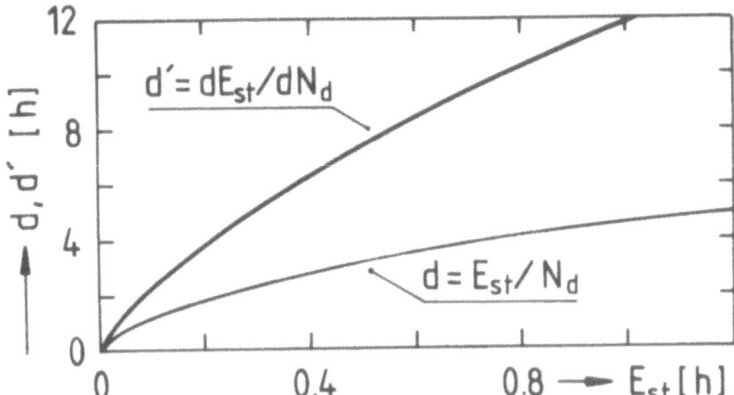

Fig. 6   Average discharge duration d and incremental discharge
         duration d' vs. E_{st}

From N_d one can find the average discharge hours d:

$$d = E_{st}/N_d$$

and also the incremental discharge hours d':

$$d' = dE_{st}/dN_d$$

Both values are shown in Fig. 6.

   Also from Fig. 5 an average ratio

$$R = N_s/N_d$$

and an incremental ratio

$$R' = dN_s/dN_d$$

may be found which express the substituted power of peaking
plants in a dimensionless way. R and R' are plotted in Fig. 7
against E_{st}.

   Now a typical annual cost diagram may be plotted (Figs. 8
and 9). Contrary to conventional base load or peaking plants
where investment cost is only power related, and utilization
is only limited by technical availibility, with storage plants
investment cost is power and energy related, and utilization is
also limited by the demand conditions in the grid.

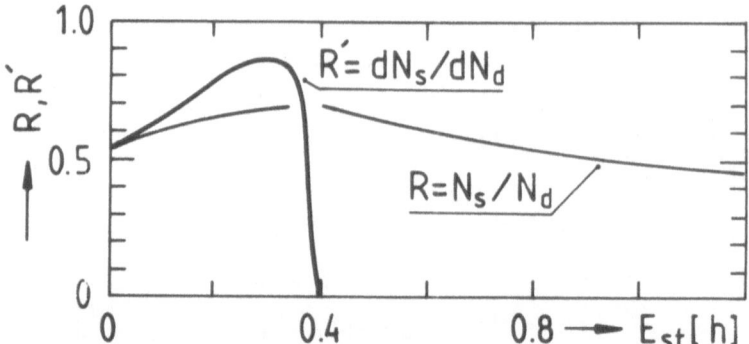

Fig. 7   Average ratio R and incremental R' vs. E_{st}

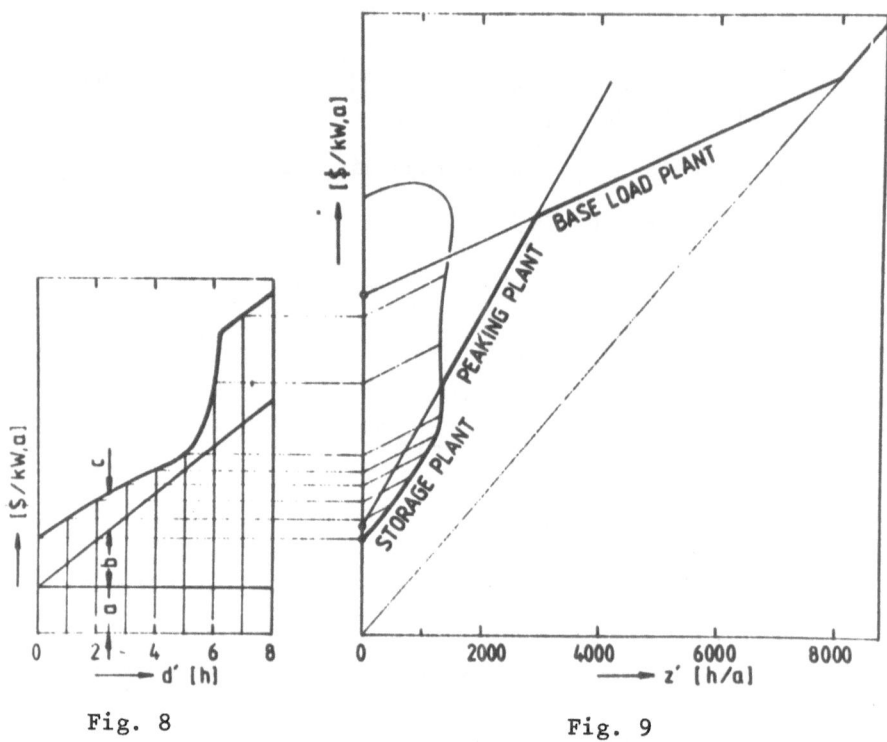

Fig. 8   Fig. 9

Fig. 8   Annual fixed cost of storage plant vs. incremental
         discharge duration d'
         a...Power related portion
         b...Energy related portion
         c...Cost of nonsubstituted peaking plant

Fig. 9   Annual cost diagram

In Fig. 8, showing fixed cost of a storage plant, the power related portion of fixed cost consists of the cost of peaking turbine etc.; the energy related portion consists of the cost of storage vessels etc. and an additional cost of the portion (1-R') of the peaking plant that has not been substituted by the storage plant. This portion, in the particular case according to Fig.8, decreases slowly first, reaches a minimum at d' = 5 hours and increases again up to the installation cost of the peaking plant.

Fig. 9 consists of the annual cost lines (currency per kW, per year) vs. the annual load factor $z = n'd'$ (h/a). The base load , e.g. a nuclear plant, shows high first cost, low fuel cost; the peaking plant, e.g. a gas turbine plant, low first cost, high fuel cost. There may be other plants in between, e.g. coal-fired plants.

The line of a storage plant is found by adding to the d'-depending fixed cost its fuel cost (oblique lines), i.e. the fuel cost of the base load plant, corrected for efficiencies (turnaround efficiency of the storage plant, part load efficiency of the base load plant). The actual curve of the storage plant is found by terminating the lines for a given d' at $z = n'd'$, taking n' from Fig. 4 and d' from Fig. 6. It turns out that the curve for the storage plant is bent upward and returns to zero hours for long discharge durations (of about 12 hrs.). For the case considered, it intersects the peaking plant line at about $z = 1300$ h/a.

It should to noted that the straight lines of fuelled plants are independent of the grid whereas the line of a storage plant depends on the grid and has to be assessed for each case.

3. THERMODYNAMICS

3.1 Storage media

Thermal energy storage media and their relevant data are given in Table I /2, 11, 15...20/. Of the liquid media storing sensible heat, water exhibits the highest volumetric heat capacity; contrary to oil, molten salt, and liquid metals, it requires, however, pressure vessels for temperatures above 100 °C. High volumetric capacities can only be achieved with large temperature swings. Thermal energy storage media with phase change liquid-solid, on the other hand, enable the storage of heat at a constant temperature (the melting temperature) without temperature swings. Many salts, metals and eutectic salt and metal alloys have been proposed.

Table I: <u>Thermal Data of Storage Media</u>

Medium	Density ρ kg/m^3	Specific Heat Capacity c kJ/kgK	Volumetric Heat Capacity ρc kJ/m^{3}K	Melting (or Reaction) Temperature $^{\circ}$C	Latent (Reaction) Heat kJ/kg	Volumetric Storage Capacity (Enthalpy density) kWh/m^3
(a) Liquids						
H$_2$O (1.0133 bar)	958	4.19	4010	0	-	-
(10 bar)	887	4.19	3720	-	-	-
(100 bar)	688	4.29	2950	-	-	-
Oil	750-800	2.1	1500-1700	-	-	-
Molten Salt (53KNO$_3$, 40NaNO$_2$, 7NaNO$_3$, HITEC)^2	1780	1.3	2300	142	-	-
Na	750	1.26	1200	98	-	-
(b) Liquid-solid						
LiNO$_3$	1776	-	-	252	530	261
NaNH	1760	-	-	318	160	78
LiOH	1385	-	-	471	1080	416
37LiOH, 63LiCl	1530	-	-	262	485	206
88Al, 12Si	2445	-	-	579	515	350
(c) Solid						
MgO	3000	1.13	3390	2600	-	-
Cast Iron	7300	0.84$^{+)}$	6130	1150-1300	-	-
Granite	2800	0.84	2350	-	-	-
(d) Thermochemical						
SO$_3$ $\gtrless$ SO$_2$+1/2 O$_2$	-	-	-	816	1235	160
CH$_4$+H$_2$O $\gtrless$ CO+3H$_2$	-	-	-	680	7350	58
NH$_4$HSO$_4$ $\gtrless$ NH$_3$+H$_2$O+SO$_3$	-	-	-	500	2930	860

$^{+)}$ higher for 700 to 800 $^{\circ}$C range

3.2 Energy densities

Storage of sensible heat in solid matter is state of the art. Cast iron shows the highest volumetric heat capacity of all media, followed by magnesia. For temperature swings including the Curie point (768 $^{\circ}$C), the average heat capacity of cast iron is even higher than shown /2/. If reversible chemical reactions are used and the reaction media are gases, they are to be stored under pressure or - after cooling down and condensing - as liquids. The volumetric storage capacities (energy densities) shown in Table I refer to the total volume required.

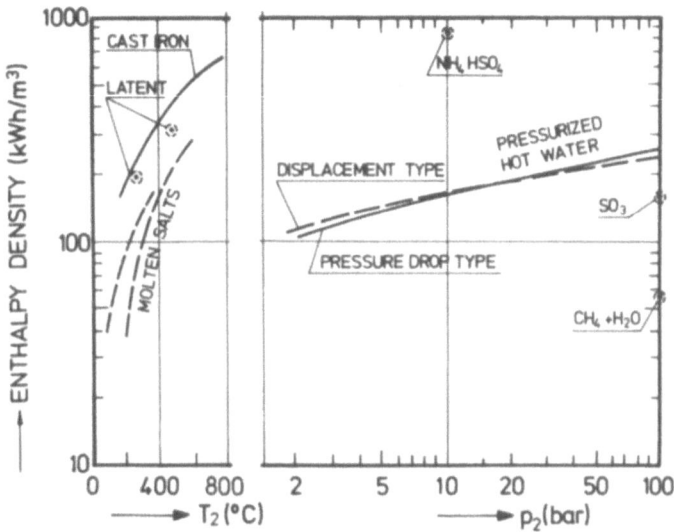

Fig. 10 Enthalpy densities of storage media

Energy densities are shown in **Fig. 10**. For sensible heat storage media, energy densities are plotted against storage temperature. For solids, the void necessary for the heat transfer agent was accounted for. For media with saturation pressures above atmospheric, enthalpy density is shown vs. storage pressure in the right hand part of Fig. 10.

The enthalpy density is a proper measure of the storage capacity only for the storage of heat without subsequent energy conversion. For power plants, the exergy (available energy) density is more appropiate; it is shown in **Fig. 11** for an ambient temperature of 32 °C. The line of pressurized hot water is now steeper than in Fig. 10, a consequence of the higher Carnot efficiencies at higher water temperatures.

3.3 Storage systems

Fig. 12 shows the basic flow sheet of the steam pumped storage case. There is an upper and a lower sliding pressure storage vessel, with a steam turbine and a steam compressor in between /27/. Such plant is well suited for siting near demand centers since it produces no emissions and requires relatively little space.

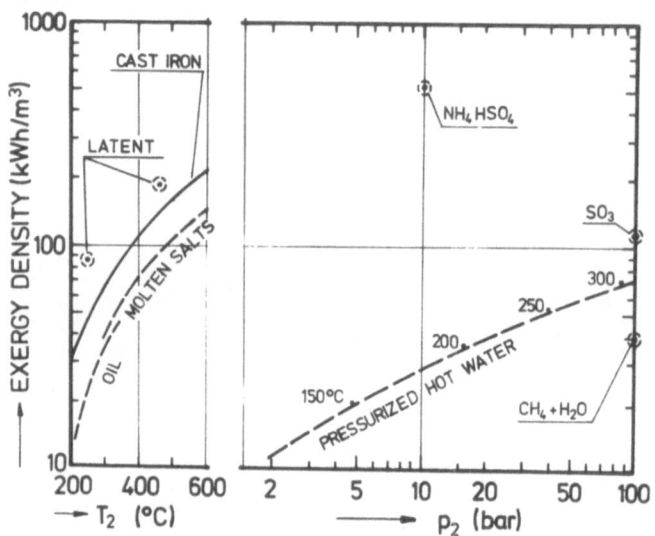

Fig. 11   Exergy (available energy) density of storage media

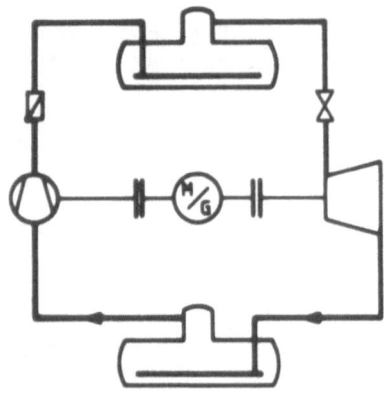

Fig. 12   Thermodynamic steam storage

Compressed air storage may be used for storing mechanical
or electrical energy in much the same way for air or gas turbines
as TES is used for steam pwer plants /11, 13, 16/. However, since
compressed air is   - contrary to steam - to be produced by
mechanical rather than thermal energy, compressed air storage
is not based on the storage of internal energy but rather on

the storage of pressure energy, similar to the storage of energy
in the steam cushion of a sliding pressure steam storage system.

3.3.4 Specific storage capacity

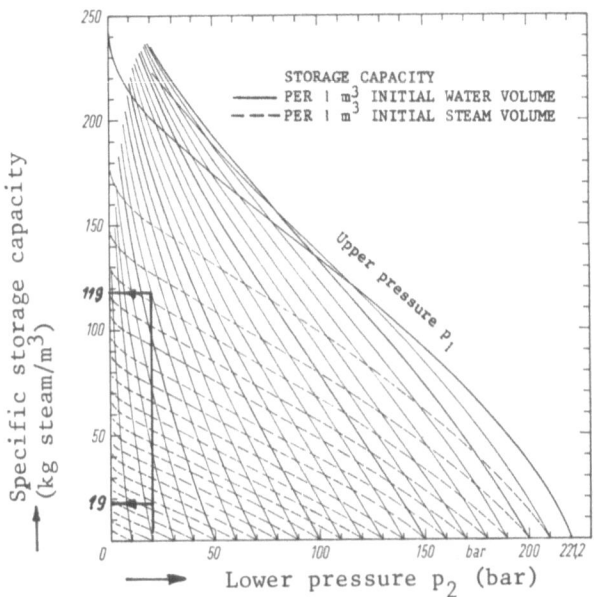

Fig. 13 Specific storage capacity (kg steam/m³) of pressurized
hot water storage tanks. Capacity to be found by
drawing a vertical line from discharge pressure to
charge pressure

Fig. 13 shows the specific storage capacity of sliding
pressure steam storage expressed as kg steam produced for a
given pressure reduction per m³ initial water volume and per
m³ initial steam volume /7/.

Fig. 14 shows the specific storage capacity of expansion-
type storage, expressed as kg hot water per m³ initial water
content, as a function of upper pressure. In the lower portion
of the diagram, the lower pressure at discharged condition (full
water discharge, no steam discharge) and the corresponding
saturation temperature are given.

Fig. 15 shows the storage capacity of a pressurized air
storage vessel /11/ as a function of pressure ratio p/p_o and
the ratio of thermal capacities of vessel and of air volume
where W = 0 is the adiabatic case (vessel of zero mass, or
vessel with internal insulation, yielding maximum temperature
drops), and W = ∞ is the isothermal case. The dimensionless

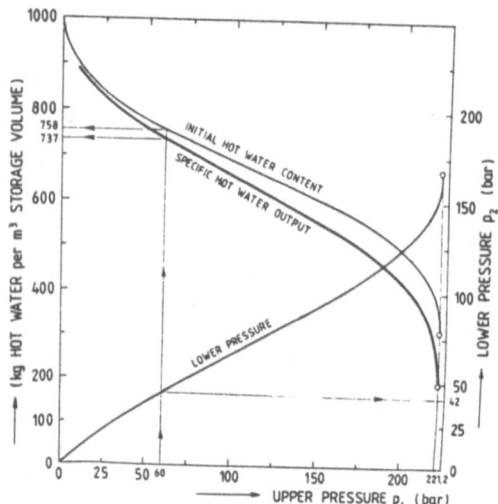

Fig. 14   Specific storage capacity (kg hot water/m³ storage
volume) of expansion-type TES systems /11/

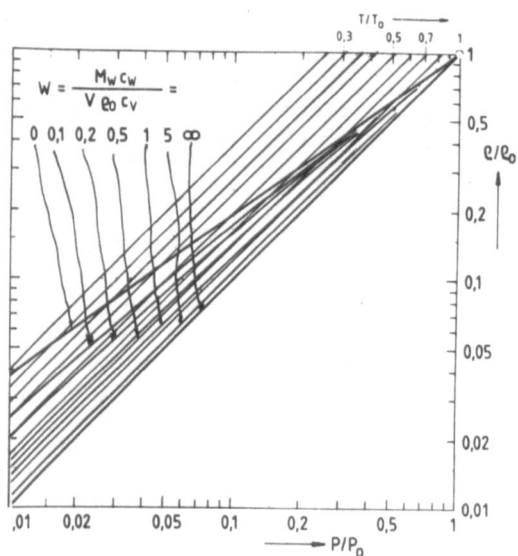

Fig. 15   Storage capacity of pressurized air storage vessel

storage capacity is plotted as ratio of density of air after discharge to air density in the charged condition, ρ/ρ_o. The

temperature drop of the pressurized air during discharge is shown as ratio of absolute temperatures T/T_o.

The use of this diagram is best be explained by an example. Assuming
p_o = 150 bar, p = 9 bar, p/p_o = 9/150 = 0.06 and W^o = 5.0
from the diagram one finds:
T/T_o = 0.94; and ρ/ρ_o = 0.064
Assuming further an initial vessel and air temperature of
T^o = 20 ^{o}C = 293 K
T^o = (0.94)(293) = 275 K = 2 ^{o}C
is obtained. Initial density at 150 bar, 20 ^{o}C is 175 kg/m^3.
Therefore,
ρ = (0.064)(175) = 11 kg/m^3,
and the specific storage capacity becomes
$\Delta\rho = \rho_o - \rho$ = 175 - 11 = 164 kg/m^3

3.5 Storage Efficiency

Approximate values of the turn-around efficiencies as defined before are given in Table II.

Table II: Turn-around Effeciency of Thermal Energy Storage Systems for Power Generation

(i) INTEGRATED STORAGE SYSTEMS	
A. Steam plants, direct storage	
(A.1) Sliding Pressure type (Internal evaporation, Ruths)	0.5...0.75
(A.2) Expansion type (external cascading flash evaporators)	0.7...0.85
A.3 Pressurized hot water displacement type	
(A.3.1) Flash evaporator + feed water storage (cascading)	0.7...0.95
(A.3.2) Feed water storage only	0.75...0.95
(A.4) High boiling liquids (heat exchangers)	0.8...0.95
B. Steam plants, indirect storage	
(B.1) High boiling liquids, feed water storage	0.6...0.75
(B.2) Solids (mainly for superheating)	0.68...0.74
(B.3) Latent heat	0.6...0.64
(B.4) Sorption and Thermochemical	0.56...0.58
C. Gas turbine plants, direct storage	
(C.1) Pressurized plants, direct storage	0.4...0.5
(ii) PUMPED STORAGE SYSTEMS	
(D) Hydraulic pumped storage	0.5...0.75
E. Air pumped storage	
(E.1) Pneumatic pumped storage	0.67
(E.2) Air pumped storage gas turbine	0.55
(F) Steam pumped storage	0.45...0.5

4. STORAGE VESSELS

For low and medium storage pressure, and small to medium sizes, the welded steel vessel is the proper choice.

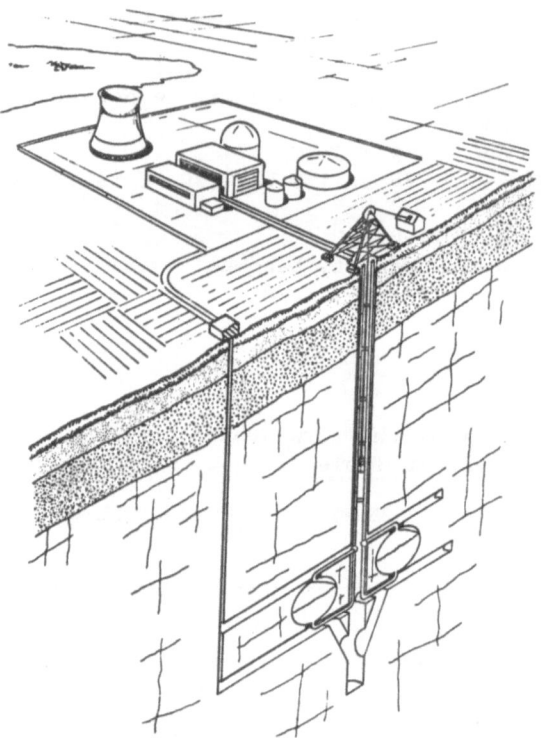

Fig. 16   Underground steam storage /4/

With suitable geological conditions at the site of the power station, underground TES in rock caverns /4, 28/ may be considered (Fig. 16).

For high pressures and large sizes, the use of the Prestressed Cast Iron Vessel (PCIV) has been proposed /1...15, 20, 23/. The PCIV is presently being developed by Siempelkamp Guss- und Reaktortechnik GmbH of Krefeld, FRG, in cooperation with GHT, HRB and L. & E. Steinmüller, as a reactor vessel for the HTR and for other applications such as TES. A PCIV vessel was built in 1972 and a nuclear coded PCIV control rod drive pressure vessel of 18 m^3 for 230 bar Helium has been built for the German THTR Schmehausen /11/.

The PCIV consists of cast iron blocks, lined on the inside
if necessary, thermally insulated on the outside or the inside,
and prestressed by cirumferential as well as axial high-strength
tendons. Its main advantages are the economic possibility of
high pressure and large unit sizes, without transportation or
site welding problems, and outstanding safety features due to
the redundancy of the prestressed design.

An assessment of transient thermal stresses in the PCIV wall
has shown that displacement storage - where sudden large temperature change occur - requires an internally insulated ("cold
going") PCIV. The same applies to the sliding pressure storage
system unless the pressure variation, and therefore the temperature variation, between loaded and unloaded condition is kept
to rather low values /11/. An inexpensive internal insulation of a
PCIV for hot pressurized water duty is under development. For
expansion-type storage, where only small temperature changes
occur during normal operation, and for sliding pressure storage
with small pressure and temperature variations, the externally
insulated ("hot going") PCIV is another possibility.

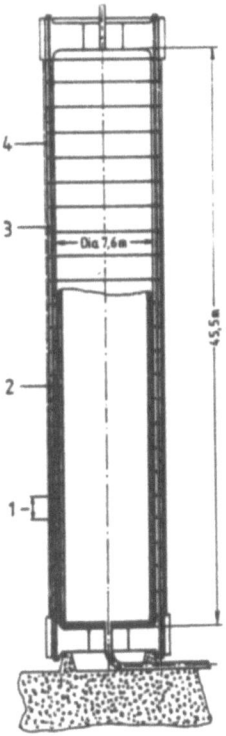

1   cast iron segments
2   cylindrical wall with insulation
    and liner at the inner surface
3   axial tendon
4   circumferential prestressing
    system

Fig. 17   Design of a Prestressed Cast Iron Vessel (PCIV) of
          2000 m³ /12/

- 283 -

Fig. 17 shows the design of a (cold going) pilot PCIV unit, designed for an internal pressure of 60 bar /14/. Internal diameter is 7.6 m, internal height is 45.5 m, total internal volume is 2000 m^3. Such a vessel is able to store between 50 and 75 MW-hours of electric energy, depending on the flow sheet. Much larger vessels, of the order of 8000 m^3 or more, are possible and should be economic /11/.

5. APPLICATIONS

5.1 Coal-fired power plants

Thermal energy storage in coal-fired plants was used already in the Twenties. The most famous plant is the one in Berlin-Charlottenburg /29/ which went in operation in 1929, was in operation decade by decade in peace, war, and peace again, and as of today is still on duty under the special conditions of the Berlin island grid. It is not a large plant by todays standards, but it certainly was in 1929. It consists of a special dual-admission steam turbine of 50 MW electric power, using 13 bar saturated steam, and 16 vertical storage vessels.

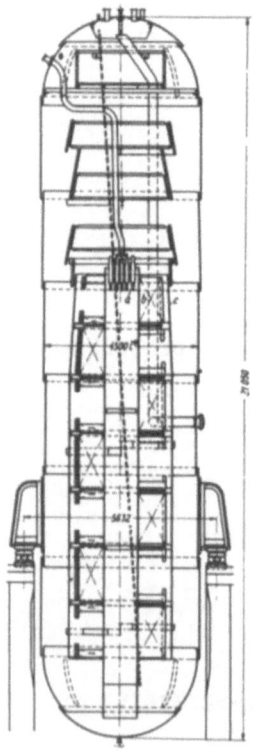

Fig. 18   One of 16 riveted steel
          vessels of the Berlin
          Charlottenburg plant /30/

Fig. 18 shows one of the riveted storage vessels of 4.5 m diameter and 21 m height. The vessels are filled almost to the top with hot saturated water, and steam is generated internally by sliding pressure according to the Ruths principle.

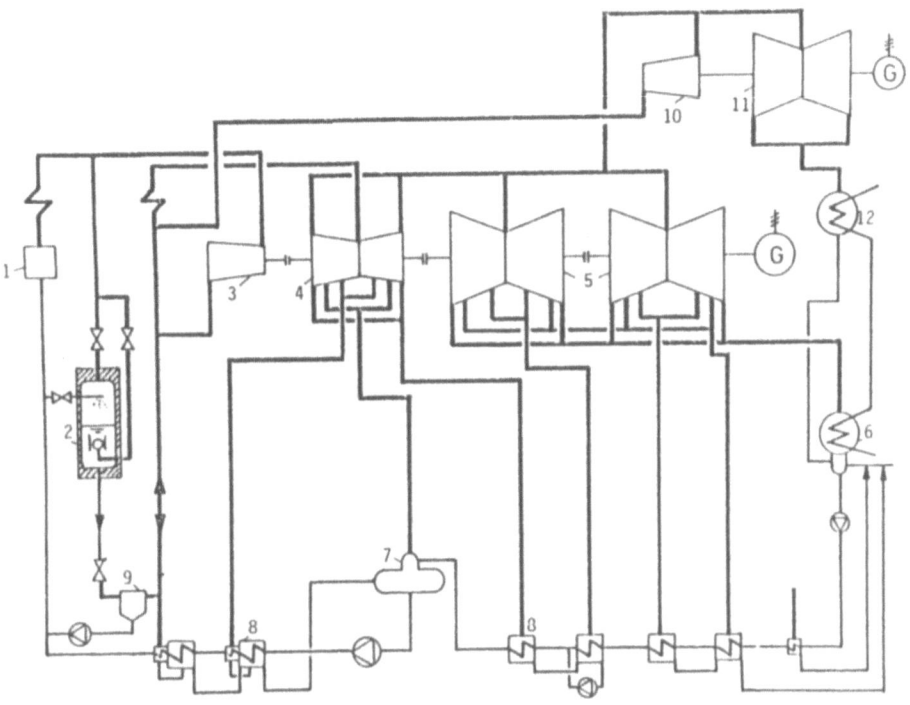

1	Heat source	7	Feed water tank (cold storage)
2	Expansion-type PCIV	8	Feed water heater
3	Main turbine, HP	9	Flash evaporator
4	Main turbine, IP	10	Peak load turbine, HP
5	Main turbine, LP	11	Peak load turbine, LP
6	Main condenser	12	Peak load turbine, condenser

Fig. 19 Proposal for integrated thermal energy storage in a large coal-fired steam power plant, with separate peaking turbine /11/

Fig. 19 demonstrates a more recent proposal for integrated TES in a large coal-fired steam power plant according to the "cascading" principle, i.e. combination of feed water storage and expansion-type steam storage that was proposed during the last years /11, 20, 23, 25/.

For the particular case /11/, an arrangement with separate peak turbine is shown, but flow sheets are similar for main turbines taking large overloads.

5.2 District heating with CHP

TES has also application in district heating systems with combined heat and power generation (CHP). One advantage of TES in large heating grids is that peak power of the grid and the systems load factor may be increased by TES near the demand centers. A further advantage is that the generation of power and of heat may be decoupled by TES, permitting power peaks and heating peaks at the same time with bleeding plants and daily heating peaks at constant full power with topper plants.

5.3 Compressed Air Energy Storage (CAES) plants

Underground compressed air energy storage (CAES) for gas turbines is a relatively new system. A first plant has been built at Huntorf, FRG (Table III, Fig. 20). This method requires special geological conditions (large salt desposits or possibly monolithic rock). Besides, the present air storage gas turbine does require additional gas- or oil-firing. It is, therefore, not a pure storage plant, but rather a symbiosis of a storage plant and a fossil-fired peak load plant /32/.

Table III: Main Data of the Compressed Air Storage Plant Huntdorf /32/

(a) Gas Turbine	
Power	290 MW
Air Flow	417 kg/s
Inlet Conditions, HP	40 bar, 550 $^{\circ}$C
" " , LP	10 bar, 825 $^{\circ}$C
(b) Compressor	
Power	60 MW
Air Flow	108 kg/s
(c) Storage	
Volume	2 x 300.000 = 600.000 m^3
Depth	650/800 m
max. dia.	60 m
Pressure	65/45 bar
(d) Operational Conditions	
Charging time	8 hrs.
Charging energy	468 MWh
Discharge power	290 MW
Discharge time	2 hrs.
Discharge energy (incl. oil firing)	580 MWh
Storage efficiency	55 %

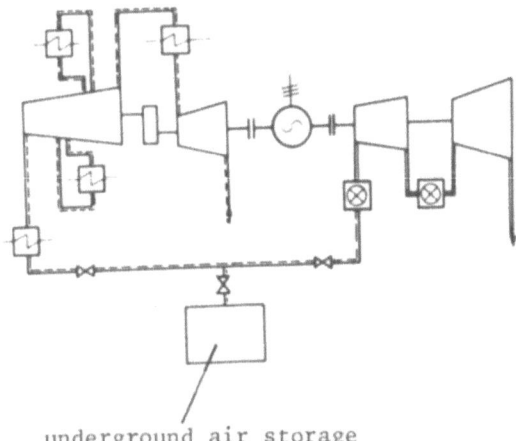

underground air storage

Fig. 20   Compressed Air Energy Storage (CAES) plant Huntorf, FRG
/32/

In order to avoid the use of the valuable oil and gas, a
coal-fired gas turbine plant with indirect heating of air as the
working medium in an Athmospheric Fluidized Bed Combuster
(AFBC) in combination with CAES has recently been proposed /34/.

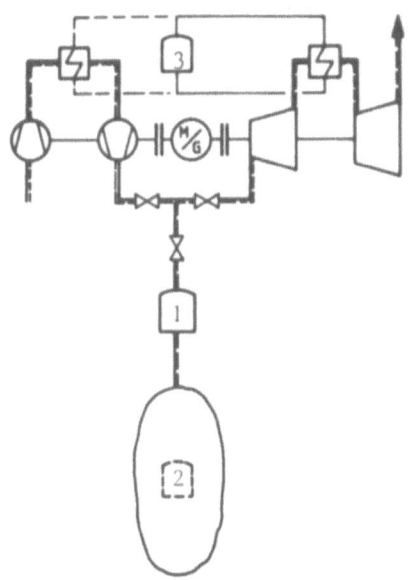

1   TES in air piping

2   TES in air storage

3   TES for intercooling/
    reheating

Fig. 21   Pneumatic energy storage plant with separate storage
of thermal energy

Another possibility of avoiding the use of gas and oil (and, indeed, all fuel) in a CAES plant is the use of an additional separate TES system (<u>Fig. 21</u>) /33/, extracting heat from the compressed air during the charging operation, storing it, and adding it to the air taken from the compressed air storage during discharge.

A further application of compressed air storage in power plants is its use for the quick start-up of open-cycle turbines for reserve (stop-gap) duty. Even without air storage, the gas turbine may be started-up rather quickly (minimum start-up time: about one minute /11, 33/). However, its lifetime is severely reduced by such quick start ups. Therefore, the quick start-up is avoided if at all possible and is practically eliminated. The gas turbine thus loses its stop-gap properties and stop-gap power must be found somewhere else. That is where high-pressure air storage using the PCIV may come in.

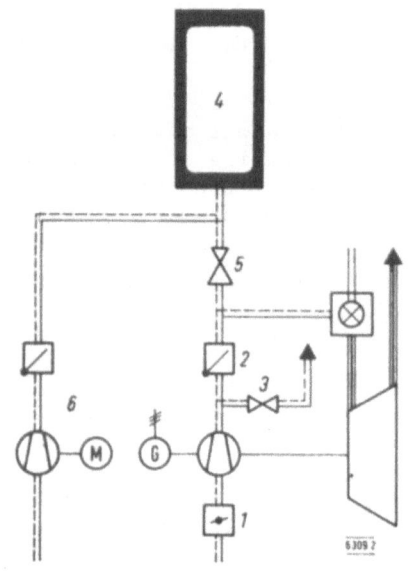

1 Start-up damper
2 Non-return valve
3 Vent
4 Air storage PCIV
5 Throttling valve
6 Charging compressor

Fig. 22   Quick start-up of a gas turbine with PCIV air storage /11, 35/

<u>Fig. 22</u> shows the arrangement. The basic principle of the method consists in a temperature increase as slow as necessary in order to avoid any marked reduction of life time and, at the same time, an immediate availability of the power output, i.e. a rate of power output increase as fast as the generator can take it. It should be noted that without the storage system, in the most critical phase power must be taken from the grid for the start-up motor /11, 33/.

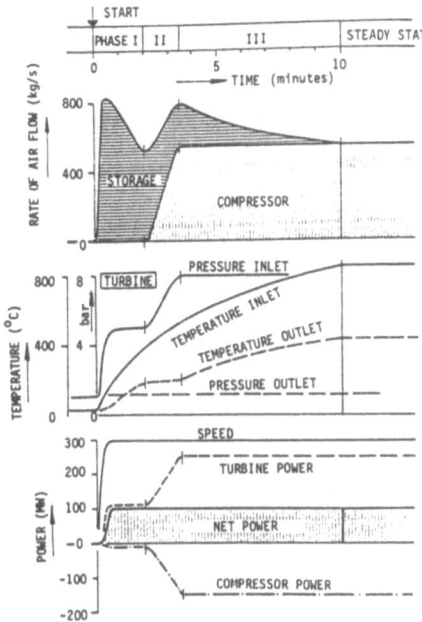

Fig. 23 Start-up procedure of gas turbine plant with PCIV air
storage /11, 35/

The procedure is shown in Fig. 23 for a 100 MW gas turbine
plant. Before start-up throttling valve and start-up damper
(Fig. 22) are closed; the vent is open. Start-up occurs in the
following steps:
- the throttling valve opens; set accelerates; fuel pipe opens;
  turbine inlet temperature is controlled by flow rate of fuel;
- generator is synchronized;
- throttling valve opens further until rated or maximum power
  is reached; pressure before turbine is lower than rated
  because compressor is idling;
- vent closes;
- start up damper opens slowly until turbine inlet pressure
  approaches rated value; throttling valve closes slowly,
  keeping output power constant.

After 10 minutes, full turbine inlet temperature is reached;
the start-up damper is fully open, the throttling valve is fully
closed; the combustion chamber is on full power; start-up
procedure is completed.

Supply pressure at the end of discharge is 8 bar in the
example shown. Storage pressure, therefore, is to be much higher.
It should be noted that the air storage vessel takes more air

mass at higher pressure. Therefore, high or very high pressure should be used. With 150 bar, a volume of about 900 m^3 is required in the example. Also, the vessel should have a large mass, in order to avoid too low air temperatures during discharge. Both effects favour the use of the PCIV.

5.4 Solar Power Plants

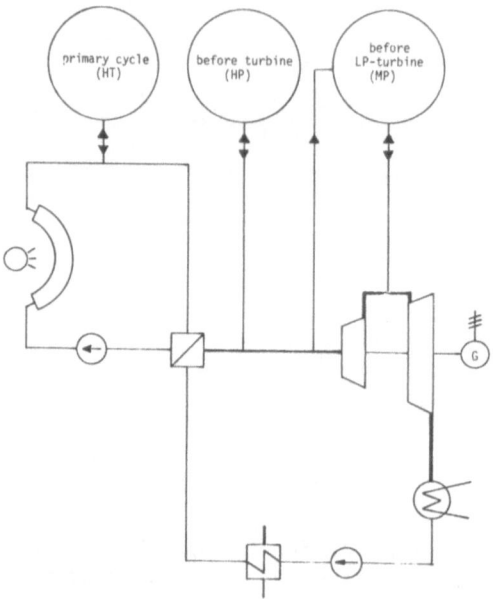

Fig. 24 Possibilities of load balancing in a solar power plant
         /23/

The solar thermal power plant (Fig. 24) exhibits several special features that make integrated thermal energy storage even more necessary /6, 7, 8, 17/:
(a) Fuel storage is not possible. Fuel control would be possible, e.g. by defocussing the heliostats, but - contrary to the fossil-fired plant - entails a loss of energy and is therefore avoided.

(b) Similar to the conditions of wind, tidal, and hydro plants, the input of primary energy keeps changing in a diurnal and annual pattern and, in addition, is depending on wheather conditions. This means that not only there is a demand pattern but also a supply pattern. The need for energy storage generally is increased except in the rare case that the two patterns should happen to coincide.

(c) The load factor of solar plants without thermal energy
    storage is rather low: about 3000 hrs./yr. in arid climates,
    about 2000 hrs./yr. in mediterranean climates and about
    1000 hrs./yr. under central European conditions. By means of
    thermal energy storage, the load factor of all components
    of the plants located after the storage system will be in-
    creased. The layout of heliostat field, turbine and storage
    unit therefore is an interesting optimization problem /6, 19/.

(d) In many practical cases not only the power input but also
    the waste heat disposal system will show a cyclic or stocha-
    stic behaviour. This will apply in particular to air cooled
    plants (cooling tower operation of a steam plant). Generally,
    this will be a disadvantages for the fossil plant in which
    peaks will occur during the day, i.e. high ambient air
    temperatures; it will, however, be an advantages for the
    solar plant where part of the power is shifted towards
    evening and night, i.e. to lower air temperature and thus
    higher efficiency.

(e) In many climatic zones a short-time energy system becomes
    necessary in order to cope with the effects on the plant
    control of sudden meteorological changes such as clouds
    passing-by. A storage capacity of about half an hour full-
    load is typical for such cases.

(f) Whereas in fossil-fired plants heat transfer occurs directly
    from flame and flue gas to the working medium, a separate
    primary heat transfer cycle is employed in some types of
    solar power plant. The primary medium may be heat transfer
    oil, molten salt (HITEC) or liquid alkali metal (Na); they
    all are suitable as high temperature storage media, too.

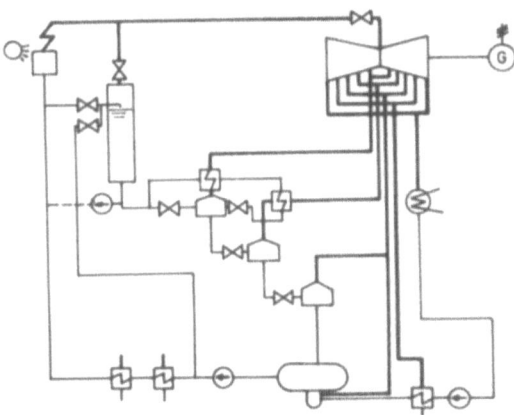

Fig. 25 Single cycle solar steam power plant with saturated
         steam storage /23/

Fig. 25 shows a flow sheet with H_2O saturated steam storage in a solar steam plant with direct steam generation in the receiver (single cycle). Charging is done by means of live steam and feed water, discharging via a series of flash evaporators. In Fig. 26, a double cycle solar steam power plant with TES in the primary cycle (heat transfer cycle) is shown. Usually, the heat transfer medium (oil, molten, salt, sodium) acts also as storage medium. There is a hot and a cold storage vessel.

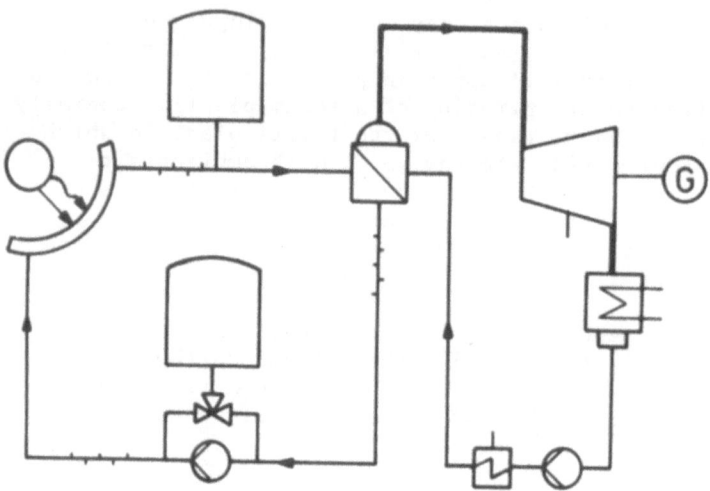

Fig. 26   Double cycle solar steam power plant with thermal energy storage in the primary (heat transfer) cycle

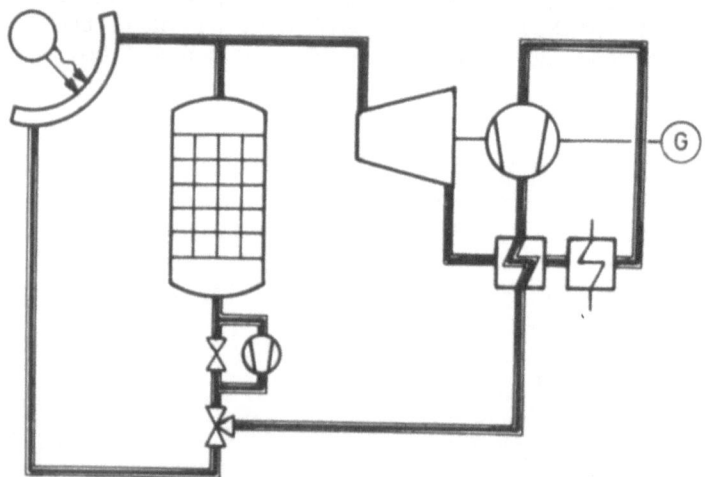

Fig. 27   Closed cycle solar gas turbine power plant with solid heat storage

A flow sheet of a solar closed-cycle gas turbine power plant is shown in <u>Fig. 27</u> . A matrix of solid bodies in a pressure vessel is in contact with the gas and acts as thermocline storage system. The upper-most layer of the matrix is always at the upper temperature, the lower-most layer at the lower temperature. The matrix may consist of MgO or cast iron bricks in a checkerwork arrangement. Cast iron, due to its higher density, needs a considerably smaller pressure vessel volume /8, 9/.

5.5 <u>Nuclear Power Plants</u>

Integrated thermal energy storage within the nuclear power plant is one of the most promising possibilities to achieve base load operation of the reactor and, at the same time, to cover daily demand peaks by means of the nuclear power plant: Base load operation of the ractor island minimizes temperature cycling of the fuel, and makes maximum use of this capital cost intensive plant item; it is, therefore, the safest and most economic way of reactor operation /1, 2, 3/. Using the nuclear power plant for peaking by means of energy storage also reduces the need for separate peak load plants and the need to use scare and expensive peak load fuel (light oil, kerosene, and natural gas) with low efficiency.

Several systems of integrated TES for nuclear power plants have already been suggested:
- displacement storage with external steam generation via heat exchangers /11/ or by flashing /20/;

- the pure feed water storage system /12, 13, 14/ that had already been proposed by Marguerre /36/ and by Margen /28/;

- the pure expansion-type TES system /3/;

- the sliding pressure system with internal steam generation (the classical Ruths principle) used in the coal-fired Berlin-Charlottenburg plant (Fig. 19 and 20) and nowadays suggested for underground storage (Fig. 16, /4/);

- the combined feed water and steam storage ("cascading") system /11, 20, 23, 25/.

The latter system warrants closer consideration. <u>Fig. 28</u> gives an example of such a cascading flow sheet, applied to a PWR with overloadable turbine set /20, 23/. In this case, the only additions to the base load plant are the PCIV, a flash evaporator, an enlargement of the feed water tank (or a separate cold storage tank), and some piping. Charging of the PCIV is performed by means of live steam plus hot feed water

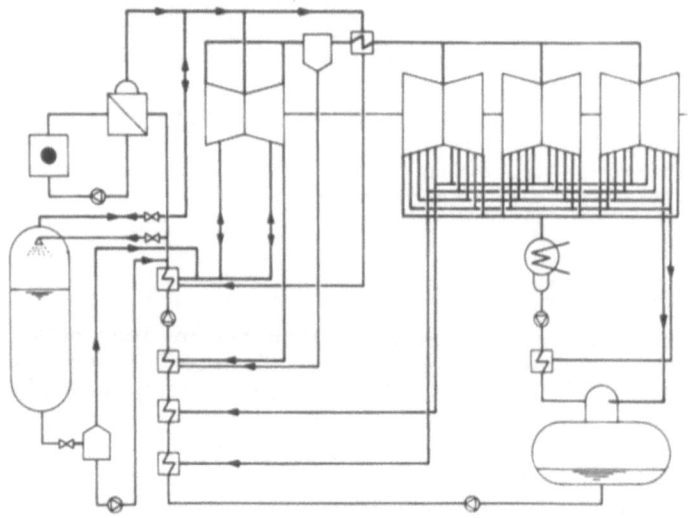

Fig. 28   Flow sheet of a nuclear power plant (PWR) with inte-
           grated thermal energy storage of the cascading type
           and overload-taking turbine set /20, 23/

being mixed in the PCIV. During discharge, the water content
of the PCIV is flashed. The steam generated in this way is
piped to the first extraction point, possibly after some super-
heating. Part of this steam is, however, used to finally heat
up some feed water that cannot be replaced by water from the
PCIV. This is because there is moisture from the moisture separator and hot condensate from the reheater that has to be used,
and there is also some feed water that has to be heated up by
the minimum amount of extraction from lower bleed points,
necessary for reason of moisture control at the cold end of the
turbine.

     If reducing of the bleed flow is considered to be unsuit-
able for a nuclear plant or overloading of the turbine set is
not possible, a separate peak load turbine is required /1, 2,
3/. The arrangement may be similar to the one of the coal-
fired station shown in Fig. 19.

6. PEAKING CAPABILITY

     The question arises whether and to what extent the peaking
capability of a TES system is limited. Obviously, there is a
fundamental limitation of stored energy in that discharged energy
during the peaking period must be less (due to storage efficiency)
than charging energy during the low load period. For the steam

storage systems this is the only limitation. However, for the
feed water storage systems there is also a thermodynamic limitation of peaking power in that only the actual feed water flow
can be replaced by stored hot water. The same limitation applies
to the cascading systems where there is only one (hot water) discharge line. It does not apply to the combined systems with a
hot water and a steam discharge line (as in Fig. 28).

The ratio of peaking power to base load power, $\Delta N/N$, depends
on the following parameters:
- Feed water temperature:
  The higher it is, the more thermal energy can be replaced by
  stored thermal energy. Feed water temperature is practically
  limited by boiling heat transfer for water reactors, and by
  blower power in the case of the HTR and GCFR.

- Cycle efficiency of the base load process:
  The higher the efficiency, the lower is $\Delta N/N$, because the flow
  rate is smaller.

- Portion of feed water that may be replayced by stored hot water:
  Depends on fixed rates such as minimum amount of bleed to
  ensure wetness control, and of hot condensate from moisture
  separator and from the reheater heated by condensing live steam.

- Storage pressure or more correctly storage temperature:
  The higher it is, the more steam will be flashed in cascading
  system by throttling down to the pressure of the first bleed
  which during peaking acts as secondary admission.

Table IV: Typical Data of Steam Cycle /20, 23/

Type of Plant	Live Steam		Reheat Steam		Feed Water Temperature	Ratio of Feed flow Replacement to Live Steam Flow Rate
	Pressure bar	Temperature $^\circ$C	Pressure bar	Temperature $^\circ$C	$^\circ$C	
CANDU	55	sat.	10	250	180	0.82
LWR	70	sat.	12	265	210	0.79
HTR	170	540	40	540	206	1.0
LMFBR	170	540	40	540	240	1.0
Fossil, subcritical	170	540	40	540	240	1.0
" supercritical	250	540	55	540	260	1.0

The value of the ratio $\Delta N/N$ has been assessed for several
reactor types as defined in Table IV and for various storage pressures P. Results of the cycle analyses are shown in Fig. 29 and
Table V versus upper storage temperature t or pressure p,

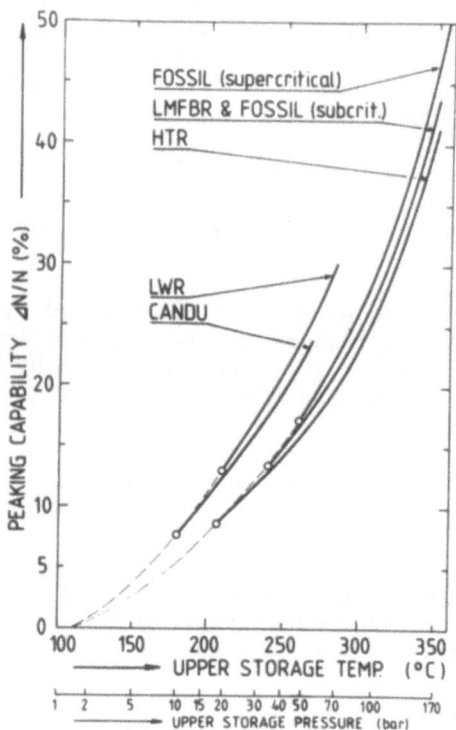

Fig. 29 Peaking capacity of steam power plant with cascadin¡
(feed water + steam) storage /22, 23/

Table V: Peaking Capability with Feed Water and Cascading Storage /20, 23/

Type of Plant	Peaking Capability $\Delta N/N$ (% of base load)					
	Feed Water Storage	Cascading (Feed Water + Steam) Storage at an upper pressure of (bar)				
		20	55	70	100	170
CANDU	7.6	12.4	23.9	-	-	-
LWR	13.0	13.3	26.0	30.2		
HTR	8.6	9.3	17.1	20.0	26.7	42.0
LMFBR	13.4	-	18.0	21.1	27.1	44.0
Fossil, subcritical	13.4	-	18.0	21.1	17.1	44.0
" supercritical	17.2	-	19.0	22.2	28.2	47.0

respectively. A lower storage temperature of 110 °C (1.5 bar)
was assumed. It turns out that in the case of pure feed water
storage, $\Delta N/N$ is only 7.6 % for the CANDU, 8.6 % for the HTR,

and about 13 % for the LWR, LMFBR, and fossil plant. An increase of $\Delta N/N$ is possible with the cascading arrangement, e.g. up to about 30 % in case of the LWR with a storage pressure of 70 bar /20, 23/.

7. CONCLUSIONS

- Energy storage in optimized electric energy supply systems leads to a larger proportion and to a better utilization of low fuel cost, high investment cost ("base load") power stations as opposed to high fuel cost, low investment cost peaking plants and, therefore, to the substitution of valuable oil and gas by coal and nuclear fuel or by solar energy.

- In optimized systems, daily, weekly and annual energy storage will, in many cases, lead to substantially lower system cost. This will depend on the load curves in the grid, the relative investment cost and the relative fuel cost of base load, peaking and storage plants. The importance and the economics of energy storage in electric grids will increase with the proportion of conventional nuclear, breeder, fusion and solar power plants. Also operational advantages will prevail.

- Hydraulic pumped storage plants have been built since many decades; they do require suitable sites. Compressed air storage plants - that are to be considered combinations of a pure energy storage plant with a gas turbine peaking plant - are in the prototype stage; they also require suitable geological conditions. Thermal energy storage (TES) plants, integrated in the steam power plant itself, based on pressurized hot water, were built decades ago and should regain importance under present and future conditions regarding power plant technology, pressure vessel technology and relative fuel cost.

- Pressurized hot water storage may be built as "steam storage", "feed water storage" or combined ("cascading") plants; they may be built underground, if suitable geological conditions prevail, or above ground; the additional power may be gained by overloading the main turbine, if this is possible, or by a separate peaking turbine.

- Advantages of integrated TES plants include: no additional energy conversions; practically no limitation of charging power; high turnaround efficiency, in particular with feed water storage - where, however, the discharge power is limited - or with cascading storage.

REFERENCES

/1/ Gilli. P.V., Beckmann, G.: The Nuclear Steam Storage Plant –
An Economic Method of Peak Power Generation, Proc. 9th WEC
(Detroit, 1974), U.S. National Committee of the World
Energy Conf., New York 1975, Vol. V, pp. 162–179, Vol.IX,
pp. 652–654

/2/ Beckmann, G., Fritz, K., Gilli. P.V.: Thermal energy
storage for economic peak power generation from nuclear
power plants, VDI-Berichte 223 (Stuttgart) 1974, ISBN 3-18-
090233-X, VDI-Verlag, Essen, pp. 21–31

/3/ Gilli, P.V., Beckmann, G.: Covering peak load means of
thermal energy, VDI-Berichte 236 (Düsseldorf, 1975), ISBN
3-18-09236-1, VDI-Verlag, Essen. pp. 121–127

/4/ Ridway, S.L., Dooley, J.L.: Underground Storage of Off-Peak
Power, Proc. 11th IECEC (1976), Vol. I, pp. 586–590

/5/ Kovach, E.G. (Ed.): Thermal Energy Storage, Report of a
NATO Science Committee Conference, Turnberry, Scotland,
1-5 March 1976, Pergamon

/6/ Gilli, P.V., Beckmann, G.: Design and Economy of Solar Power
Plants with Integrated Thermal Energy Storage, UNESCO/OMM
Symposium on the Problems of Solar Energy Utilization,
Geneva, August 30 – September 3, 1976

/7/ Beckmann, G.: Die Entladekapazität von Gefällespeichern,
Die Wärme 82 (1976), pp. 110–117

/8/ Boeing Engineering & Construction: Advanced Thermal Energy
Storage Concept Definition Study for Solar Brayton Power
Plants, Vol. 1: Final Technical Report, July 1 – December 31,
1976, SAN/1300-1, NTIS, Springfield, Va., 1977

/9/ Beverly, W.D., Engle, W.W., Mahony, F.O.: Integration of
High Temperature Thermal Energy Storage into a Solar
Thermal Brayton Cycle Power Plant, paper 779192, 12th
IECEC (1977) Washington, pp. 1195–1220

/10/ Pierce, B.L., Spunsier, F.R., Wright, M.K.: Thermal Energy
Storage, paper 779191, 12th IECEC (1977), Washington, pp.
1189–1194

/11/ Gilli, P.V., Beckmann, G., Schilling, F.E.: Thermal Energy
Storage using Prestressed Cast Iron Vessels (PCIV), COO-
2886-1, NTIS, Springfield, Va., 1977

/12/ Bitterlich, E.: Bildung von Speicherwärme mittels Entnahme-
dampf in Lastsenken und deren Nutzung in Spitzenbelastungs-
zeiten, Brennstoff-Wärme-Kraft 29 (1977), pp. 51-58

/13/ Bitterlich, E., Brandes, H.: Kraftwerksprojekt zur Heiß-
wasserspeicherung für elektrische Spitzenlast und Fernwärme,
VGB-Kongress Kraftwerk 1977, Kopenhagen

/14/ Brandes, H., Gülicher, L.: Einsatz von Heisswasserspeichern
in Kraftwerken zur Deckung der elektrischen Spitzenlast und
des Fernwärmebedarfes, VDI Bericht 288 (Düsseldorf, 1977),
ISBN 3-18090288-4, VDI-Verlag, Essen, 1978, pp. 145-151

/15/ Gilli, P.V.: Energy Storage by Means of Prestressed Cast
Iron Vessels, 12th Mid Atlantic Regional Meeting, American
Chemical Society (ACS), Hunt Valley, Maryland, April 5 - 7,
1978

/16/ Vrable, D.L., Quade, R.W.: High Efficiency Thermal Energy
Storage Systems for Utility Application, paper 789073, 13th
IECEC (1978), San Diego, pp. 917-922

/17/ Bramlette, T.T.: Recent Advanes in Thermochemical Energy
Storage and Transport, paper 789151, 13th IECEC (1978) San
Diego, pp. 928-934

/18/ Calogeras, J.E.: Storage Systems for Solar Thermal Power,
paper 789681, 13th IECEC (1978) San Diego, pp. 970-976

/19/ Fujita, T.: Comparative Evaluation of Distributed Collector
Solar Thermal Electric Power Plants, paper 789346, 13th
IECEC (1978), San Diego, pp. 1600-1607

/20/ Gilli, P.V., Beckmann, G., Schilling, F.E.: Nuclear Peak
Load Plant with PCIV Thermal Energy Storage, paper No.
C 1/10, NUCLEX 78, Basel, October 3-7, 1978

/21/ Hausz, W., Berkowitz, B.J., Hare, R.G.: Conceptual Design
of Thermal Energy Storage Systems for Near Term Utility
Applications, Vol. 1: Screening of Concepts, Vol. 2:
Appendices-Screening of Concepts, NASA CR-159411, NTIS,
Springfield, Va., 1978

/22/ J. Silverman(Ed.): Energy Storage, Trans. First Int. Assembly
(Dubrovnik, 1979), Pergamon Press, 1980

/23/ Gilli , P.V., Beckmann, G.: Thermal Energy Storage for Non-
Fossil Power Plants; /22/, pp. 264-278

/24/ Oplatka, G., Schmid, G.: Deckung der Lastspitzen mit
thermisch gespeicherter Energie, Brown Boveri Mitt. 67
(1980), pp. 457-464

/25/ Goding, P., Sammon, D.: Energy Storage in an Electric Power
Grid, IEA Conference on New Energy Conservation Technologies
and their Commercialisation, Berlin, 6-10 April 1981

/26/ Fernandes, R.A.: Hydrogen Cycle Peak-Shaving for Electric
Utilities, 9th IECEC (1974), pp. 413-422

/27/ Marguerre, F.: Das thermodynamische Speicherverfahren von
Marguerre, Escher Weyss Mitteilungen, Jg. VI, Nr. Mai-
Juni 1933, pp. 67-76

/28/ Margen, P.H.: Thermal Energy Storage in Rock Chambers - A
Complement to Nuclear Power, Peaceful Uses of Atomic Energy,
IAEA, Vienna, 1972, Vol.4, pp. 177-194

/29/ Halle, K.:, Schmid, J.: Die Dampfspeicher-Anlage im Kraftwerk Charlottenburg, Die Wärme 54 (1931), 920-925

/30/ Goldstern, W.:Dampfspeicheranlagen, 1. Ed. 1933, 2. Ed. 1963,
Springer-Verlag, Berlin/Göttingen/Heidelberg: English
Ed.: Steam Storage Installations, Pergamon Press, 1970
(International Series of Monographs in Mechanical Engineering,
Vol. 4)

/31/ Schäck, R.: Betrieb von Fernheiznetzen mit und ohne Speicher,
Energie 25 (1973) 157

/32/ Herbst, H.Ch., Maass, P.: Das 290-MW-Luftspeicher-Gas-
turbinenkraftwerk Huntorf, VGB Kraftwerkstechnik 60, Heft 3
(März 1980) 174-187

/33/ Schüller, K.H., Haaland, M.: Pneumatische Speicherkraftwerke
ohne Brènnstoffzufuhr, VGB-Tagung Kraftwerke 1977

/34/ Giramontir, A.J.: Fluid Bed Augmented CAES Systems /22/
509-518

/35/ Gilli, P.V., Beckmann, G.: Schnellstart von Gasturbinen mit
Hilfe von VGD-Luftspeichern, Elektrizitätswirtschaft Jg. 75
(1976) Heft 6, 116-118

/36/ Marguerre, E., Koch, J.: Gleichdruckspeicher als Ausgleich
in Vorschalt- und Heizkraftanlagen, Die Wärme 52 (1929)
334-340

TRANSMISSION OF HEAT USING HOT WATER PIPES

PART I

Sven Hadvig

Laboratory of Heating and Air Conditioning
Technical University of Denmark, Lyngby

INTRODUCTION

In consequence of recent years' energy crises, more will be staked on power-heating plants than on conventional power plants. In the future, district heating plants proper, i.e. plants whose only purpose is to supply heating, will more and more have to switch over to coal-burning, thus using the same cheap fuel as do the power plants.

Natural gas, however, which is to be conducted to consumers may also be a possibility if it can be obtained at a reasonable price.

The energy crises reversed the Danish energy policy to a quite considerable degree and an extension of the power-heating supply will take place within the next few years.

In proportion to its size, Denmark probably has the most developed district heating network with at present 16,000 km district heating pipes which have been laid and extended in the course of many years over an area of 43,000 km^2.

Accordingly, owing to our early start within the field of district heating we now possess wide experience as regards district heating networks.

Consequently, the Laboratory of Heating and Air Conditioning at the Technical University of Denmark is frequently visited by foreign guests who want to be informed about the lessons we have learned, good as well as bad ones.

It should be added that part of the material to be presented in the following has not been published before.

PROBLEMS TO BE SOLVED

We shall confine ourselves to the problems arising in connection with district heating networks. There is no concealing the fact that extensive district heating networks do present certain problems unless a number of details have been taken care of in advance.

The problems exist at various levels.

The essential purpose of this paper is to exhibit the problems and to give some guide-lines for their avoidance.

DESIGN

Already in the design phase it should be fully realised whether the area contemplated lends itself to district heating at all. Several aspects should be brought up, such as:-

Is the area in question only sparsely built up?

Are the houses in the area likely to be connected in adequate numbers? If not, it is no good idea to lay out a large and costly network. In this connection I assume the price of the heat supply to the network to be acceptable.

Are the soil conditions suitable for district heating pipes, or must allowance be made for considerable parts of the network to be laid on the ground?. Experience shows that this involves considerable difficulties.

Once it has been decided to design a district heating network it is a great advantage to have a thorough knowledge of the heat demand of the individual buildings in question, whether industrial or dwelling.

Now the question of how to dimension the pipe system comes into the picture. In this connection a number of questions arise:-

How do I design the pipe system so as to obtain a reasonable quarantee of supply in the case of damage to the pipes?

What pipe diameter must be used to dimension the network

sufficiently close to an economical optimum?

Is a well worked-out EDP system available for the calculation of complex networks with ring connections (i.e. shunt connections in which the direction of the flow is not known in advance), by-passes, and a large number of joints?

Is the EDP system of such a character that it can go over the calculation of different proposals without the financial expenses in connection with the calculations becoming too high?

What type of pipe and protective casing should be used?

After the design phase follow the invitation to submit tenders and the accomplishment of the work.

An important aspect is the way the pipes are laid and joined. A poor quality of the preceding design work often turns out to result in a far too short life of the pipes owing to corrosion of the steel.

A short life of the pipe system causes thermal losses above normal from parts of the network, both

1) loss of heat owing to the fact that the heat insulating material is wet, and
2) loss of heat owing to the fact that new, cold water is added and then led into the earth at a relatively high temperature.

That the quantities of water involved according to 2) are not unimportant can be seen from the fact that some plants consume between 50 and 100 m^3 fresh water per day, so that in some places the district heating plant is the main consumer of water.

It is obvious that the time allotted for this presentation does not allow a thorouth treatment of all these aspects. Therefore, it has been chosen to take a closer look at the utilization of the thermal energy induced and to make some reflections on the economy in connection with the design of networks.

Finally, it should be added that all these difficulties were not mentioned in order to deter engineers from designing extensive networks for district heating, but merely to call attention to some of the problems we have had to solve in Denmark and thus to let others benefit from the lessons we have learnt and to make it possible for others to have district heating networks which are just as well functioning as the ones we now have.

THE UTILIZATION OF THERMAL ENERGY

As a consequence of the high and ever-rising prices of energy it has been necessary to take the question of energy prices into account when calculating the thermal loss from dwellings.

This means, among other things, that - anyway in the North European countries - we have had to effect a rather high degree of heat insulation in both industrial buildings and dwellings. This has no unimportant influence on the energy economy of district heating plants.

HEAT LOSS FROM DISTRICT HEATING PIPES

How much thermal energy is being lost from a network under ground? Or, to put it another way, to what extent do we exploit the thermal energy for, e.g., heating including domestic water in dwellings provided with district heating?

This question will be treated in greater detail, partly by a discussion of a pipe system which has been measured through and through with regard to thermal losses and partly by going over the calculations of different pipe diameters. This was done at the Laboratory of Heating and Air Conditioning at the Technical University of Denmark.

Before presenting these data it is necessary to mention a number of expressions indicating thermal interchange between the district heating pipes as well as heat loss from the pipelines. [1]

These expressions are based on theoretic considerations, and a considerable number of cases have been gone over in accordance with the Finite Element Method.

HEAT LOSS FROM PIPELINES IN DISTRICT HEATING SYSTEMS

Absolute resistance of earth. Temperature difference.

It is useful to start with these two quantities since they recur in all the expressions.

Although the corrected laying depth will be treated later, the expression is given here in advance.

The laying depth is h. It is corrected so as to include the earth resistance at the surface.

Figure 1 shows a pipe at laying
depth h and with external diameter D_u.

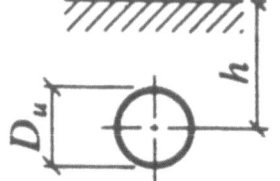

$$H = h + 0.0685 \cdot \lambda_J \quad (m)$$

for $\dfrac{h}{D_i} = \gtrless 1$

Figure 1.

$$R_J = \frac{1}{2\pi\lambda_J} \ln \frac{4H}{D_u} \quad (m^\circ C/W) \tag{1}$$

for $\dfrac{h}{D_i} < 1$

$$R_J = \frac{1}{2\pi\lambda_J} \ln \left(\frac{2H}{D_u} + \sqrt{\left(\frac{2H}{D_u}\right)^2 - 1} \right) \quad (m^\circ C/W) \tag{2}$$

Temperature difference

$$\Delta t = t_r - t_h \quad (^\circ C) \tag{3}$$

where
t_r = temperature of pipe (water temperature) $(^\circ C)$
t_h = mean earth temperature of month under consideration
at depth h for undisturbed earth.

For normal Danish conditions, t_h can be found in Table 1. For a
more accurate adaptation of t_h to local conditions, see equations
(16) and (17).

THEORY

Heat loss from single pipes in earth

Heat loss from a single pipe under ground can be written

$$q = \frac{1}{R_o + R_J' + R_I} \cdot \Delta t \quad (W/m) \tag{4}$$

where
R_o = absolute earth resistance at surface $(^\circ Cm/W)$
R_J' = absolute resistance of earth $(^\circ Cm/W)$
R_I = absolute resistance of pipe insulation $(^\circ Cm/W)$

Δt = temperature difference $(^\circ C)$

In most cases R_O and R_J' are combined into one resistance

$$R_J = R_O + R_J' \quad (^\circ Cm/W)$$

Table 1.

t_O	t_h for different values of h (m)									
$^\circ C$	0.2	0.4	0.6	0.8	1.0	1.2	1.4	1.6	1.8	2.0
Jan. 0.3	1.1	1.9	2.6	3.2	3.8	4.3	4.8	5.2	5.6	5.9
Feb. 0.0	0.7	1.3	1.8	2.3	2.8	3.3	3.7	4.1	4.6	5.0
Mar. 1.5	1.8	2.1	2.4	2.7	3.0	3.3	3.5	3.9	4.2	4.6
Apr. 5.4	5.2	5.0	4.8	4.7	4.6	4.6	4.6	4.7	4.8	5.0
May 11.5	10.6	9.8	9.0	8.4	7.9	7.4	7.0	6.8	6.7	6.6
Jun. 17.3	16.0	14.9	13.8	12.9	12.1	11.4	10.8	10.2	9.7	9.2
Jul. 21.0	19.7	18.6	17.5	16.6	15.7	14.9	14.2	13.5	12.8	12.1
Aug. 18.1	17.9	17.6	17.3	17.1	16.8	16.5	16.2	15.7	15.0	14.3
Sep. 14.4	14.5	14.6	14.6	14.6	14.6	14.5	14.4	14.3	14.2	14.0
Oct. 9.0	9.7	10.3	10.8	11.2	11.6	11.9	12.1	12.2	12.3	12.3
Nov. 4.1	5.0	5.9	6.6	7.3	7.9	8.4	8.8	9.2	9.6	9.9
Dec. 2.2	2.9	3.5	4.1	4.6	5.1	5.5	5.9	6.4	6.9	7.4

Corresponding tables can be set up for other climatic conditions.

Absolute_resistance_of_earth

Figure 2 shows a pipe placed
in an infinite speace of earth.

In the following calculations a
slice of earth is considered
which measures 1 m in the direction of the pipe axis. The heat
flux through a comparison surface at distance x from the
centre can be written

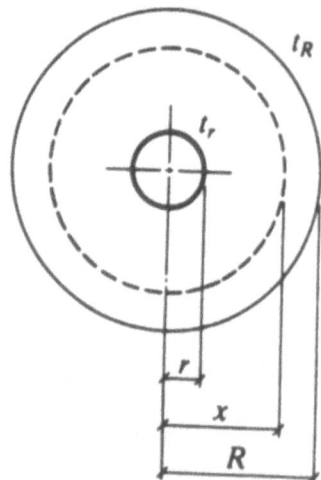

$$q = \int_0^{2\pi} \lambda_J \cdot -\frac{dt}{dx} \cdot x \; d0 = -2\pi \cdot \lambda_J \; x \; \frac{dt}{dx}$$

Figure 2.

q is taken to be positive in the case of flux from the centre

$$dt = -\frac{q}{2\pi\lambda_J \cdot x}\, dx$$

Integration yields

$$t = -\frac{q}{2\pi\lambda_J} \cdot \ln x + C$$

for $x = 1$, $t = t_r$

$$t = t_r - \frac{q}{2\pi\lambda_J} \cdot \ln \frac{x}{r} \tag{5}$$

for $x = R$, $t = t_R$.

q can now be written

$$q = (t_R - t_r)\frac{2\pi\lambda_J}{\ln(\frac{r}{R})}$$

We now consider a pipe placed at depth h under the surface of the ground.

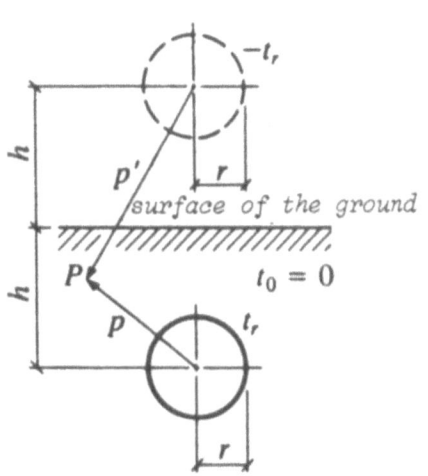

Figure 3.

The earth is considered as a semi-infinite space. In order to find t we introduce a fictitious pipe with radius r and temperature $-t$ placed at height h above the surface.

The temperature of point p is found by superposition of the thermal exposure from the two pipes. Because of symmetry, the heat loss from the fictitious pipe will be $-q$

$$t_p = t_r - \frac{q}{2\pi\lambda_J}\ln\frac{p}{r} - t_r + \frac{q}{2\pi\lambda_J}\ln\frac{p'}{r}$$

$$= \frac{q}{2\pi\lambda_J}\ln\frac{p'}{p} \tag{6}$$

For p placed on the surface of the ground, $p = p' \Rightarrow t_p = 0$
The boundary condition is thus fulfilled. If p leaves the pipe

$$\frac{p'}{p} \to 1 \quad \text{and} \quad t_p \to 0$$

Heat loss

For the determination of q, point p is placed on the brim of the

pipe. As will be seen from Equation (6) it is not unimportant
exactly where on the brim p is placed. The position will affect
p'/p and hence q. By letting p' be equal to 2h we get the simplest
equation

$$q = \frac{2\pi\lambda_J}{\ln(\frac{2h}{r})} \cdot t_r = \frac{1}{R'_J} \cdot t_r$$

$$R'_J = \frac{1}{2\pi\lambda_J} \ln(\frac{2h}{r}) \quad (^\circ Cm/W) \tag{7}$$

The thermal loss from the pipe can also be determined by a
conformal picture [2]. The solution then becomes

$$R'_J = \frac{1}{2\pi\lambda_J} \ln \left(\frac{\sqrt{h^2 - r^2} + (h - r)}{\sqrt{h^2 - r^2} - (h - r)} \right)$$

This solution is identical with the formula used in [3] and [4]

$$R'_J = \frac{1}{2\pi\lambda_J} \ln \left(\frac{h}{r} + \sqrt{\left(\frac{h}{r}\right)^2 - 1} \right) \quad (^\circ Cm/W) \tag{8}$$

R_J determined by the "Finite Element Method"

The program "ICES STRUDL-II IUG VERSION V3MI, MAY 1977", available at the regional EDP centre NEUCC, was used. By means of the
finite element method this program is able to compute two and
three-dimensional heat conduction in the stationary as well as
the instationary case.

Comparison of methods for the calculation of R_J

In order to get an idea of the differences between the equations
and the numerical method they are shown in Figure 4 as functions
of h/r.

From the figure it will be seen that there is good agreement
between STRUDL and Equation (8). For small values of h/r, Equation
(7) gives too large values of $R'_J \cdot \lambda_J$. In practice it will be
possible to use Equation (7) for h/r > 2. For smaller values of
h/r Equation (8) should be used.

Absolute earth resistance at surface of the ground

Heat transfer at the surface can be written

$$W_o = \alpha_o(t_o - t_\ell) + \epsilon_j \cdot \sigma(T_o^4 - T_s^4) \quad (W/m^2) \tag{9}$$

where

T_o = absolute temperature of surface of the ground (K)
t_o = temperature of surface of the gound ($^{\circ}$C)
t_ℓ = air temperature ($^{\circ}$C)
T_s = absolute temperature of firmament (K)
α_o = convective heat transfer coefficient at surface of the ground (W/m$^2\,^{\circ}$C)
ε_j = emissivity of surface of the ground

The first term of Equation (9) describes the convective heat transfer. In accordance with [4] α_o is taken to be equal to 9.7 W/m$^2\,^{\circ}$C. The second term of Equation (9) gives the radiation, the firmament being regarded as a black surface with temperature T_s. If the radiation term is linearised, Equation (9) can be written

$$W_o = \alpha_o(t_o - t_\ell) + \varepsilon_j \cdot \sigma \cdot 4 \cdot T_o^3 (T_o - T_s)$$

The temperature t_s is defined by

$$t_a = \frac{\alpha_o t_\ell + \varepsilon_j \cdot \sigma \cdot 4 \cdot T_o^3 (T_s - 273)}{(\alpha_o + \varepsilon_j \cdot \sigma \cdot 4 \cdot T_o^3)}$$

By insertion of t_a, W_o becomes

$$W_o = (\alpha_o + \varepsilon_j \cdot \sigma \cdot 4 \cdot T_o^3) \cdot (t_o - t_a)$$

If T_o is fixed at 283°C and ε_j at 0.95, W_o becomes

$$W_o = 14.6 \cdot (t_o - t_a) \tag{10}$$

From Equation (10) we get the earth resistance

$$M_o = 0.0685 \quad (m^{\circ}C/W)$$

R_o cannot be calculated directly from M_o as the area by which we have to divide in order to get from M_o to R_o is difficult to calculate. In order to include the earth resistance into the total resistance in a simple way, R_o is combined with R_j'. R_o is included into R_j' as a stratum with thickness δ.

$$\delta = \lambda_J \cdot M_o \quad (m)$$

Consequently, Equations (7) and (8) change into

$$R_j = \frac{1}{2\pi\lambda_j} \ln\left(\frac{2h+2\delta}{r}\right) \quad (^{\circ}Cm/W) \tag{1}$$

and

$$R_J = \frac{1}{2\pi\lambda_J} \ln\left(\frac{h+\delta}{r} + \sqrt{(\frac{h+\delta}{r})^2 - 1}\right) \quad (^\circ Cm/W) \qquad (2)$$

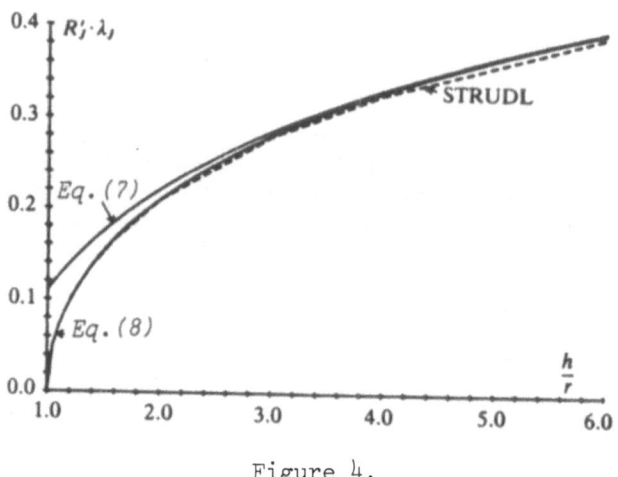

Figure 4.

Absolute resistance of insulation

Where possible, measured values of R_I are used for prefabricated pipes. For pipes of simple geometry, R_I can be determined by calculations.

Insulation of circular cross section

$$R_I = \frac{1}{\alpha_r \pi D_i'} + \frac{1}{2\pi\lambda_{ri}} \ln \frac{D_i}{D_i'} + \frac{1}{2\pi\lambda_I} \ln \frac{D_u}{D_i} + \frac{1}{2\pi\lambda_{ru}} \ln \frac{D_u'}{D_u} \qquad (11)$$

In most cases the terms describing the internal earth resistance, the absolute resistance of the internal pipe, and the absolute resistance of the jacket pipe will constitute only a small part of the total resistance of the pipe. In these cases, Equation (10) is reduced into

$$R_I = \frac{1}{2\pi\lambda_I} \ln \frac{D_u}{D_i} \qquad (12)$$

Where Equation (12) is used, D_u will normally be taken as the external diameter of the pipe.

Insulation of rectangular cross section

For the calculation of R_I, Equation (12) is used, introducing a fictitious external diameter of the insulation

$$D_u = \frac{1.1 \cdot 8}{\dfrac{1}{L_1} + \dfrac{1}{L_2} + \dfrac{1}{L_3} + \dfrac{1}{L_4}} \qquad (13)$$

If $L_1 = L_2 = L_3 = L_4$ we get

$$D_u = 1.1 \cdot 2 \cdot L_1 \qquad (14)$$

The factor 1.1 is a correction for the cross section not being circular. Brauer [4] uses the correction factor 1.073, whereas Harris & Fitzgerald [5] uses the factor 1.08. In the present case the factor 1.1 was chosen which corresponds better with the calculations made by means of STRUDL.

Equation (13) is applicable only so long as there are not too large differences between maximum and minimum insulation thicknesses. This is shown in Figure 7.

The calculations the results of which are shown in the figure were carried out on the basis of a pipe with $D_i = 0.15$ m. The shape of the insulation is rectangular and the pipe is placed in the centre of the insulation. Calculated in accordance with Equation (13), $D_u = 0.4$ m. The thermal loss is shown in the figure as a function of the ratio of the thickness of the lateral insulation to that of the top insulation.

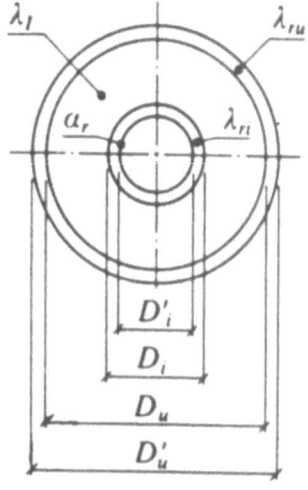

Figure 5.

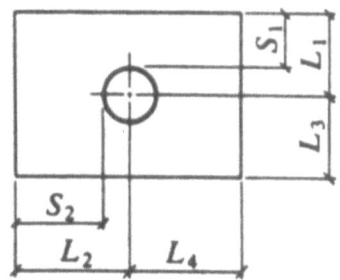

Figure 6.

Determination of temperature difference

In the case of stationarity

$$\Delta t = t_r - t_a \qquad (^\circ C)$$

where

t_r = the temperature of the heat conducting medium ($^\circ$C)
t_a = oriented temperature ($^\circ$C)

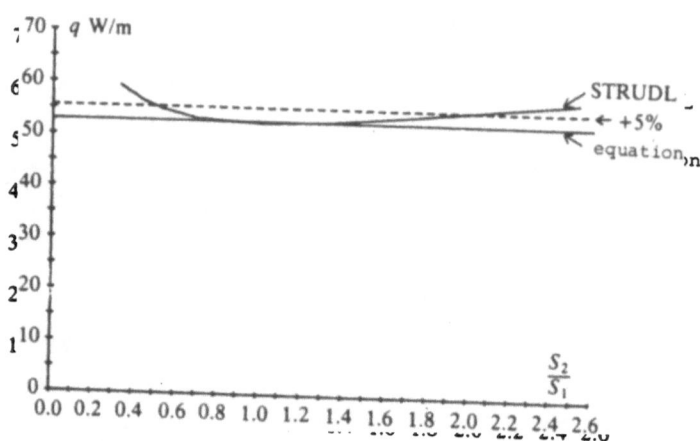

Figure 7a.

As earth has a very high heat capacity, stationarity can be expected only for the determination of the annual mean temperature difference.

For the calculation of the mean temperature difference of a given month, the heat capacity of earth must be taken into account. The mean temperature difference of a given month is determined as follows:-

$$\Delta t = t_r - t_h \quad (^\circ C) \qquad\qquad (3)$$

where

t_h = mean earth temperature of the month in question at depth h for undisturbed earth ($^\circ$C)

h = laying depth of pipe (m)

Calculating Δt in this way, the result will be that a pipe at depth h and with the temperature of earth will have a thermal loss of 0.

Calculation of t_h

Regarding the earth as a semi-infinite space and assuming the annual variation of the surface temperature to form a sinus curve we get the following expression for t_o

$$t_o = t_{om} + A \cos\left(\frac{2\pi}{\tau_o} (\tau - \tau_{max})\right)$$

where

t_{om} = annual mean temperature of surface of the ground

A = amplitude (°C)

τ = time (s)

τ_o = period (annual) (s)

τ_{max} = moment of maximum temperature (s)

According to [5] t_h becomes

$$t_h = t_{om} + A\, e^{-h\sqrt{\frac{\pi}{a\tau_o}}} \cos\left(\frac{2\pi}{\tau_o} (\tau - \tau_{max}) - h\sqrt{\frac{\pi}{a\tau_o}}\right) \qquad (15)$$

where

a = thermal diffusivity (m²/s)

From Equation (15) it will be seen that the amplitude of t_h follows that of t_o except for a damping and a time lag. Consequently, t_h at time τ can be written

$$t_h = t_{om} + (t_{of} - t_{om})\, e^{-h\sqrt{\frac{\pi}{a\tau_o}}} \qquad (16)$$

where

$t_{of} = t_o$ at time $\tau - h\sqrt{\dfrac{\tau_o}{4a\pi}}$

Weather data
============

Weather data were taken from Knud Schlosser [7] who, on the basis of the reference year and using K.J. Kristensen "Temperature and heat balance of soil" calculated values of t_ℓ, t_o, and W_o.

As an approximate sinus curve for t_o, a curve was used with mean temperature = 9°C, amplitude=11°C, and maximum temperature of 1 August.

From Table 2 the sinus approximation to t_o appears to deviate somewhat from t_o. Consequently, for the calculation of t_h, Equation (16) and the table values of t_o are used, not Equation (15). The following parameters are used:-

$$\lambda_J = 2 \text{ W/m}^\circ\text{C}$$
$$a_J = 8 \cdot 10^{-7} \quad \text{m}^2/\text{s}$$
$$t_{om} = 9^\circ\text{C}$$
$$\tau_o = 3.15 \cdot 10^7 \quad \text{s}$$

Hence, Equation (16) becomes

$$t_h = 9 + (t_{of} - 9) \, e^{-0.35h}$$

where

$$t_{of} = t_o \text{ at time } 0.67 \cdot h \text{ month earlier}$$

The variations in t_h as regards time and depth according to this equation are shown in Table 1.

Table 2.

	t_ℓ $^\circ$C	t_o $^\circ$C	W_o $^\circ$C	t_a $^\circ$C	sinus approximation to t_o
Jan.	0.0	0.3	8.30	-0.3	-1.5
Feb.	-0.4	0.0	5.70	-0.4	-1.5
Mar.	0.9	1.5	1.40	1.4	1.3
Apr.	4.4	5.4	-5.00	5.7	6.2
May	9.7	11.5	-11.10	12.3	11.8
Jun.	13.2	17.3	-12.40	18.1	16.7
Jul.	16.5	21.0	-10.40	21.7	19.5
Aug.	15.3	18.1	-5.00	18.4	19.5
Sep.	12.9	14.4	1.40	14.3	16.7
Oct.	8.4	9.0	7.70	8.5	11.8
Nov.	3.9	4.1	9.90	3.4	6.2
Dec.	2.0	2.2	9.90	1.5	1.3

Pairs of pipes can be treated in a similar way as single pipes, but to save space this part of the theory will not be included here.

Characteristically, an additional resistance R_h enters into the calculations in the case of pairs of pipes. This resistance takes the thermal interchange between the two pipes into account thus reducing the total heat loss.

Let us now proceed to the final practical expressions for the heat loss from pipes.

PRACTICAL EXPRESSIONS

Pairs of separately insulated pipes

Heat loss

$$q_1 = \frac{\Delta t_1 (R_{J2}+R_{I2})-\Delta t_2 R_h}{(R_{J2}+R_{I2})(R_{J1}+R_{I1})-R_h^2} \qquad (17)$$

Indices 1 and 2 refer to pipes 1 and 2.

$$H = h+0.0685 \; \lambda_J \quad (m)$$

$$R_{I1} = \frac{1}{2\pi\lambda_I} \ln \frac{D_{u1}}{D_{i1}} \quad (m^\circ C/W) \qquad (13)$$

$$R_{I2} = \frac{1}{2\pi\lambda_I} \ln \frac{D_{u2}}{D_{i2}} \quad (m^\circ C/W) \qquad (13)$$

$$R_h = \frac{1}{2\pi\lambda_J} \ln \sqrt{1 + (\frac{2H}{C})^2} \quad (m^\circ C/W) \qquad (18)$$

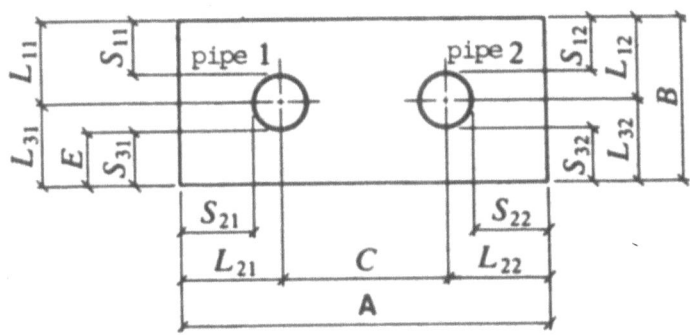

Figure 7

Jointly isulated pairs of pipes

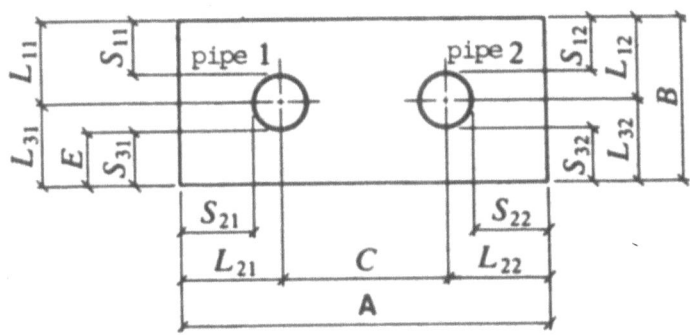

Figure 8

Heat loss

$$q_1 = \frac{\Delta t_1 (R_{J2}+R_{I2})-\Delta t_2 \cdot R_{h2}}{(R_{J2}+R_{I2})(R_{J1}+R_{I1})-R_{h1} \cdot R_{h2}} \qquad W/m \qquad (19)$$

Indices 1 and 2 refer to pipes 1 and 2

S_{max1} = maximum of S_{11}, S_{21}, and S_{31}

S_{min1} = minimum of S_{11}, S_{21}, and S_{31}

For $\dfrac{S_{max1}}{S_{min1}} < 2$ we have

$$D_{ul} = \frac{35.2}{\dfrac{5}{L_{11}} + \dfrac{4}{L_{21}} + \dfrac{5}{L_{31}} + \dfrac{2}{C}} \quad (m) \tag{13}$$

$$R_{I1} = \frac{1}{2\pi\lambda_I} \ln \frac{D_{ul}}{D_{il}} \quad (m\,^{\circ}C/W) \tag{12}$$

$$R_{h1} = R_{J1} + R_{I1} - \left(\frac{2\pi\,\lambda_I}{\ln\left(\dfrac{C}{D_{il}} + \sqrt{(\dfrac{C}{D_{il}})^2 - 1}\right)} + \frac{0.4 \cdot C \cdot (\lambda_J - \lambda_I)}{B \cdot R_{I1} \cdot \lambda_J} \right)^{-1}$$

S_{max2} = maximum of S_{12}, S_{22}, and S_{32}

S_{min2} = minimum of S_{12}, S_{22}, and S_{32}

For $\dfrac{S_{max2}}{S_{min2}} < 2$ we have

$$D_{u2} = \frac{35.2}{\dfrac{5}{L_{12}} + \dfrac{4}{L_{22}} + \dfrac{5}{L_{32}} + \dfrac{2}{C}} \quad (m) \tag{13}$$

$$R_{I2} = \frac{1}{2\pi\lambda_I} \ln \frac{D_{u2}}{D_{i2}} \quad (m\,^{\circ}C/W \tag{12}$$

$$R_{h2} = R_{J2} + R_{I2} - \left(\frac{2\pi\lambda_I}{\ln\left(\dfrac{C}{D_{i2}} + \sqrt{(\dfrac{C}{D_{i2}})^2 - 1}\right)} + \frac{0.4 \cdot C \cdot (\lambda_J - \lambda_I)}{} \right)^{-1}$$

EXAMPLES

A couple of examples will show the applicability of the equations.

Pairs of separately insulated pipes

The pair consists of two identical prefabricated pipes.

External diameter of insulation (D_u) = 0.220 m
Internal diameter of insulation (D_i) = 0.160 m
Thermal conductivity of earth (λ_J) = 1.5   W/m$^{\circ}$C
Thermal conductivity of insulation (λ_I) = 0.027 W/m$^{\circ}$C
Laying depth (h)   = 1.2   m
Pipe centre distance (C)   = 0.34 m
Pipe temperature forward   = 70°C
Pipe temperature return   = 45°C

The average thermal loss from the pipes in the month of July is to be determined.

Resistance

$$R_{I1} = R_{I2} = \frac{1}{2\pi \cdot 0.027} \ln \frac{0.22}{0.16} = 1.88 \text{ m}^{\circ}\text{C/W}$$

$$H = 1.2 + 0.0685 \cdot 1.5 = 1.30 \text{ m}$$

$$R_{J1} = R_{J2} = \frac{1}{2\pi \cdot 1.5} \ln \frac{4 \cdot 1.30}{0.22} = 0.34 \text{ m}^{\circ}\text{C/W}$$

$$R_h = \frac{1}{2\pi \cdot 1.5} \ln \sqrt{1+(\frac{2 \cdot 1.30}{0.34})^2} = 0.22 \text{ m}^{\circ}\text{C/W}$$

Heat loss forward

According to Table 1, $t_h = 14.9^{\circ}$C

$$\Delta t_1 = 70^{\circ}\text{C} - 14.9^{\circ}\text{C} = 55.1^{\circ}\text{C}$$
$$\Delta t_2 = 45^{\circ}\text{C} - 14.9^{\circ}\text{C} = 30.1^{\circ}\text{C}$$
$$q_1 = \frac{55.1(1.88+0.34)-30.1 \cdot 0.22}{(1.88+0.34)(1.88+0.34)-(0.22)^2} = 23.7 \text{ W/m}$$

Heat loss return

$$q_2 = \frac{\Delta t_2(R_{J1}+R_{I1})-\Delta t_1 R_h}{(R_{J1}+R_{I1})(R_{J2}+R_{I2})-R_h^2}$$

$$q_2 = \frac{30.1(1.88+0.34)-55.1 \cdot 0.22}{(1.88+0.34)(1.88+0.34)-(0.22)^2} = 11.2 \text{ W/m}$$

Total heat loss

$$q_1 + q_2 = 34.9 \text{ W/m}$$

Jointly insulated pairs of pipes

Steam pipe in a tunnel with cellular concrete insulation.

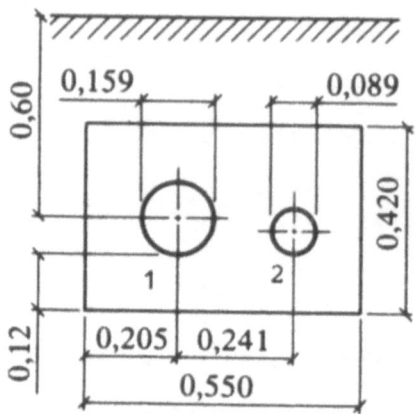

Figure 9.

Heat conductivity insulation (λ_I) = 0.1 W/m°C
Heat conductivity earth (λ_J) = 2.0 W/m°C
Temperature pipe 1 (t_{r1}) = 270°C
Temperature pipe 2 (t_{r2}) = 70°C

The average thermal loss from the pipes in the month of February
is to be determined.

$$S_{max1} = 0.141 \text{ m}$$

$$S_{min1} = 0.120 \text{ m}$$

$$\frac{S_{max1}}{S_{min1}} = 1.18 < 2$$

$$D_{ul} = \frac{35.2}{\dfrac{5}{0.221} + \dfrac{5}{0.205} + \dfrac{5}{0.200} + \dfrac{2}{0.241}} = 0.467 \text{ m}$$

$$R_{I1} = \frac{1}{2\pi \cdot 0.1} \ln \frac{0.467}{0.159} = 1.71 \text{ m°C/W}$$

$$H = 0.6 + 0.0685 \quad 2.0 = 0.74$$

$$R_{J1} = \frac{1}{2\pi \cdot 2.0} \ln \frac{4 \quad 0.74}{0.467} = 0.15 \text{ m°C/W}$$

$$R_{hl} = 0.15+1.71$$

$$= -\left(\frac{2\pi \cdot 0.1}{\ln\left(\dfrac{0.241}{0.159}\right) + \sqrt{\left(\dfrac{0.241}{0.159}\right)^2 - 1}} + \frac{0.4 \cdot 0.241(2.0-0.1)}{0.420 \quad 1.71 \quad 2.0} \right)^{-1}$$

$$= 0.56 \text{ m}^\circ\text{C/W}$$

$$S_{max2} = 0.211 \text{ m}$$

$$S_{min2} = 0.060 \text{ m}$$

$$\frac{S_{max2}}{S_{min2}} = 3.5 > 2$$

The condition of S_{max2}/S_{min2} not being fulfilled, the error in the result cannot be expected to be less than 5%.

$$D_{u2} = \frac{35.2}{\dfrac{5}{0.256} + \dfrac{4}{0.104} + \dfrac{5}{0.165} + \dfrac{2}{0.241}} = 0.364 \text{ m}$$

$$R_{I2} = \frac{1}{2\pi \cdot 0.1} \ln \frac{0.364}{0.089} = 2.24 \text{ m}^\circ\text{C/W}$$

$$R_{J2} = \frac{1}{2\pi \cdot 2.0} \ln \frac{4 \quad 0.74}{0.364} = 0.17 \text{ m}^\circ\text{C/W}$$

$$R_{h2} = 0.17+2.24$$

$$= -\left(\frac{2\pi \cdot 0.1}{\ln\left(\dfrac{0.241}{0.089}\right) + \sqrt{\left(\dfrac{0.241}{0.089}\right)^2 - 1}} + \frac{0.4 \cdot 0.241(2.0-1)}{0.42 \cdot 2.24 \cdot 2.0} \right)^{-1}$$

$$= 0.32 \text{ m}^\circ\text{C/W}$$

Heat loss forward

According to Table 1, $t_h = 1.8^\circ\text{C}$

$$t_1 = 270^\circ\text{C} - 1.8^\circ\text{C} = 268.2^\circ\text{C}$$
$$t_2 = 70^\circ\text{C} - 1.8^\circ\text{C} = 68.2^\circ\text{C}$$

$$q_2 = \frac{68.2(0.15+1.71)-268.2 \cdot 0.56}{(0.15+1.71)(0.17+2.24)-0.32 \cdot 0.56} = -5 \text{ W/m}$$

Total heat loss

$$q_1 + q_2 = 140 \text{ W/m}$$

By calculations using the finite element method STRUDL, the heat losses are defined as

$$q_1 = 139 \text{ W/m}$$
$$q_2 = 1 \text{ W/m}$$

STRUDL calculations have shown that the error in the expressions in this case is less than 5% although

$$\frac{S_{max}}{S_{min}} > 2$$

SIMPLIFÍED CALCULATIONS

If one is only interested in the heat loss from district heating pipelines not wanting to know the amount of heat passing from one pipe to the other, the expressions can be simplified by the introduction of the following three limitations:-

1) Forward and return current pipes must be identical and identically insulated.
2) Vertically, the pipes must be placed in the middle of the cellular concrete insulation.
3) The ratio λ_I/λ_J must be approx. 0.05.

Item 3) is not a very strong condition as it has only little influence on the final result.

On the basis of the preceding equation we thus derive:-

Two separately insulated pipes

$$q_1 + q_2 = \frac{t_1 + t_2}{R_J + R_I + R_h} \quad (\text{W/m})$$

$$R_I = \frac{1}{2\pi\lambda_I} \ln \frac{D_u}{D_i}$$

$$R_h = \frac{1}{2\pi\lambda_I} \ln \sqrt{1+\left(\frac{2H}{C}\right)^2}$$

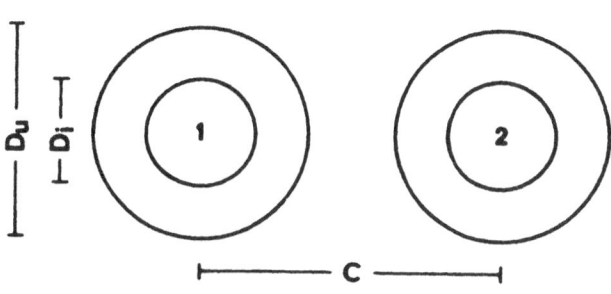

Figure 10.

Two jointly insulated pipes

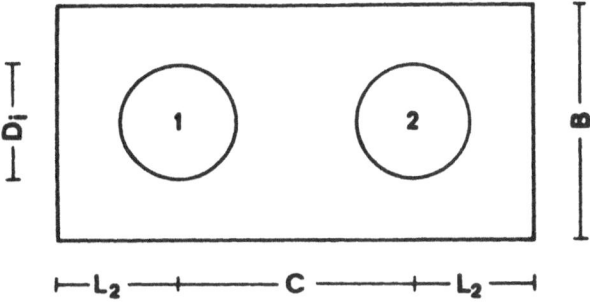

Figure 11.

$$q_1 + q_2 = \frac{\Delta t_1 + \Delta t_2}{R_J + R_I + R_h} \quad (W/m)$$

$$D_u = \frac{17.6}{\frac{10}{B} + \frac{2}{L_2} + \frac{1}{C}}$$

$$R_I = \frac{1}{R\pi\lambda_I} \ln \frac{D_u}{D_i}$$

$$R_h = R_J + R_I - \left(\frac{2\pi\lambda_I}{\ln\left(\dfrac{C}{D_i} + \sqrt{\left(\dfrac{C}{D_i}\right)^2 - 1}\right)} + \frac{0.38\ C}{B\ R_I} \right)^{-1}$$

To find the heat loss of a given section, $q_1 + q_2$ has to be multiplied by the length of the forward (or return) current, i.e. the distance in metres.

THE MEASURING OF PIPELINE LOSSES

In the above the theoretical expressions were given. Below follow some measuring results obtained from an existing district heating network.

In all cases measurements of the pipelines were taken after the pipes had been laid in earth and covered. In no case consumers were connected to the measuring section, either had they been disconnected or no connection at all existed at the time of the measurements. All pipelines measured were new and preinsulated. Consequently, it does not appear from these measurements whether or to what extent an increase of the thermal loss occurs with the years caused by, e.g. leaking joints, ageing of insulation materials, or changes in the compaction of the ground.

Measuring method

For the measuring of thermal losses, a section of a district heating network was cut off. In an appropriate point of this section, usually at draw-off and air relief tubes in connection with main cocks, a pumping and heating unit was connected which circulated and heated the water. At the end of any secondary pipes, by-passes (as large as possible) were mounted for a quick circulation.

The water in the measuring sections was heated by electric heating elements. In this way the definition of the heat loss becomes very simple: For a given electric power induced, the stagnation temperature of the section in question is measured. This method of measurements is applicable since, as mentioned before, no consumers were connected to the section measured. At stagnation temperature the power lost equals the power induced.

The quantities used for the measurement of thermal loss, viz. electric power and temperature, can both be measured with great accuracy and by means of simple and reliable instruments: kWh-meter, clock, and thermometer.

Another characteristic feature of this measuring method is that the temperatures of forward and return currents are identical within the range of a few grades, i.e. in practice the section to be measured can be assumed to be isothermal with a temperature corresponding to the arithmetic mean value of forward and return current temperatures.

Concurrently with the above measurements also earth temperature was measured. The thermal loss in the sections measured is shown as a function of the difference between water and earth temperatures and it is thus valid also for other times of the year than that at which the measurements were carried out.

DESCRIPTION OF NETWORK

The network measured was designed to supply a residential quarter comprising 47 lots with single-family houses. The network included a $\phi114.3/200^x$ mm main and intersections of dimensions from $\phi42.4/110$ to $\phi76.1/140$ totalling 1797 metres of pipe as measured from the plan. It should be noted that service pipes to the individual lots were not included in the network these having not yet been laid at the time of the measurements. All pipes, stop cocks, sockets, etc. were pre-insulated.
The design of the network is shown in Figure 12.

All pipes were laid in accordance with the instructions of the manufacturer, i.e. with 10 cm vibrated sand free from stones under the pipes and not less than 10 cm hand-tamped sand free from stones on either side. The cover was also sand. The surrounding ground consisted of clay.

MEASURING RESULTS - CALCULATIONS

A total of four measuring series was carried out with a rated power induced of 31.5 kW, 27 kW, 18 kW, and 9 kW respectively. Each measuring series covered a period of 48-63 hours. During the measuring period, measurements of surface temperatures currently checked whether the water was circulating in all intersections. [8]

The calculations were carried out assuming $\lambda_I = 0.027$ W/m^{o}C. This value is based on an estimate, the heat conductivity being a function of specific weight, temperature of insulation, and

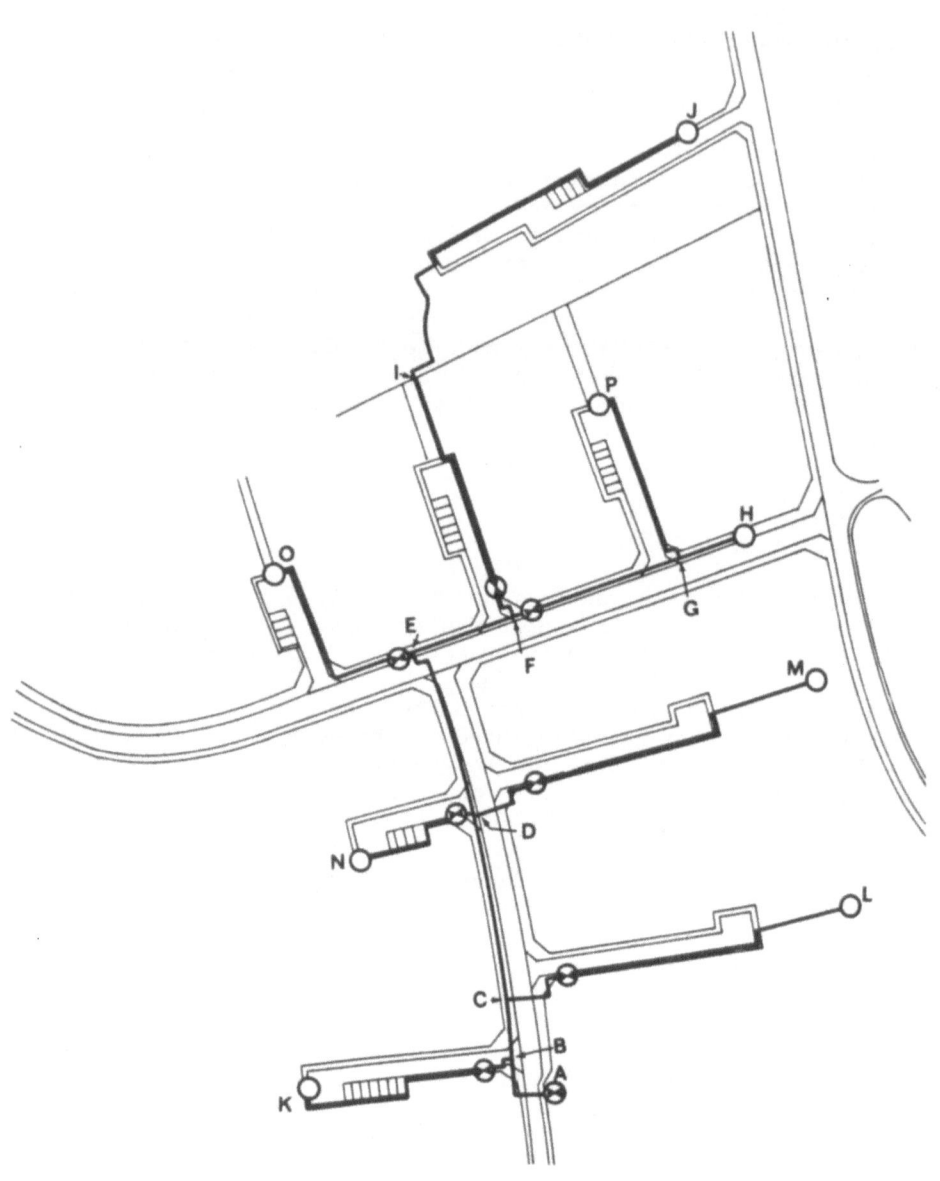

Figure 12. Measuring unit connected in point A.
By-passes installed in points K, N, O, J, P, H, M, and L.

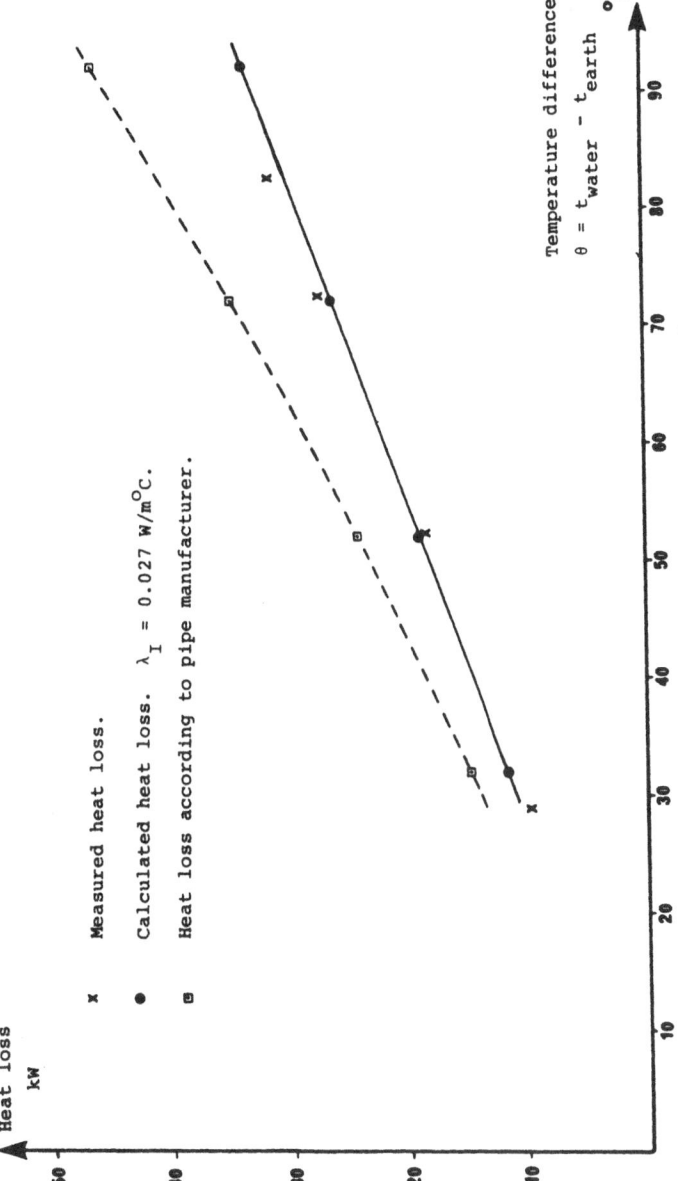

Figure 13. Comparison between measured and calculated heat losses.

degree of ageing.

Correspondingly, $\lambda_J = 1.2$ W/m^{o}C was estimated for earth. Kvisgaard & Hadvig[1] states $\lambda = 2$ W/m^{o}C for earth and $\lambda = 1$ W/m^{o}C for sand. $\lambda_J = 1.2$ W/m^{o}C seems to be a reasonable value considering that the pipes are laid in sand surrounded by earth.

COMPARISON OF MEASURED AND THEORETICAL THERMAL LOSSES

The thermal loss is shown in Figure 13 as a function of the difference between the temperatures of earth and water.

It should be added that the thermal losses measured have a good agreement with the theoretical ones calculated in accordance with Hadvig's method, whereas the manufacturer states a thermal loss which exceeds the measured one by 20%. One of the reasons is that the statements of the manufacturer are based on a distance from top of insulation to surface of 0.4 m whereas the actual distance is considerably larger, typically 0.8 m. The laying depth, however, influences the thermal loss only to a small degree, in this case about 2%.

REFERENCES

[1] Kvisgaard, B. & Hadvig, S.: Heat loss from pipelines in
        district heating systems, 1980. (In Danish).
[2] Krischer, O.: Gesundheits-Ingenieur 37, 1936.
[3] Weber, A.P.: Sanitär und Heizungstechnik 33, II, 1968.
[4] Brauer, H.: Energia, vol. 15, no. 9, Sep. 1963.
[5] Harris, B.R. & Fitzgerald, D.: Estimations of heat losses
        from buried insulated hot pipes. J.I.H.V.E., vol. 38,
        Sep. 1970.
[6] Gröber, Erk, & Grigull: Wärmeübertragung. Springer-Verlag,
        1963.
[7] Schlosser, K.: En simpel model for varmepumpens indflydelse
        på temperaturfelterne i jorden. Laboratoriet for Køle-
        teknik, Danmarks tekniske Højskole, 1975.
[8] The measurements were carried out by P. Tørslev.

TRANSMISSION OF HEAT USING HOT WATER PIPES

PART II

Sven Hadvig

Laboratory of Heating and Air Conditioning
Technical University of Denmark, Lyngby

INTRODUCTION

Part I was a theoretical contribution to the determination of heat
loss from pipes under ground and by means of a number of measurements a reasonably good agreement between theory and reality was
demonstrated to exist.

On this basis the determination of heat loss from underground
district heating piping systems is further developed. First,
attention is drawn to the results, several of which give food for
thought. All contingencies should be provided for where district
heating is planned for extensive piping systems supplying sub-
scribers who need only small amounts of heat.

Part II has another aim too, however, and that is to provide an
introduction into the calculation of the distribution of water
in complex district heating piping systems.

This introduction is useful for the design of small systems, but
it does not lend itself to the calculation of extensive, complex
district heating networks; in the latter case it may be an
advantage to supplement with other computation systems to reduce
the computing time and thus the expenses for computation.

The Laboratory of Heating and Air Conditioning at the Technical
University of Denmark has developed such a program. On application
to me this program can be used for the computation of designs.

EFFICIENCY

The efficiency of a district heating piping system could be
defined as that share of the total heat volume produced by a
heating plant, power station, or other heat producing source
which is of benefit to the consumers.

The introduction of the following notations will make it easy to
understand this simple concept which shall be used here both for
periods of one month and for periods of 12 months:-

q_H is the heat consumption for the heating of all
houses

q_W is the heat consumption for the heating of domestic
water in all houses

q_L is the heat loss from all pipes under ground

$$\eta = \frac{(q_H + q_W)\ 100}{q_H + q_W + q_L}$$

Consequently, if the heat loss from the pipes under ground is
zero we get $q_L = 0$ and hence $\eta = 100\%$, which means that all
consumers receive the heat volume produced by the power station
of heating plant.

SOME EXAMPLES OF THE EFFICIENCY

The piping system shown in Part I, Figure 12, is now provided
with the missing service pipes to the 47 houses planned for this
area. These pipes were also made by I.C. Møller.

The length of each service pipe is estimated at 15 m. The pipes
are preinsulated 26.9/90 mm, depth 0.60 cm, centre distance 30 cm.

During the period September - May the forward and return temperatures are 80°C and 50°C respectively. During the summer
period, i.e. the months of June, July, and August, the forward
temperature is reduced to 70°C whereas the return temperature
remains unchanged.

For the calculations λ_I is put equal to 0.027 W/m°C and λ_J to
1.2 W/m°C.

The calculations include 47 houses of approx. 120 m² each.

1) The BR-77 houses are heat insulated in accordance
with the Danish Building Regulation of 1977 which

applies to all single-family houses [1]. The annual
energy consumption for heating (21°C) amounts to
14000 kWh per house.
2) The heat insulation of low-energy houses is stronger
than that prescribed by the BR-77 and consequently
the annual energy consumption for heating (21°C)
amounts to 6200 kWh per house.

For systematic reasons, 1) and 2) are subdivided to show the consumption of domestic water (240 l water per day heated from 10°C
to 50°C) which amounts to 4400 kWh and 3000 kWh respectively [2].

Figure 1   gives a symbolic illustration of the cases A, A*, B,
and B* for BR-77. In the cases with a * there is no space heating
in the months of June, July, and August.

Figure 2 shows the same symbolism for low-energy houses while
Figure 3 shows the efficiency for 47 low-energy houses and Figure
4 shows the efficiency for 47 BR-77 houses.

A comparison of the *-values shows that η has the same numerical
value in the months of June, July, and August.

Figure 5 shows the annual efficiency for both BR-77 and low-energy
houses.

In the following the piping system thus examined will be called
the original or the existing system.

A number of questions now arise:-

1) Is the original piping system the most economical
system when the commercial prefabricated pipes are
taken into account?
2) Are there methods for an economical design of the
piping systems and, if so, which η-value would in
that case be obtained?
3) An increase of the insulation of the pipes beyond the
normal thickness would probably cause η to increase,
but how much?
4) For the calculation of the efficiencies the power
station was implicitly assumed to be located in point
A of Figure 12 in Part I. This situation will hardly
occur, however. What will a pipe line of e.g. 100 m
or 1000 m from the heating plant to point A (Figure
12, Part I) mean in terms of efficiency?
5) Where fossil fuels are used there will furthermore be
a heat loss through the chimney. What will the total
degree of utilisation of the fuel be, if this loss is
estimated at 10% thus assuming the fuel to be

Low energy house

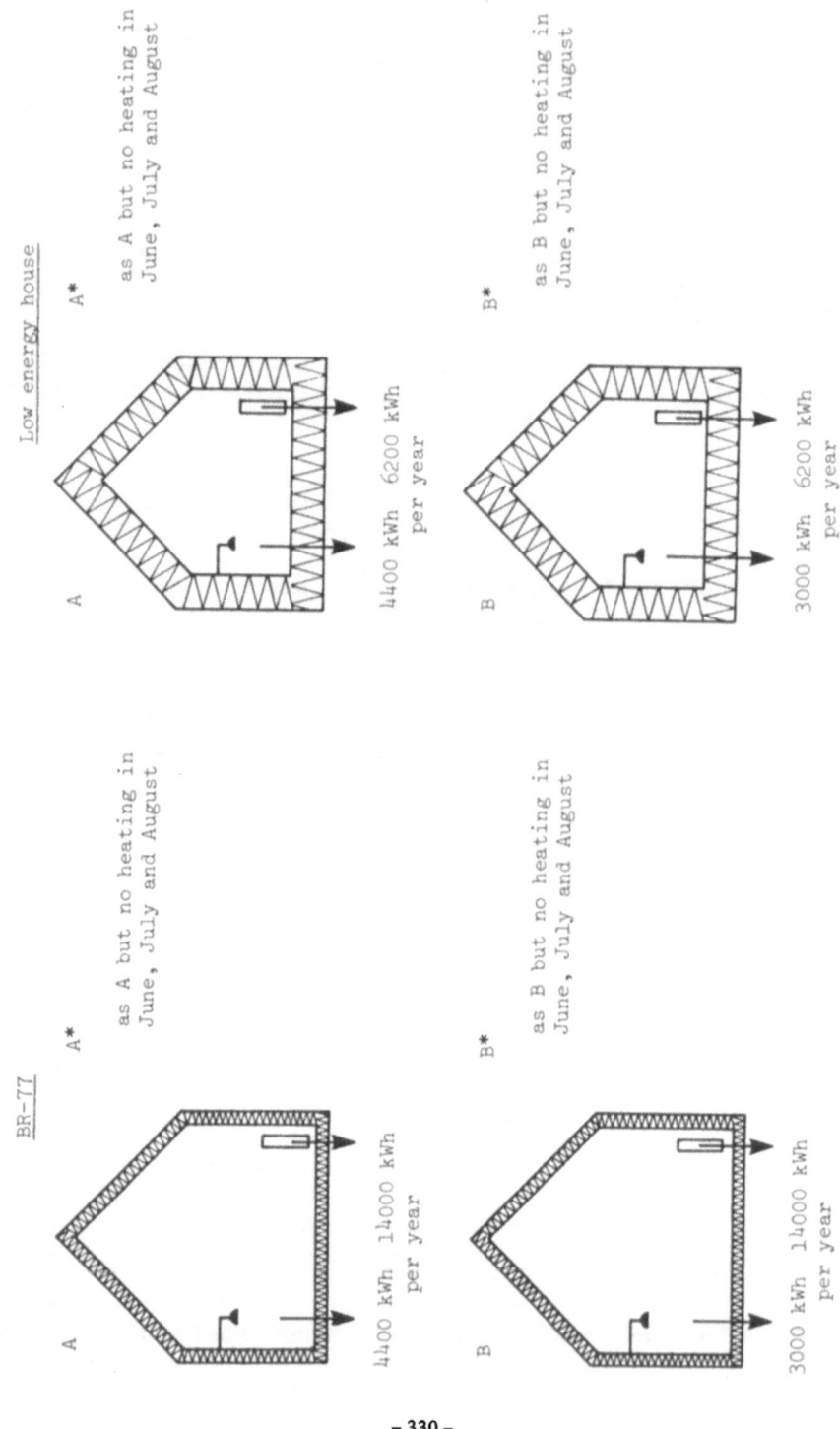

A

A*

as A but no heating in
June, July and August

4400 kWh   14000 kWh
            per year

B

B*

as B but no heating in
June, July and August

3000 kWh   14000 kWh
            per year

Figure 1.

A

A*

as A but no heating in
June, July and August

4400 kWh   6200 kWh
            per year

B

B*

as B but no heating in
June, July and August

3000 kWh   6200 kWh
            per year

Figure 2.

Low energy houses   6200 kWh per year

		Heating q_H	Hot water supply q_w	Loss q_L	Efficiency η % $\dfrac{(q_H+q_w)100}{q_H+q_w+q_L}$
	47 houses kW				
Jan.	A=A*	68.2	23.6	34.0	73
	B=B*		16.0		71
Feb.	A=A*	67.7	23.6	34.5	73
	B=B*		16.0		71
Mar.	A=A*	52.6	23.6	34.2	69
	B=B*		16.0		67
Apr.	A=A*	37.6	23.6	33.1	65
	B=B*		16.0		62
May	A=A*	16.0	23.6	33.1	56
	B=B*		16.0		51
June	A	0.9	23.6	25.9	49
	A*	0	23.6		48
	B	0.9	16.0		40
	B*	0	16.0		38
July	A	0.9	23.6	23.9	51
	A*	0	23.6		50
	B	0.9	16.0		42
	B*	0	16.0		40
Aug.	A	0.9	23.6	23.6	51
	A*	0	23.6		50
	B	0.9	16.0		42
	B*	0	16.0		40
Sep.	A=A*	7.5	23.6	27.7	53
	B=B*		16.0		46
Oct.	A=A*	29.6	23.6	29.6	64
	B=B*		16.0		61
Nov.	A=A*	47.5	23.6	31.7	69
	B=B*		16.0		67
Dec.	A=A*	72.4	23.6	33.2	74
	B=B*		16.0		73

Figure 3.

		Heating q_H	Hot water supply q_w	Loss q_L	Efficiency η % $\dfrac{(q_H+q_w)100}{q_H+q_w+q_L}$
	47 houses kW				
Jan.	A=A* B=B*	149.9	23.6 16.0	34.0	83 82
Feb.	A=A* B=B*	141.9	23.6 16.0	34.5	83 82
Mar.	A=A* B=B*	116.1	23.6 16.0	34.2	80 80
Apr.	A=A* B=B*	82.8	23.6 16.0	33.1	76 75
May	A=A* B=B*	41.8	23.6 16.0	33.1	67 65
June	A A* B B*	10.8 0 10.0 0	23.6 23.6 16.0 16.0	25.9	57 48 51 38
July	A A* B B*	9.9 0 9.9 0	23.6 23.6 16.0 16.0	23.9	58 50 52 40
Aug.	A A* B B*	12.7 0 12.7 0	23.6 23.6 16.0 16.0	23.6	61 50 55 40
Sep.	A=A* B=B*	26.8	23.6 16.0	27.7	65 61
Oct.	A=A* B=B*	67.1	23.6 16.0	29.6	76 74
Nov.	A=A* B=B*	105.3	23.6 16.0	31.7	80 79
Dec.	A=A* B=B*	148.5	23.6 16.0	33.2	84 83

Figure 4.

utilised by 90% in the furnace?

These problems are essential as far as the utilisation of our resources is concerned and, consequently, the above questions will be answered in the following. But first it might be appropriate to discuss the economical design.

ECONOMICAL DESIGN

The economical design of district heating pipelines is a very complex matter involving a number of aspects such as:-

        Cost of construction
        Interest
        Provision for depreciation
        Fuel prices
        Heat loss from pipes
        Electricity prices
        Working time for pumps
        etc.

A thorough treatment has been given in [3] both theoretically and from practical points of view. In the following some details about the practical design will be given only very briefly.

Figure 6 shows a design diagram for preinsulated pipes of commercial standard. The abscissa gives the total quantity of water passing through the pumps (m^3/s). On the curves is found the flow desired in the individual pipe lines and from this point it is possible to read on ordinates to the left the economical dimension (inner diameter in metres) assuming the 1974 price level (Denmark) still to be valid which, of course, it is not. All prices have risen, especially fuel prices. Consequently the actual dimension has to be defined on the basis of the expression

$$d_{e(actual)} = \gamma^{-020} \cdot d_{e(1974)} \tag{1}$$

$d_{e(1974)}$ has already been defined
γ is defined by means of the expression in Figure 7.

An example of the definition of γ:-

        The cost of construction has risen by 100% since 1974
        The fuel price has risen by 50% since 1974
        The electricity price has risen by 50% since 1974
        The working time is 30% less than in 1974

BR. 77   14000 kWh per year

Efficiency (year)
A : η = 76.7
A* : η = 76.2
B : η = 75.2
B* : η = 74.6

Low energy houses   6200 kWh per year

Efficiency (year)
A : η = 65.4
A* : η = 65.4
B : η = 62.1
B* : η = 62.0

Figure 5.

47 BR. 77   14000 kWh per year

	Efficiency (year) %								
	Original system			Economic design			More insulation		
Extra tube length, m	0	100	1000	0	100	1000	0	100	1000
A	76.7	67.6	56.7	78.3	69.0	58.3	79.1	69.9	60.0
A*	76.2	67.1	56.1	77.8	68.5	57.7	78.6	69.4	59.4
B	75.2	66.2	55.0	76.9	67.7	56.6	77.7	68.6	58.3
B*	74.6	65.6	54.4	76.3	67.2	56.0	77.2	68.1	57.7

Figure 9.

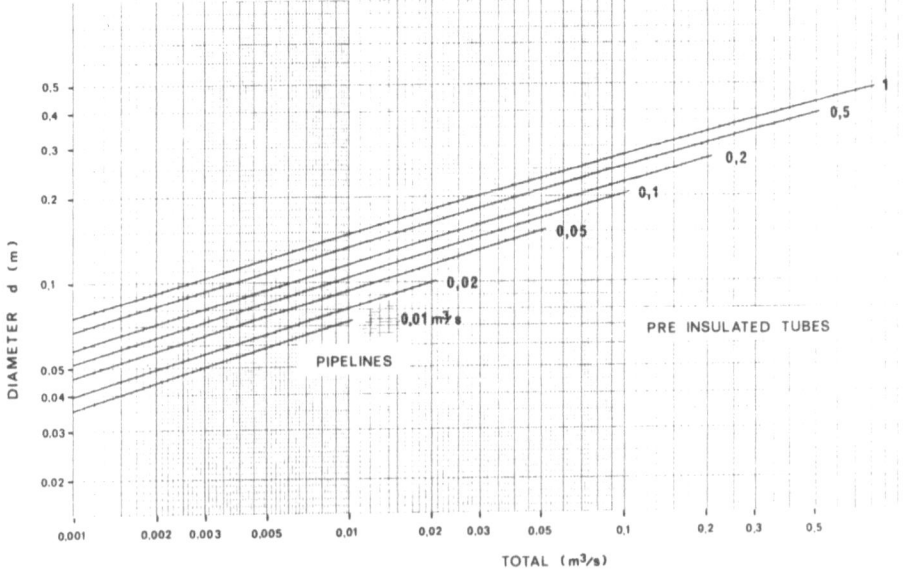

Figure 6.

$$\chi = \cfrac{0.9 \times \cfrac{\begin{pmatrix} \text{Cost of construction} \\ \text{Interest} \\ \text{Provision for depreciation} \end{pmatrix}}{\begin{pmatrix} \text{Cost of construction} \\ \text{Interest} \\ \text{Provision for depreciation}\,(1974) \end{pmatrix}} + 0.1 \cfrac{\begin{pmatrix} \text{Fuel} \\ \text{prices} \end{pmatrix}}{\begin{pmatrix} \text{Fuel} \\ \text{prices}\,(1974) \end{pmatrix}}}{\cfrac{\begin{pmatrix} \text{Electricity} \\ \text{prices} \end{pmatrix}}{\begin{pmatrix} \text{Electricity} \\ \text{prices}\,(1974) \end{pmatrix}} \times \cfrac{\begin{pmatrix} \text{Pumps work-} \\ \text{ing time} \end{pmatrix}}{\begin{pmatrix} \text{Pumps work-} \\ \text{ing time}\,(1974) \end{pmatrix}}}$$

$$d_e \,(\text{actual}) = \chi^{-0.2} \cdot d_e \,(1974)$$

Figure 7.

$$\gamma = \frac{0.9 \cdot 2 + 0.1 \cdot 1.5}{1.5 \cdot (1 - 0.30)} \approx 1.9$$

Accordingly, $d_{e(actual)}$ is defined by means of (1), $d_{e(1974)}$ being known from the diagram.

Giving (as well as reading) so brief a description of such complex matters is most unsatisfactory, but the reader is referred to [3] for further details.

Figure 8 gives the pipe dimensions of I.C. Møller pipes, calculated on the basis of the above economical design.

MORE INSULATION

The manufacturer has made an effort to obtain a more effective heat insulation of the preinsulated pipes. For all diameters the jacket dimension was one size larger than the standard dimension. A 26.9/90 pipe thus became a 26.9/110 pipe, a 42.4/110 became a 42.4/125, etc.

COMPARATIVE CALCULATIONS

The answers to questions 4) and 5) are now combined into one answer, which means that the efficiency values for the 47 BR-77 houses shown in Figure 9 valid for the distances 100 and 1000 m from the heating plant to point A include a boiler efficiency of 90%. Naturally, η_{year} for the distance of 0 m in Figure 9 is the same as in Figure 5.

Figure 10 gives the corresponding values for the 47 low-energy houses.

We now clearly see the dilemma of district heating, viz. the poor utilisation of the heat in the piping system where the system is supposed to supply detached, well insulated houses. The assumption has been made of a 100% connection to the piping system, which means that all houses in the area that have the possibility of being supplied with district heating do receive district heating. If fewer houses had been connected, the efficiency would have been lower.

A comparison with individual heating with oil or natural gas shows that for good individual plants the efficiency is 85% and thus somewhat higher than that in the examples shown. It should be remembered, however, that the examples concern detached houses with small heat losses (according to Danish conditions). In built-

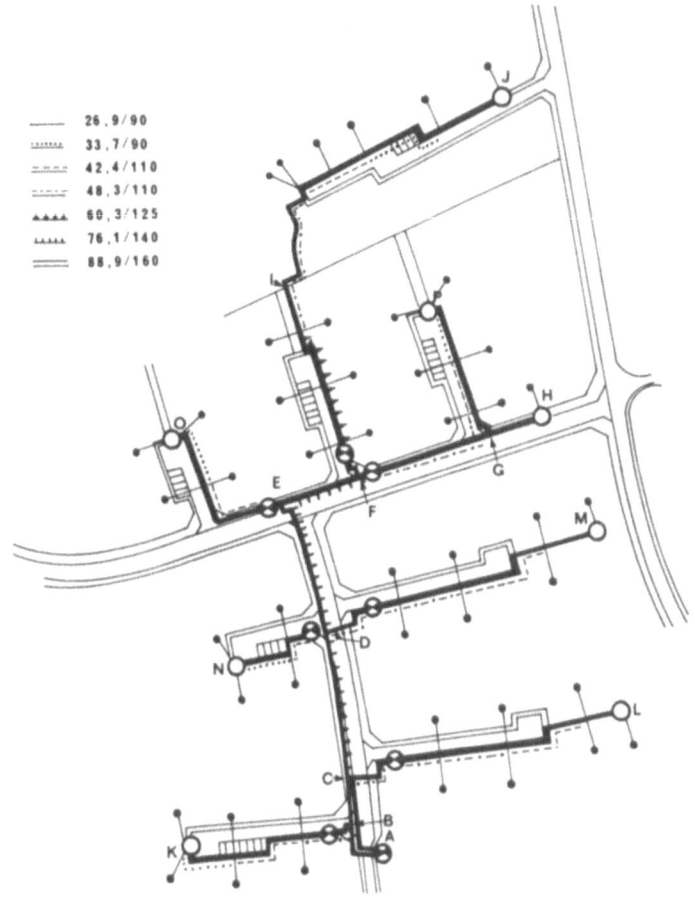

Figure 8.

47 Low energy houses 6200 kWh per year

	Efficiency (year) %								
	Original system			Economic design			More insulation		
Extra tube length, m	0	100	1000	0	100	1000	0	100	1000
A	65.4	57.1	44.6	67.5	58.9	46.3	68.5	60.0	48.1
A*	65.4	57.0	44.5	67.4	58.8	46.2	68.4	59.9	48.1
B	62.1	54.0	41.4	64.2	55.9	43.1	65.3	57.1	44.9
B*	62.0	53.9	41.3	64.1	55.8	43.0	65.2	57.0	44.8

Figure 10.

up areas, if only the degree of connection is reasonable.

Our considerations have been focussed on the energy utilisation, but where the fuel has to be paid for they also apply to the financial aspect, i.e. the fuel consumption. Where it is a matter of production of heat, however, that would normally be carried off by water or air, not being wanted for any purpose, a different view might be applied. In that case it might be better to lead it through district heating pipes, even if a considerable part goes to heat loss from the piping system.

BETTER UTILISATION

It seems natural to investigate how much the efficiency is increased if the water temperature in the district heating pipes is lowered.

The results of such a computation are shown in Figure 11 for "existing system" and "economical design" as well as for "more insulation" as mentioned above. In the figure the abscissa is the mean temperature of forward and return flow.

It appears from Figure 11 that in the case of a mean temperature of 50°C (e.g. forward flow 65°C and return flow 35°C) the system supplying BR-77 houses is able to compete with individual heating systems.

A reduction of the mean temperature, i.e. the application of low-temperature district heating is a method which may render district heating reasonably economical in sparsely built-up areas as far as energy is concerned.

Some countries do quite the opposite thing, i.e. they increase the mean temperature far beyond what is shown in the examples given here. This procedure must be strongly deprecated where the heat supply of detached single-family houses is concerned.

COMPUTATION OF COMPLEX DISTRICT HEATING PIPING SYSTEMS

Danish authorities want a rather drastic extension of the national district heating piping systems. This was the incentive for our laboratory to develop a

General Computer Program for District Heating Systems.

The program includes up to 800 joints and is also able to include complex loops in the piping system. At the top of Figure 11 a part of such a system is shown schematically (one stroke re-

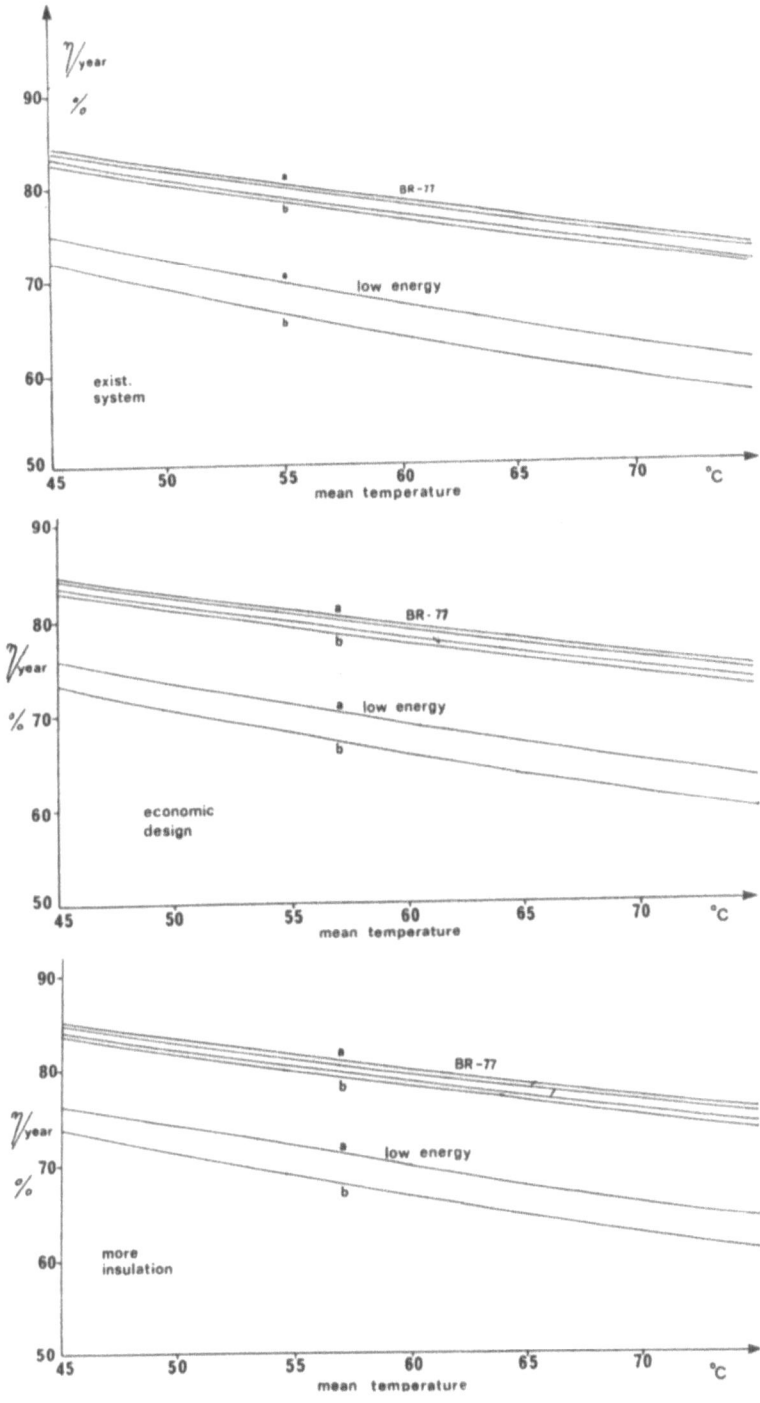

Figure 11.

presents two pipes). At the bottom of the figure a loop is shown which results from shunting at each house.

For the above-mentioned program fairly sophisticated computation methods have been applied in order to make the computer time as short as possible and thus to make a computation economically feasible, consequently making it possible to analyse quite a large number of combinations. In the present paper, however, some fundamental methods will be discussed which can, of course, always be used but they require longer computer time than our program. The best known is probably the Hardy Cross method [4] which appeared in the thirties. It is not very practical to use this method for the initial computations as in that case too many conversions are required.

The BOP method [5], on the other hand, which was developed in Denmark is applicable as a first approximate computation.

First the BOP method and then the Hardy Cross method will be discussed below and finally an example of calculation will be shown.

Simple hydraulic relations are taken as known such as the pressure drop p in the pipe length ℓ that can be expressed by

$$\Delta p = \left(\lambda \frac{\ell}{d} + \Sigma \zeta \right) \frac{1}{2} \rho w^2 \tag{2}$$

where

Δp is the pressure drop having the unit Pa
λ is the friction factor, dimensionless
ℓ is the pipe length, m
d is the pipe diameter (inner), m
ζ is the single resistance, dimensionless
ℓ is the density of water, kg/m^3
w is the mean velocity in the pipe, m/s

For most "practical pipes" Colebrook and White's equation for the definition of λ appeared to be a reasonable approximation.

With volumetric flow v m^3/s we have

$$\Delta p = r \cdot v^2 \tag{3}$$

where we have the resistance r from

$$r = \left(\lambda \frac{\ell}{d} + \Sigma \zeta \right) \frac{\rho}{2 \left(\frac{\pi}{4} d^2 \right)^2}$$

Within a suitable sphere of application it is possible to write

$$\Delta p = r \cdot v^z$$

where z is close to 2.

THE BOP METHOD

In each joint of a piping system the sum of the volumetric flows taken with signs must be zero. In brief:-

$$\Sigma v = 0 \qquad (5)$$

This yields the linear continuity equations.

The non-linear energy equations require the total pressure loss in a ring to be equal to the pressure available in the ring. In brief:-

$$\Sigma \Delta p = p_p \qquad (6)$$

If there is no pump in the circuit, p_p is equal to zero.

By means of these two conditions just as many equations are obtained as there are pipings in the rings (circuits).

In the case of the BOP method it is useful to use the equations with v^2, this making it possible to split up Δp_i in the following way:-

$$\Delta p_i = |r_i \cdot v_i| \cdot v_i \qquad (7)$$

and in the summation to let this be equal to zero or to the pump pressure P depending on whether a pump is inserted in the ring or not.

In order to be able to make the summation it is necessary in advance to have marked the flow directions in the individual pipings between the joints and thus to introduce a calculation with signs for v_i. $|r_i \cdot v_i|$ is only a numerical quantity.

The conditions (5) and (6) are both put into a matrix and the equations with the energy losses are divided by v_1. For the i'th we then have

$$r_i \frac{v_i}{v_1} = r_i \frac{w_i \frac{\pi}{4} d_i^2}{w_1 \frac{\pi}{4} d_1^2}$$

Assuming the same flow velocity in all pipes ($w_i = w_1$) we then have

$$r_i = \frac{v_i}{v_1} = r_i \frac{d_i^2}{d_1^2} \tag{8}$$

which can now be introduced into the equations for energy losses. The system of equations can now be solved and v_1, v_2, ... v_i, ... can be defined. The BOP method will be shown later in an example.

THE HARDY CROSS METHOD

First the positive direction in each ring is indicated by an arrow. We now use the equation (4), here written as

$$\Delta p = r \cdot v^z = r \cdot \left| v^{z-1} \right| \cdot v \tag{9}$$

the last v containing the sign according to the positive direction decided on.

On the basis of the volumetric flows defined by the BOP method we are now in a position to calculate the pressure drop in a ring (to sum up the pressure losses). As mentioned above this must be equal to the pressure available which may be zero or euqal to the pumping pressure p_p.

The result of the summation is not zero, however, among other reasons because the numerical value of v determined by the previous method is not the correct numerical value v_o. A volumetric flow Δv thus has to be added to v in order to obtain the correct numerical value

$$v_{io} = v_i + \Delta v$$

The summation in (6) then becomes

$$\Sigma \ r_i(v_i + \Delta v) \left| v_i + \Delta v \right|^{z-1} = p_p$$

and hence

$$\Sigma \ r_i \frac{(v_i + \Delta v)}{\left| v_i + \Delta v \right|} \left| v_i + \Delta v \right|^z = p_p$$

Assuming

$$\left| \Delta v \right| \ll \left| v_i \right| \tag{10}$$

we can develop in series the last term on the left-hand side. As an approximation we write

$$\Sigma \; r_i \; \frac{v_i}{|v_i|} \; \left|\frac{v_i}{v_i}\right|^z \; \left(1 \, + \, z \; \frac{\Delta v}{v_i}\right) = p_p$$

By multiplication we get

$$\Sigma \; r_i v_i |v_i|^{z-1} \, + \, \Delta v \; \Sigma \; z \cdot r_i |v_i|^{z-1} = p_p$$

and hence

$$\Delta v \; = \; \frac{p_p \, - \, \Sigma \; r_i |v_i|^{z-1} \cdot v_i}{\Sigma \; z \cdot r_i |v_i|^{z-1}} \tag{11}$$

It should be remembered that v_i is taken with signs.

We made the assumption (10), but often this is not fulfilled. Still, the method is applicable and the calculation will generally be convergent.

The correction Δv is used for the correction of all partial flows in the circuit in question and these corrected values are now used for the pipelines forming part of the next circuit to be computed so that the "best values" are always used.

Below is given an example of the application of the BOP and Hardy Cross methods.

EXAMPLE

A simple joint can be described by an equivalent resistance and the volumetric flows. The figure below shows a ring system to which two systems are connected characterised only by their equivalent resistances.

For convenience, the individual pipelines are numbered.

First we use the BOP method and, consequently, flow directions have to be indicated by arrows for the sake of the signs in the calculations. Before the arrows are sketched in, the piping system is drawn in the following way:-

First the forward flow is split up into 1 and 3, then 1 is split up into 2 and 4 while 3 is split up into 2 and 6 and thus the first ring has been formed. The figure below shows the final result.

Pipes 1, 2, and 3 represent the internal forward flow and pipes 5, 7, and 8 represent the internal return flow of the system.

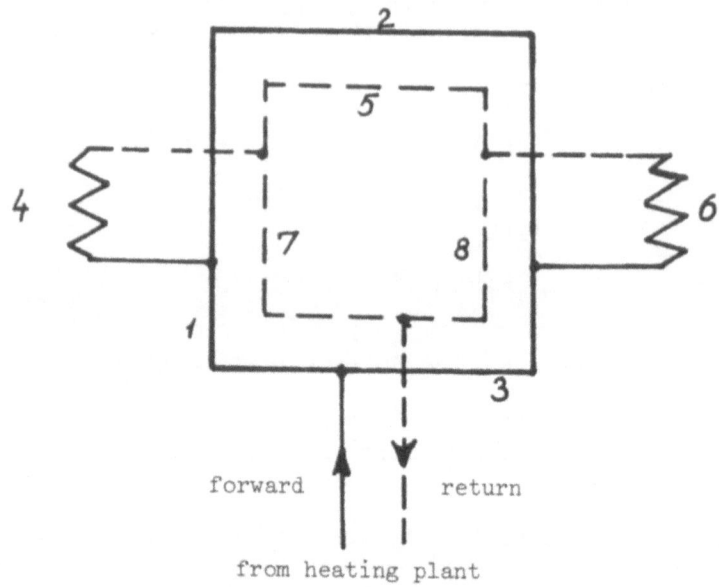

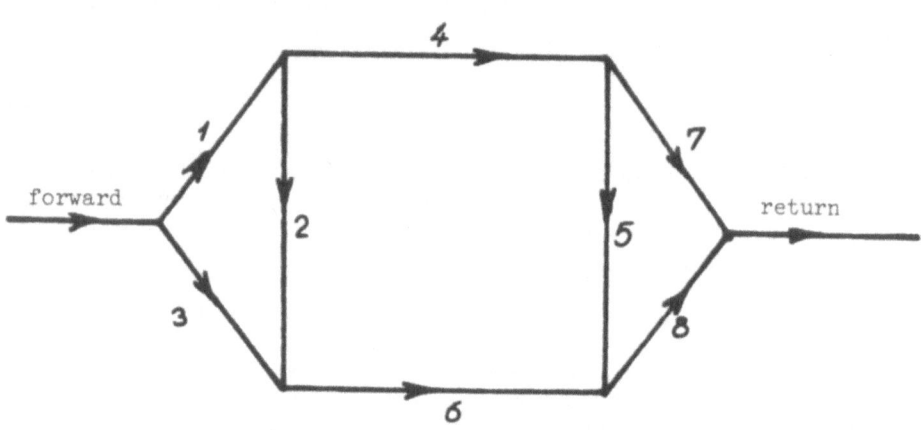

For the sake of convenience the following round numbers for the
resistance were chosen for the example. These numbers are not
realistic, but practical for the illustration of the method:-

$$r_1 = r_8 = 10 \qquad r_3 = r_7 = 15 \qquad r_4 = 16$$

$$r_2 = r_5 = r_6 = 20$$

This also goes for the inner pipe diameters

$$d_i = 0.1 \qquad i = 1, 2, 3, 7 \text{ and } 8$$

$$d_i = 0.05 \qquad i = 4 \text{ and } 6$$

The flow from the heating plant and back is equal to 10. We now try to define the eight unknown flows.

a) Equations of the joint.
   The volumetric flow from and to the heating plant is written v_{tot}.

$$
\begin{aligned}
v_{tot} &= v_1 + v_3 = 10 \\
v_1 &= v_2 + v_4 \\
v_6 &= v_2 + v_3 \\
v_4 &= v_5 + v_6 \\
v_8 &= v_5 + v_6 \\
(v_{tot} &= v_7 + v_8 = 10) \text{ not necessary.}
\end{aligned}
$$

b) The pressure loss in a closed circuit is zero. There is no pump in the circuits.

$$
\begin{aligned}
\Delta p_1 + \Delta p_2 - \Delta p_3 &= 0 \\
\Delta p_4 + \Delta p_5 - \Delta p_2 - \Delta p_6 &= 0 \\
\Delta p_5 + \Delta p_8 - \Delta p_7 &= 0
\end{aligned}
$$

These three equations are rewritten in the basis of $\Delta p_i = r_i \cdot v_i^2$. Accordingly, we have

$$
\begin{aligned}
r_1 v_1^2 + r_2 v_2^2 - r_3 v_3^2 &= 0 \\
r_4 v_4^2 + r_5 v_5^2 - r_2 v_2^2 - r_6 v_6^2 &= 0 \\
r_5 v_5^2 + r_8 v_8^2 - r_7 v_7^2 &= 0
\end{aligned}
$$

Then, the equations from a) and b) are written in matrix form with the three equations for pressure loss at the top.

The three top equations (the pressure-loss equations) are divided by v_1. According to (18) we thus have

1st element, top row

$$10 \, \frac{d_1^2}{d_1^2} = 10$$

$$
\begin{Bmatrix}
r_1v_1 & r_2v_2 & -r_3v_3 & 0 & 0 & 0 & 0 & 0 \\
0 & -r_2v_2 & 0 & r_4v_4 & r_5v_5 & -r_6v_6 & 0 & 0 \\
0 & 0 & 0 & 0 & r_5v_5 & 0 & -r_7v_7 & r_8v_8 \\
1 & 0 & 1 & 0 & 0 & 0 & 0 & 0 \\
-1 & 1 & 0 & 1 & 0 & 0 & 0 & 0 \\
0 & 1 & 1 & 0 & 0 & -1 & 0 & 0 \\
0 & 0 & 0 & -1 & 1 & 0 & 1 & 0 \\
0 & 0 & 0 & 0 & 1 & 1 & 0 & -1
\end{Bmatrix}
\cdot
\begin{Bmatrix}
v_1 \\ v_2 \\ v_3 \\ v_4 \\ v_5 \\ v_6 \\ v_7 \\ v_8
\end{Bmatrix}
=
\begin{Bmatrix}
0 \\ 0 \\ 0 \\ 10 \\ 0 \\ 0 \\ 0 \\ 0
\end{Bmatrix}
$$

2nd element, top row

$$20 \frac{d_d^2}{d_1^2} = 10 \quad \text{and so on,}$$

and hence

$$
\begin{Bmatrix}
10 & 20 & -15 & 0 & 0 & 0 & 0 & 0 \\
0 & -20 & 0 & 4 & 20 & -5 & 0 & 0 \\
0 & 0 & 0 & 0 & 20 & 0 & -15 & 10 \\
1 & 0 & 1 & 0 & 0 & 0 & 0 & 0 \\
-1 & 1 & 0 & 1 & 0 & 0 & 0 & 0 \\
0 & 1 & 1 & 0 & 0 & -1 & 0 & 0 \\
0 & 0 & 0 & -1 & 1 & 0 & 1 & 0 \\
0 & 0 & 0 & 0 & 1 & 1 & 0 & -1
\end{Bmatrix}
\cdot
\begin{Bmatrix}
v_1 \\ v_2 \\ v_3 \\ v_4 \\ v_5 \\ v_6 \\ v_7 \\ v_8
\end{Bmatrix}
=
\begin{Bmatrix}
0 \\ 0 \\ 0 \\ 10 \\ 0 \\ 0 \\ 0 \\ 0
\end{Bmatrix}
$$

This system of equations is solved and we have

$$v_1 = 5.63$$
$$v_2 = 0.47$$
$$v_3 = 4.37$$
$$v_4 = 5.16$$
$$v_5 = 0.64$$
$$v_6 = 4.84$$
$$v_7 = 4.52$$
$$v_8 = 5.48$$

This ends the application of the BOP method.

Then the Hardy Cross method is applied. As mentioned above, the positive direction is indicated in circuits I, II, and III using the arrows from above.

1st guess. The above-mentioned numerical values are used.

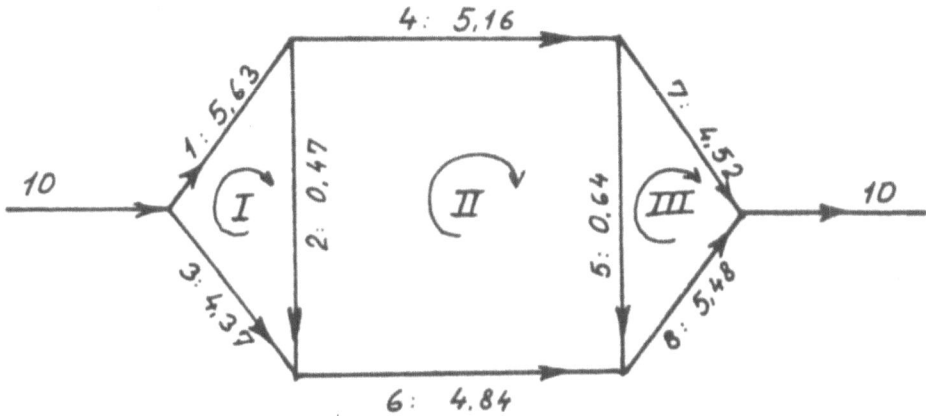

For the sake of convenience we put z = 2. We now calculate the pressure loss in circuit I.

$$\Delta p_I = 10 \cdot 5.63^2 + 20 \cdot 0.47^2 - 15 \cdot 4.37^2 = 34.9 \neq 0$$

p_I should have been zero.

$$\Sigma \, 2r_i |v_i| = 2 \cdot 10 \cdot 5.63 + 2 \cdot 0.47 + 2 \cdot 15 \cdot 4.37 = 263$$

and hence

$$\Delta v = \frac{-\Sigma \, r_i v_i^2}{\Sigma \, 2r_i |v_i|} = -\frac{34.9}{263} = -0.13$$

v_1, v_2, and v_3 are now corrected by Δv, whereas $v_4 - v_8$ are un-changed. These corrected values represent the 2nd guess.

$$v_1 = 5.63 - 0,13 = 5,50$$
$$v_2 = 0.47 - 0.13 = 0.34$$
$$v_3 = 4.37 + 0.13 = 4,50$$

Pressure loss in circuit II:-

$$\Delta p_{II} = 16 \cdot 5.16^2 + 20 \cdot 0.64^2 20 \cdot 4.84^2 - 20 \cdot 0.34^2 = -36.6 \neq 0$$

Note that we used $v_2 = 0.34$ for the 2nd guess.

$$\Sigma \, 2r_i |v_i| = 2 \cdot 16 \cdot 5.16 + 20 \cdot 2 \cdot 0.64 + 2 \cdot 20 \cdot 4.84 + 2 \cdot 20 \cdot 0,34 = 398$$

This also includes $v_2 = 0.34$ from the 2nd guess.

$$v = - \frac{-36.6}{398} = 0.09$$

3rd guess

$$\begin{aligned}
v_4 &= 5.16+0.09 = 5.25 \\
v_5 &= 0.64+0.09 = 0.73 \\
v_6 &= 4.84-0.09 = 4.75 \\
v_2 &= 0.34-0.09 = 0.25
\end{aligned}$$

Pressure loss in circuit III:-

$$\Delta p_{III} = 15 \cdot 4.52^2 - 10 \cdot 5.48^2 - 20 \cdot 0.73^2 = -4.51 \neq 0$$

$$\Sigma \; 2r_i |v_i| = 2 \cdot 15 \cdot 4.52 + 2 \cdot 10 \cdot 5.48 + 2 \cdot 20 \cdot 0.73 = 274$$

$$\Delta v = - \frac{-4.51}{274} = 0.02$$

4th guess

$$\begin{aligned}
v_7 &= 4.52+0.02 = 4.54 \\
v_8 &= 5.48-0.02 = 5.46 \\
v_5 &= 0.73-0.02 = 0.71
\end{aligned}$$

Pressure loss in circuit I. Second computation.

$$\Delta p_I = 10 \cdot 5.50^2 + 20 \cdot 0.25^2 - 15 \cdot 4.50^2 = 0$$

and hence $\Delta v = 0$.

Pressure loss in circuit II. Second computation.

$$\Delta p_{II} = 16 \cdot 5.25^2 + 20 \cdot 0.71^2 - 20 \cdot 4.75^2 - 20 \cdot 0.25^2 = -1.42 \neq 0$$

$$\Sigma \; 2r_i |v_i| = 32 \cdot 5.25 + 40 \cdot 0.71 + 40 \cdot 4.75 + 40 \cdot 0.25 = 396$$

$$\Delta v = - \frac{-1.42}{396} = 0.004$$

This error is thought to be acceptable.

Pressure loss in circuit III. Second computation.

$$\Delta p_{III} = 15 \cdot 4.54^2 - 10 \cdot 5.46^2 - 20 \cdot 0.71^2 = 0.98 \neq 0$$

$$\Sigma \; 2r_i \; v_i = 30 \cdot 4.54 + 20 \cdot 5.46 + 40 \cdot 0.71 = 274$$

$$\Delta v = - \frac{0.98}{274} = -0.004$$

This error is thought to be acceptable.

From the second computation it is seen that all pressure losses
in the circuits are below an acceptable value. Since all joint
equations are still fulfilled the solution for the ring systems
is

$$v_1 = 5.50$$
$$v_2 = 0.25$$
$$v_3 = 4.50$$
$$v_4 = 5.25$$
$$v_5 = 0.71$$
$$v_6 = 4.75$$
$$v_7 = 4.54$$
$$v_8 = 5.46$$

This result is comparable with the result obtained by the BOP
method.

I hope that the above will serve as an illustration of the simple
methods of computation used for district heating pipings.

REFERENCES

[1] Danish Building Regulations 1977. National Building Agency.
    Ministry of Housing. Copenhagen.
[2] The computations were carried out by P. Tørslev and Ole
    Quistgaard.
[3] Johansen, J. and Hadvig, S.: Economical design of district
    heating pipings
[4] Cross, Hardy: Analysis of flow in networks of conduits or
    conductors. Eng. Exp. Station, Univ. III. Bull. 286. 1936.
[5] Pedersen, B.F.: Analysis of complex water conduit networks
    (in Danish). Ingeniøren no. 2, 1965.

METAL HYDRIDES FOR HEAT STORAGE OF COMBUSTION AND REACTION
PROCESSES

O. Bernauer, H. Buchner

Abstract
Metal hydrides are heat storage means with high storage densities. By variation of the hydrides it is possible to store and
regain waste heat on nearly all temperature levels. Therefore,
such systems would be suitable for a large number of combustion
and chemical reaction processes. Different applications of
hydrides in closed systems (where hydrogen is not consumed) and
in open systems (where hydrogen is consumed) are discussed.

1   INTRODUCTION

   The shortage of fossile energy resources makes it urgent to
develop new technologies which enables us, on the one hand, to
fully utilize the available fossile energy sources and on the
other to employ new technological methods.

   One way of better utilizing energy is based on storing and
regaining the waste heat of combustion and chemical reaction processes by use of absorption and desorption processes. For these
absorption and desorption processes metal hydrides are particularly suitable, because they guarantee with respect to weight and
volume high energy densities and beside this high desorption and
absorption kinetics.

   For the technical application of the hydrides two cases must
be distinguished for heat storage:

- Hydrogen is consumed, i. e. the hydride storage tank serves as
  an intermediate hydrogen or heat storage unit.

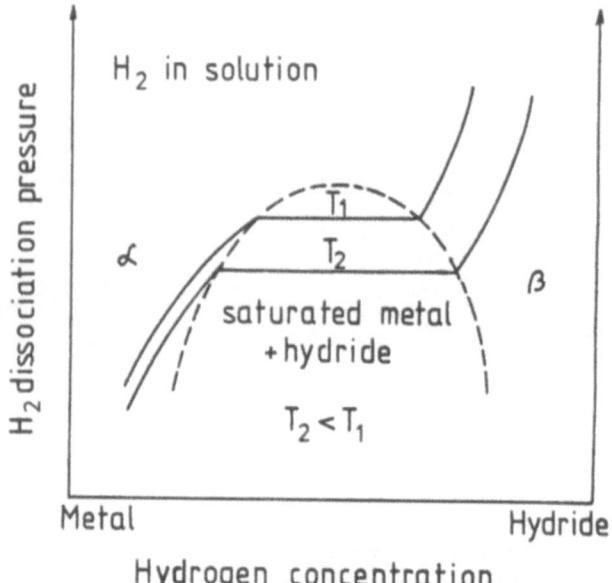

Figure 1. Schematic CPI diagram for a metal-
hydrogen system forming a hydride

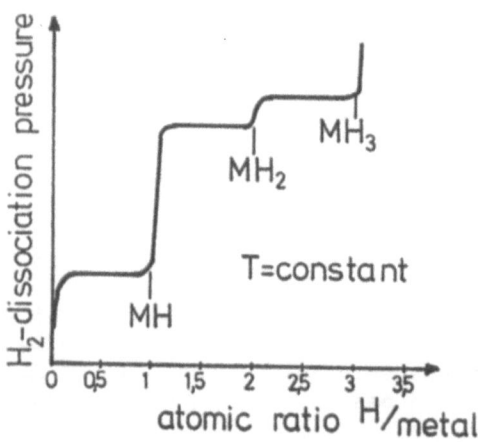

Figure 2. Pressure concentration isothermes
for hydrides with different hydride
phases

- Hydrogen is not consumed, i. e. it is pumped to and fro between (at least) two hydride storage units.

Heat is stored in both cases and (perhaps even at a higher temperature level) released again (heat pump effect).

As, in a closed hydrogen circuit 2 to 2 n (n = 1, 2, ...), storage units are required, the corresponding total storage density decreased by the factor 2 to 2 n. The total system therefore becomes less economical, the more hydride storage units have to be employed. This case then becomes economical when the absorption and desorption intervals have to be small (heat pumps and cooling systems with metal hydrides). In the open hydrogen circuit (hydrogen supply, hydrogen consumption) the heat storage densities of the hydrides can be fully utilized. Potential applications, on the one hand, could be the hydrogen processing industry and on the other hand, in the case of future hydrogen technology, households and vehicles. However, the hydride heat storage unit, in this case, will always be dependent on a hydrogen infrastructure.

2 CHARACTERISTICS OF METAL-HYDROGEN SYSTEMS

If hydrogen reacts with a metal in which a stable hydride MH_y is formed this can be described by the following equation:

$$MH_x + \frac{1}{2}(y - x) H_2 \rightleftarrows MH_y$$

where $y > x$.

The formation of MH_y is the absorption process, the reverse reaction is the desorption process. As the hydride formation of the metal hydrides which are suitable for heat storage is exothermic, the heat for the reaction must be supplied during the desorption process by external means and during absorption, the heat from the reaction must be taken away. The formation of various exothermic hydrides as a function of temperature and their stoichiometry can be established with the aid of pressure/concentration isotherms (fig. 1).

At low hydrogen concentrations there is a strong dependency on the composition of the hydrogen pressure. This is the region (α-phase) in which gaseous hydrogen is dissolved in the metal without the occurrence of a new phase. A further increase in the hydrogen concentration leads to a region where the saturated solid solution in the α-phase is in equilibrium with a hydrogen deficient hydride phase (β-phase). In this field the hydrogen pressure is independent of the concentration. After the α-phase has been completely converted into the β-phase the pressure

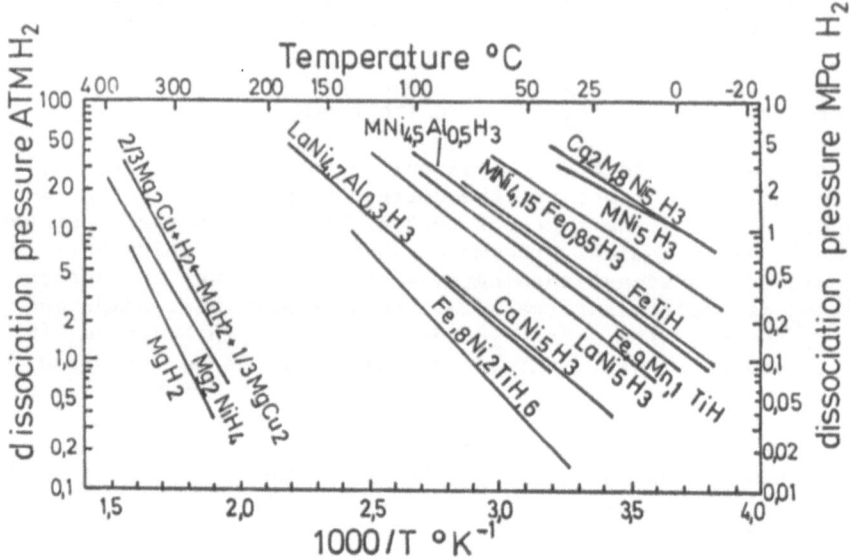

Figure 3. van't Hoff plots (desorption
for various hydrides)

LaNi$_5$H$_x$ Desorption

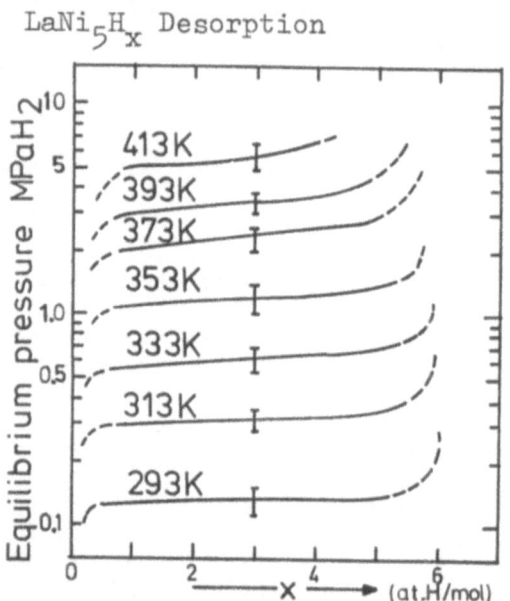

Figure 4. CPI-diagram for LaNi$_5$-H /1/

increases again very rapidly in proportion to the hydrogen load. In some cases there still exist several hydride phases, which lead to different steps in the pressure/concentration isotherms (fig. 2).

For the technical application of the hydrides the temperature at which the hydrogen pressures reach values of more than 1 bar is of special interest, although there naturally exists the possibility of extending this region by compressing the gas mechanically. Among the large number of hydrides one group (fig. 3, top right) is described as low temperature hydrides (LTH), as their dissociation pressure, even at temperatures below the freezing point of water, is above 1 bar. The remaining hydrides which have a pressure of 1 bar above the boiling point of water are called high temperature hydrides (HTH). These are shown in the left-hand part of fig. 3.

The physical properties of low temperature hydrides can be summarized as follows:

- The H_2-storage capacities of these hydrides are in the range between 1.3 and 2.0 w % H_2 with respect to the weight of the metal alloy. Well known exponents are for example: $LaNiH_6$ /1/ (fig. 4); $CaNi_5H_5$ /2/ (fig. 6); $TiFeH_1$ and $TiFeH_{1.94}$ /3/ (fig. 5); $Ti_{0.8}Zr_{0.2}Cr_{0.8}Mn_{1.2}H_3$ /4/ (fig. 7); $TiCrMnH_3$ /4/ (fig. 8); $ZrCr_2H_4$ and $ZrMn_2H_3$ /5/ and a new one $Ti_{0.9}Zr_{0.1}V_{0.2}Cr_{0.4}Mn_{1.4}H_{3.2}$ /6/ (fig. 8). Hence there follow energy densities ranging from 1,500 kJ/kg hydride to 2,500 kJ/kg hydride with respect to the lower calorific value of hydrogen.

- The heat of formation of the hydriding process of these hydrides ranges from - 15 kJ/mol H_2 ($TiCrMnH_3$) to - 35 kJ/mol H_2 ($CaNi_5H_5$). With this data we get energy densities with respect to heat storage means of about 200 kJ/kg hydride to 350 kJ/kg hydride. The heat is produced on a temperature level up to 100 °C so that it is possible to use water under common pressure as a exchanging medium.

- The gravity of the low temperature hydrides amounts to 5.4 - 6.2 kg/dm³. In a technical storage unit it is possible to bring in up to 5 kg/dm³ of the pulverized hydride material when it is compressed isostatically (5,000 - 6,000 bars). In this case one gets energy densities for concrete heat storage units of 1,000 kJ/dm³ up to 1,750 kJ/dm³.

The physical properties of the high-temperature hydrides can be described in the following way:

FeTiH Desorption

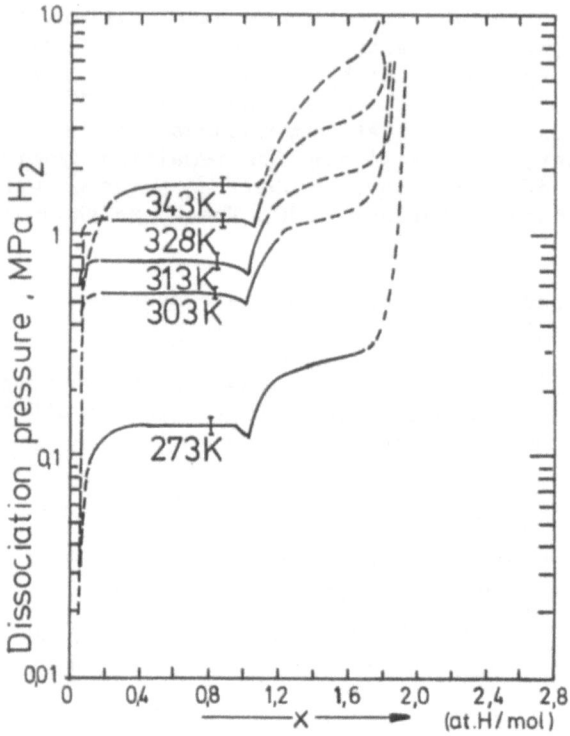

Figure 5. CPI-diagram for TiFe-H /3/

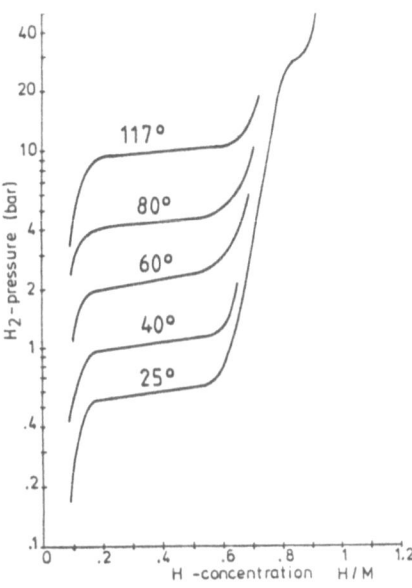

Figure 6. CPI-diagram of CaNi$_5$-H /2/

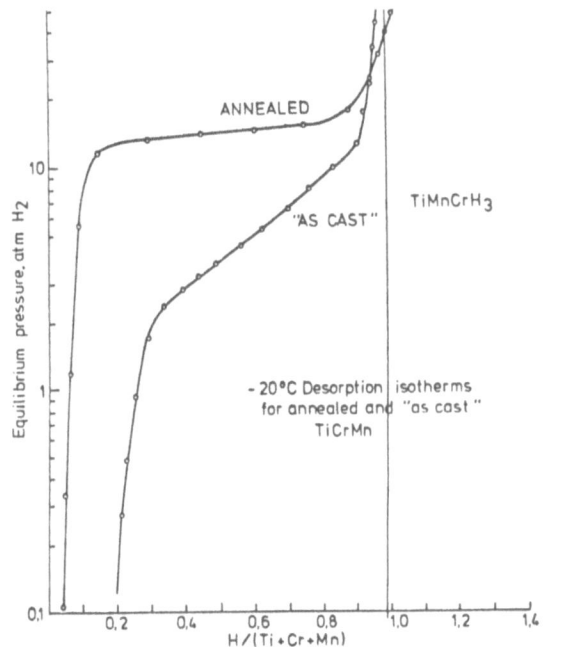

Figure 7. CPI-diagram for annealed TiCrMn-H /4/

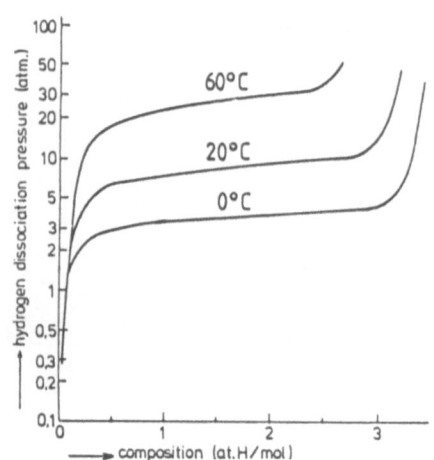

Figure 8. CPI-diagram for
$Ti_{0,9}Zr_{0,1}Cr_{0,4}V_{0,2}Mn_{1,4}$-H /6/

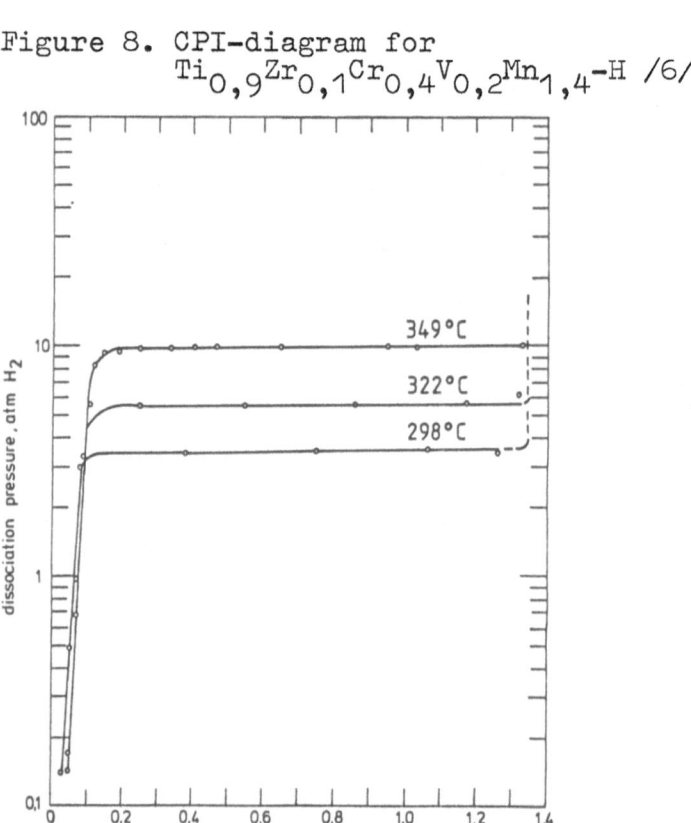

Figure 9. CPI-diagram for the MgNi-H system /7/

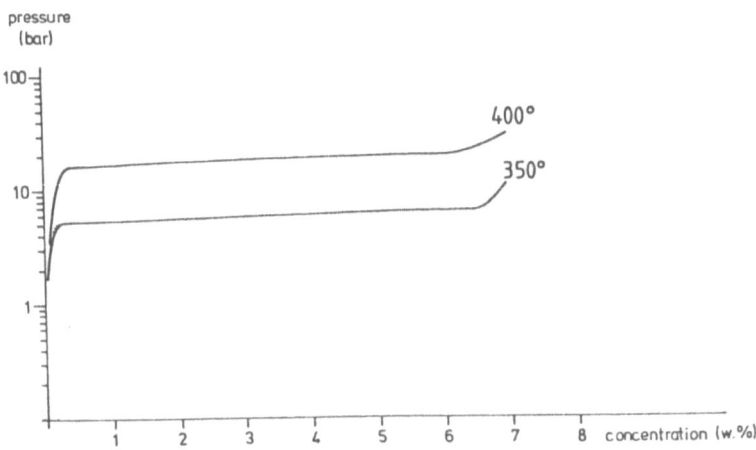

Figure 10. CPI for Mg-H /8/

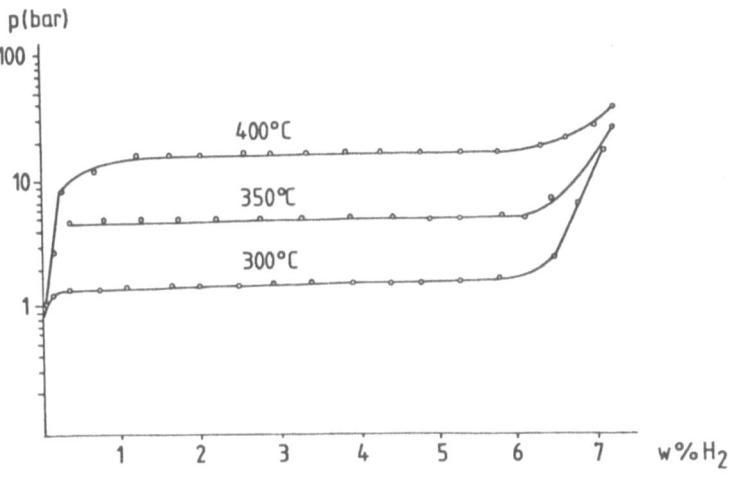

Figure 11. Pressure-concentration-isotherms
of $Mg_{95}Ni_5$ (GfE)

Table 1   Physical Properties for Metal Hydrides

Alloy composition	ΔH (kj mol^{-1}H$_2$)	A (K)*	B*	capacity (w %)	reference
TiFe	28.03	-3383	12.76	1.8	3
LaNi$_5$	30.96	-3712	12.96	1.4	1
MNi$_5$	20.92	-2539	11.64	1.4	2
CaNi$_5$	31.80	-3838	12.17	1.4	2
Ti$_{0.8}$Zr$_{0.2}$CrMn	24.66	-2968	13.07	1.8	–
Mg$_2$Ni	64.43	-7736	14.71	3.8	7
Mg$_2$Cu	72.80	-8771	17.12	2.0	7
Mg$_{0.95}$Ni$_{0.05}$	75.86	-9128	16.34	7	–
Mg	77.40	-9314	16.63	7	8
TiVMn	38.52	-4640	13.81	2.6 – 3	–

* $\ln p = \dfrac{A}{T} + B$ (plateau pressure (bar) taken in the middle of the plateau)

- The H_2-storage capacities of these hydrides reach values between 3 w % (TiVMn-H) and 7.4 w % ($Mg_{0.95}Ni_{0.05}H_2$) with respect to the metal weight. Hence there follow energy densities
  ranging from 3,600 kJ/kg hydride to 8,400 kJ/kg hydride with
  respect to the lower calorific value of hydrogen. Well known
  exponents are for example:
  Mg_2NiH_4 /7/ (fig. 9); MgH_2 /8/ (fig. 10); $Mg_{0.95}Ni_{0.05}H_{1.9}$ /7/
  (fig. 11); TiVMn (fig. 12).

- The heat of formation of the hydriding process amounts to
  ~ 40 kJ/mol H_2 (TiVMn-H) to 73 kJ/mol H_2 (MgH_2) or in the case
  of still higher temperature levels to > 160 kJ/mol H_2 (TiH_2).
  Therefore a large heat quantity on a high temperature level is
  required in order to release hydrogen. At the same time, for
  heat storage purposes energy densities between 600 kJ/kg and
  2,600 kJ/kg (and 2,600 kJ/kg) are available. The heat is produced
  on a temperature level between 150 - 200 °C (TiVMn-H),
  250 - 350 °C (Mg_2NiH_4, $Mg_{0.95}Ni_{0.05}H_{1,9}$) and $\geq$ 600 °C (TiH_2).

- The gravities of the high temperature hydrides depend on the
  collection of the containing elements. In the case of Mg-based
  hydrides one gets gravities between 1.7 - 2 kg/dm³ and with
  Ti-based hydrides 5 - 6 kg/dm³. By that the real energy densities with respect to the volume of a concrete container
  reach values ranging from 2,400 kJ/dm³ up to 4,000 - 4,500
  kJ/dm³.

Metal hydrides are therefore suited for high temperature
and low temperature heat accumulation because of their high heat
storage densities (with respect to weight and volume) and due to
the high absorption and desorption kinetics of hydrogen. In table
1 the main physical properties of some hydrides are shown.

3   GENERAL CONSIDERATION ON HEAT STORAGE WITH METAL-HYDROGEN
    SYSTEMS

The use of thermal energy at a high temperature level to
pump heat from a low temperature source to an environment at an
intermediate temperature is possible with the combination of
metal-hydrogen systems when the absorption and desorption processes involve thermal energy. These processes then can be used
for heating and cooling purposes.

The equilibrium pressure in the two-phase region of metal
hydrides depends on the temperature, according to the following
equation

$$\ln p = \Delta H/RT - \Delta S/R$$

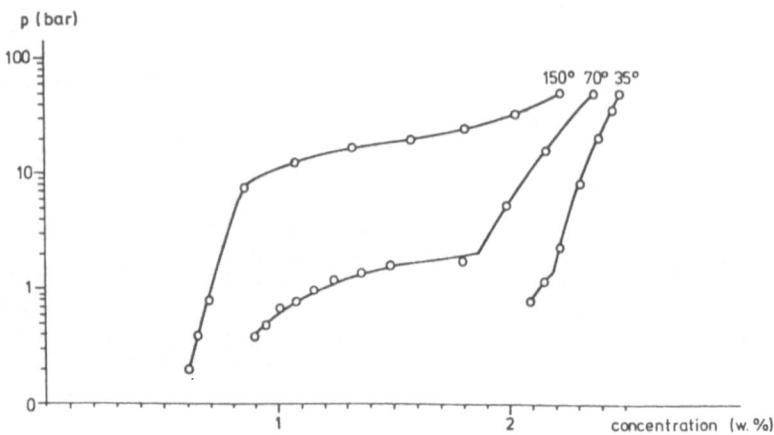

Figure 12. CPI-diagram for TiVMn-H

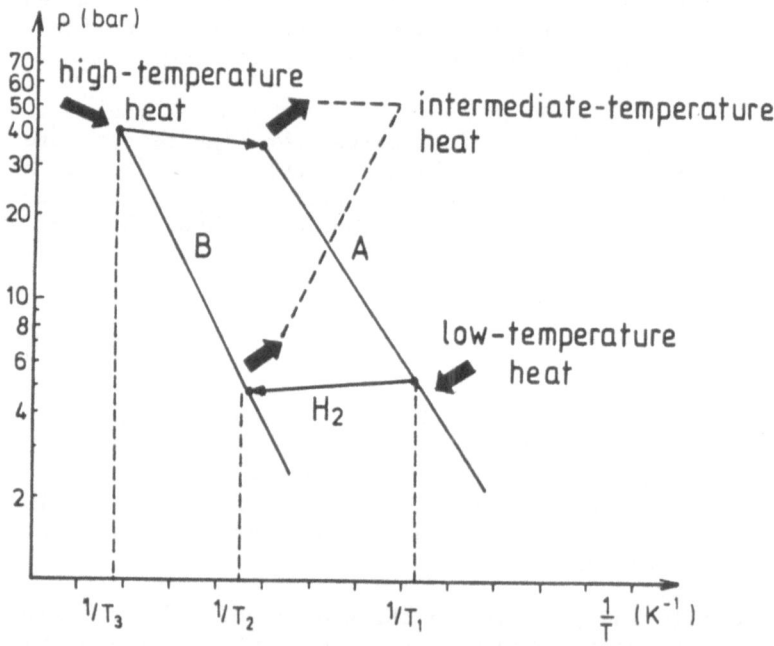

Figure 13. Hydride heat pump cycle with
two different hydrides

where ΔS is the entropy difference between gaseous and absorbed hydrogen and ΔH is the heat of formation of the hydriding process.

Fig. 13 illustrates a typical heat pump cycle for a hydride combination A and B where the hydride B is the more stable one and $T_1 > T_2 > T_3$. It can be seen that hydride A has a slightly higher and slightly lower hydrogen equilibrium pressure at temperatures T_1 and T_2, respectively, than hydride B has at temperatures T_2 and T_3, respectively.

A heat pump mode of operation for cooling or heating purposes immediately suggests itself using such a hydride pair.

In order to have a continuous cooling or heating effect it is at least necessary to have four tanks which run on two alternating cooling or heating cycles /9/. Tanks 1 and 2 (fig. 14) are filled with the metal hydride A, while tanks 3 and 4 are filled with the hydride B. The hydride B in tank 4 absorbs heat from external heating on a temperature level T_3 and desorbs hydrogen which flows to tank 2 where it reacts to form hydride A (A' is the metal alloy). The heat from this reaction on a temperature level T_2 can be used when the system works as a heat pump. At the same time, the metal hydride A in tank 1 decomposes by absorbing energy on a low temperature level into metal A' and hydrogen. The hydrogen flows to tank 3 to form in combination with metal B' the hydride B. If the system works as a heat pump the heat of formation on the temperature level T_2 can be used again. The overall efficiency of such a system depends on the heat of formation of the hydriding processes, the heat capacities of the hydride and container materials and on the temperature differences between hydriding and dehydriding modes as already shown by Gruen et.al. /10/, and van Mal /11/.

If such a system works as a heat pump the thermodynamic efficiency can be expressed by the following equation:

$$\eta_h = \frac{\Delta H_B + \Delta H_A - C_p \cdot (m_A + m_1) \cdot (T_2 - T_1)}{\Delta H_B + C_p \cdot (m_B + m_2) \cdot (T_3 - T_2)} +$$

$$+ \frac{C_p \cdot (m_B + m_2) \cdot (T_3 - T_2)}{\Delta H_B + C_p \cdot (m_B + m_2) \cdot (T_3 - T_2)}$$

The efficiency tends towards 2, when the heat of formation ΔH_A and ΔH_B are nearly equal and when the heat capacities are negligible. Under concrete conditions the efficiencies reach values between 1.3 and 1.8 and, therefore, are comparable to standard absorption heat pump cycles.

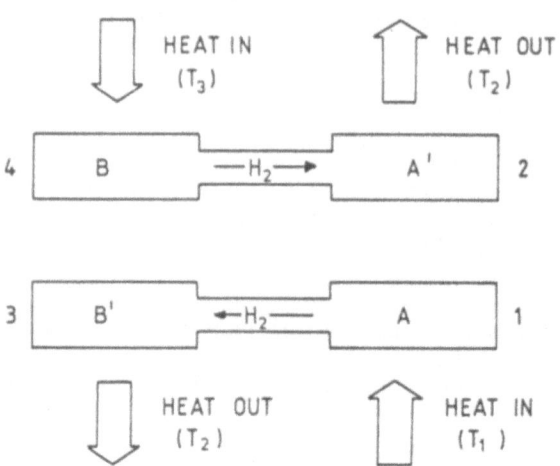

Figure 14. A schematic diagram of a
hydride heat pump

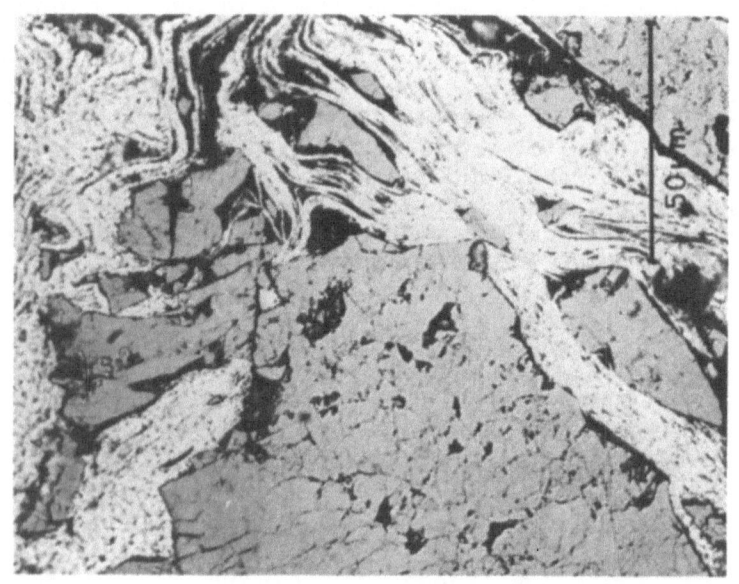

Figure 15. TiFe + 5 w% Al
(several times hydrided and dehydrided)

If such a metal hydrogen system works in a cooling mode the thermodynamic efficiency can be expressed by the following equation

$$\eta_c = \frac{\Delta H_A - C_p \, (m_A + m_1) \cdot (T_2 - T_1)}{\Delta H_B + C_p \, (m_B + m_2) \cdot (T_3 - T_2)}$$

and in a concrete case it can reach values again comparable to standard absorption refrigeration cycles ($\sim 0.4 - 0.6$).

4 APPLICATIONS FOR CLOSED METAL HYDROGEN SYSTEMS WITHOUT HYDROGEN CONSUMPTION

The rising demand for energy and the diminishing supply of high-grade fossil fuels have given rise to a number of efforts to make alternate energy sources feasible. Thermal energy from industrial processes and heat engines like internal combustion engines, steam and gas turbines etc. is available in quantities which are definitely worth considering. This energy is both low-grade and high-grade and is not utilized efficiently today.

Power plants, for example, are typically over designed to meet peak loads with the results that such plants function with lower efficiency during off-peak periods. Here, it is important to find methods and devices for storing energy and making it available upon demand for heating and/or cooling like the closed metal hydrogen systems do. An important feature of these systems in applications involving waste heat recovery is their versatility. Recoverable waste heat will range in temperature from 60 °C - 500 °C. The composition of the hydridable metallic materials used as storage media in such systems can be readily tailored to efficiently use the energy in this entire temperature range depending upon the temperature at which it is available. The result is overall energy utilization efficiencies of up to 80 % on a practical basis. Energy stored in such systems can be utilized in many ways. It can be employed for space heating, preheating of engines and cooling to supply heat to and remove heat from industrial processes, and for many other purposes which will be readily apparent to those skilled in the arts to which such systems pertain.

Another application where the principles of the hydride systems can be used to particular advantage is in air conditioning and the preheating of vehicles and their engines /12, 13/. The typical automobile air conditioner requires in general 2 to 10 kW during normal operation depending on the cooling demand. Consequently, a significant part of the energy consumed goes to operate the air conditioner. This problem can be solved by

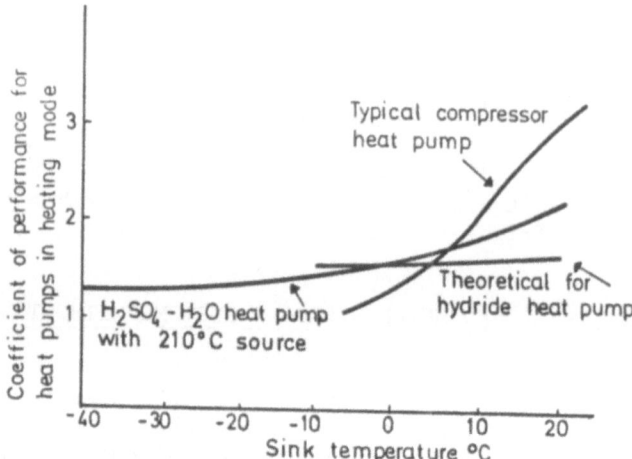

Figure 16. The operating efficiencies of
three types of heat pumps

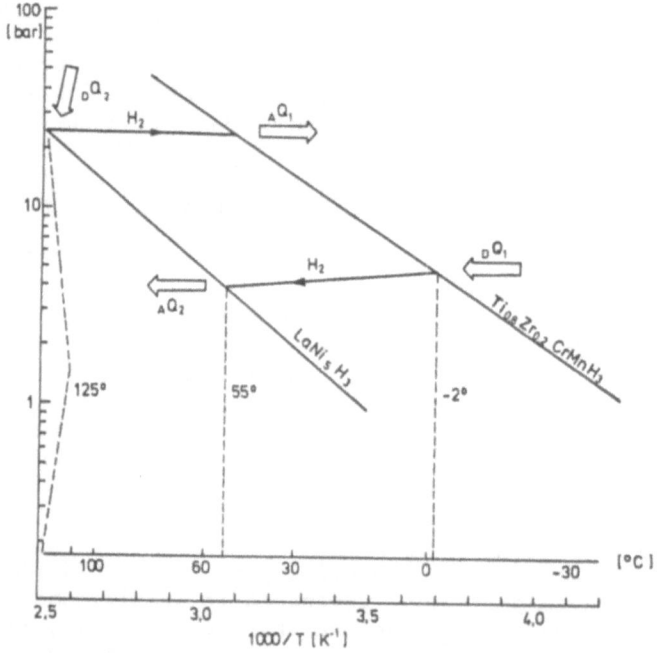

Figure 17. Plateau pressure vs 1/T for
LaNi$_5$H$_3$ and Ti$_{0,8}$Zr$_{0,2}$CrMnH$_3$

employing an air conditioning system which uses the waste heat
from the exhaust to operate the air conditioner. In this way no
mechanical energy is drained from the automobile engine; and, in
comparison to an automobile equipped with a conventional air
conditioner, a 10 to 20 % reduction /14/ in fuel consumption can
be obtained while the air conditioner is working.

The same advantage can be reached by employing a hydride
based preheating system which reutilizes the waste heat from the
exhaust gas or coolant water for operation. The effect is that
there are no cold start problems with the engines, resulting in
a reduction in consumption and an extension of the life of the
engine.

4.1 Thermal conductivity of the metal hydrides

Besides hydrogen storage capacity and hydriding enthalpy the
efficiency of closed hydrogen hydride systems depends on the
ability to transfer heat to and fro between the hydride container
and heat exchanging media. In most applications the heat transfer
of pulverized metal hydrides is a restriction on the design and
construction of closed hydrogen hydride storage systems.

The integral thermal conductivity of a total system depends
on the

- thermal conductivity of the hydride or metal powder,
- thermal conductivity of the hydrogen gas,
- thermal conductivity of the metal (tank).

Since the thermal conductivity of the hydride or metal powder is
greater than the thermal conductivity of the gas by the factor 2
– 10 (depending on the compression of the material in the pipe)
and smaller than the solid metals by the factor 100 – 200, the
thermal conductivity in the storage unit must be improved in the
latter's technical design, in order to ensure good charging and
discharging rates of these hydride storage units.

Hydrides compacted to form porous solids supported by a thin
metal matrix which does not absorb hydrogen have already been
utilized by Daimler-Benz in cooperation with Battelle/Geneva in
the course of our hydride battery development since 1969 /15/.

The improvement of the thermal conductivity which remains
stable over a long operating life is to be explained at this
point with the following process developed by Daimler-Benz, which
improves the thermal conductivity of the system by as much as the
factor 10 – 30 /16/.

The metal hydride powder is mixed with 5 – 20 per cent by

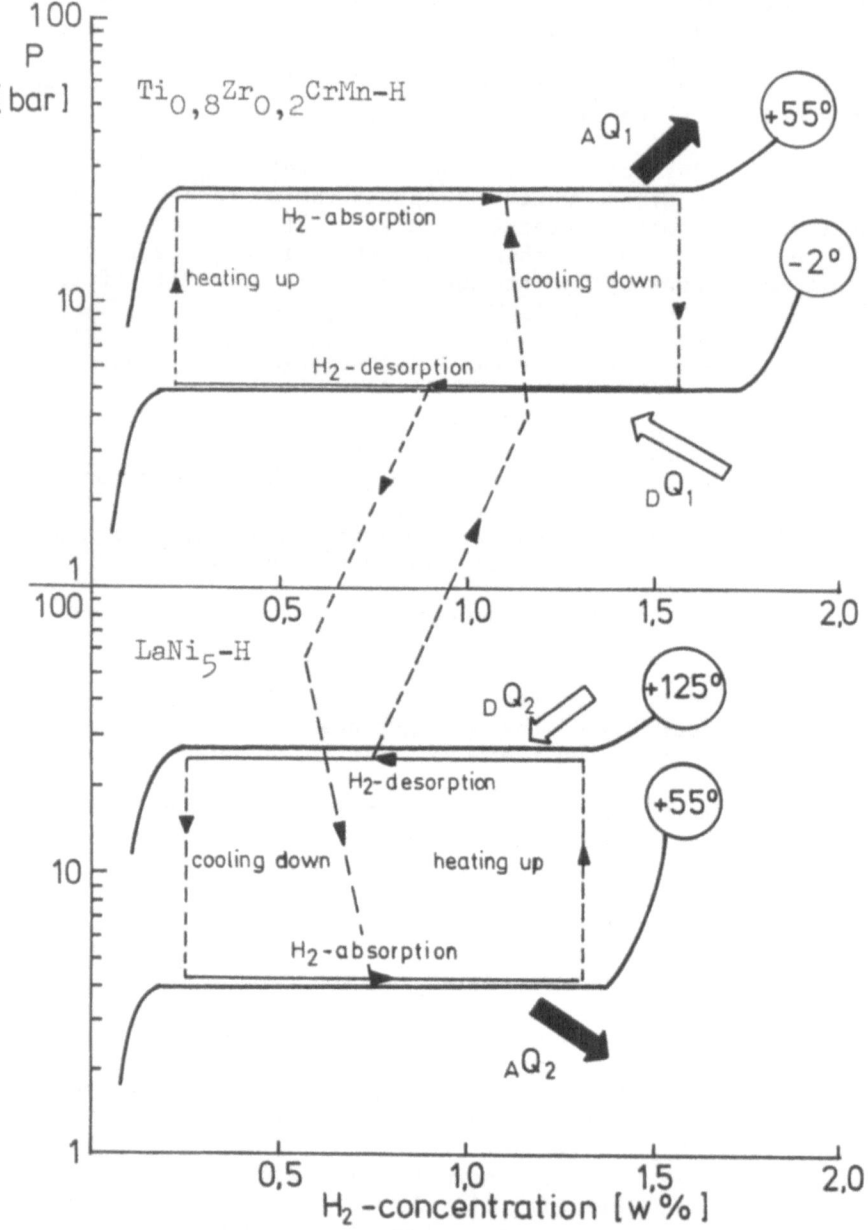

Figure 18. Hydride heat pump cycle for the system $LaNi_5-H/Ti_{0,8}Zr_{0,2}CrMn-H$

weight of an inactive metal in powder or flake form and compressed under high pressure (10 - 20 tons/cm³). If the material
is subsequently tempered at 450 °C - 500 °C under continual H_2
charging and discharging, a stable matrix body of the inactive
material is formed in which the hydride powder is embedded (figure 15). Metals suitable for this application are aluminum, magnesium, copper, i. e. metals with a high thermal conductivity
and a relatively low melting point. When adding 5 per cent by
weight of aluminum to TiFe, for instance, a thermal conducticity
is achieved in this way of 3 $\pm$ 1 W m^{-1} K^{-1}. If 10 per cent by
weight of aluminum is added, the thermal conductivity was found
to be between 7 and 10 W m^{-1}K^{-1}, as compared with 0,5 - 1 W m^{-1}
K^{-1} with loosely to lightly pressed powder. Hydride tanks designed in this way ensure high exchange rates without the necessity
for large heat exchanging surfaces. In addition to this such a
hydride storage unit offers maximum safety since the hydride material cannot get into the open should the tank break, but it is
held in the matrix.

A further advantage of such a matrix technique is that the
storage body can be of any shape whatsoever (plates, cylinders
etc.), which also remain stable during cycling. Battery building
can be mentioned here as an important field of application if the
cathode is to consist of hydrides (example: TiNi electrode).
These types of electrode have a very high energy density.

This gives the provision to store electrical energy. In
collaboration with Battelle in Geneva /15/ Daimler-Benz has developed an electrochemically reversible hydride battery system,
consisting basically of a combination of electrolysis, hydride
storage unit and fuel cell. The hydride electrodes are manufactured in the way that hydride powder is either embedded in a
matrix of copper or it is copper coated. A hydride accumulator
of this type supplies an energy density between 60 and 70 Wh/kg
during a 10 hours discharge period when titanium/nickel hydrides
are used as negative, and nickel hydroxide as positive, electrode
substances. Direct electric power storage in hydrides and direct
power generation during accumulator discharge from hydrides is
thus possible, guaranteeing a total efficiency of more than 70 %
/15/. The electrochemical and metallurgical aspects of the mentioned hydride accumulators are described in ref. /17, 18, 19/.

The technical know how from this development which requires
an electrode with excellent electrical conductivity of the hydride
bed was used for the aluminum-hydride technique described previously to increase the thermal conductivity of the storage tank.

4.2 Hydride heat pumps for stationary application

Waste heat from industrial processes and from the household

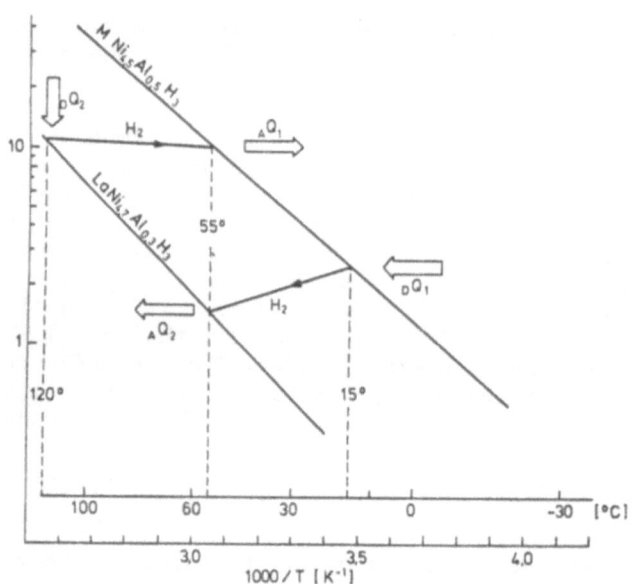

Figure 19. Plateau pressure vs 1/T for
LaNi$_{4,7}$Al$_{0,3}$H$_3$ and MNi$_{4,5}$Al$_{0,5}$H$_3$

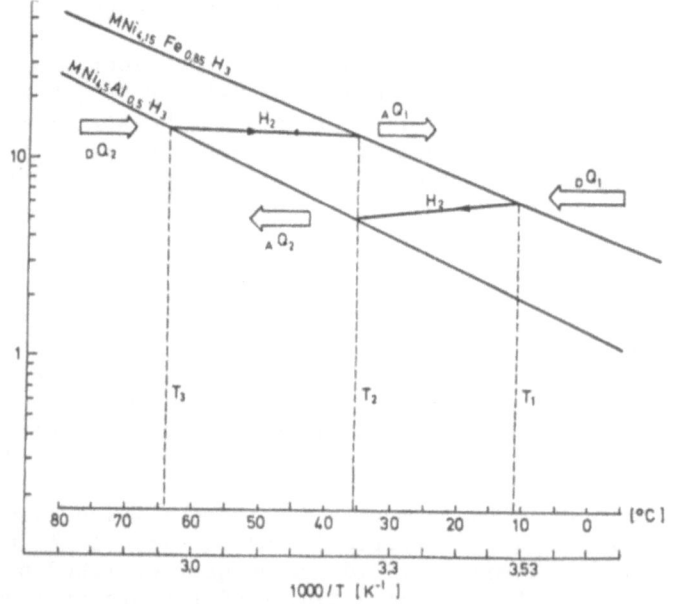

Figure 20. Plateau pressure vs 1/T for
MNi$_{4,5}$Al$_{0,5}$H$_3$ and MNi$_{4,15}$Fe$_{0,85}$H$_3$

is available on a temperature level between 500 °C and 50 °C and nowadays often not used because the energy is typically low-grade and it may not be available when the demand arises.

One important objective of hydride heat pumps and heat storage systems resides in the provision of systems which are capable of efficiently utilizing low-grade energy while retaining the capacity for utilizing high-grade thermal energy and in other cases which are capable of storing energy and making it available upon demand for heating or cooling or both heating and cooling. As shown in chapter 3, the coefficient of performance η_H for heating for the hydride heat pump does not explicitly contain the operating temperature of the system, when we neglect the heat capacities of the materials, but is determined by the enthalpies of the reactions in the heat pump and is independent of the ambient temperature. In comparison to this fact, for mechanical heat pumps utilizing a Carnot cycle, the coefficient of performance is given by the equation

$$\eta_c = \frac{T_H}{T_H - T_L}$$

where T_H is the source temperature and T_L is the sink temperature.

The η_c is totally a function of temperatures and decreases as $T_H - T_L$ increases. The typical efficiencies for a mechnical, hydride and chemical heat pump (using sulfuric acid and water) are shown in fig. 20 /20/. The efficiency for a mechnical heat pump can be seen to drop off quickly below the freezing temperature of water while the efficiency of a hydride heat pump is nearly temperature independent. Therefore, hydride heat pumps become especially advantageous in the lower temperature range. An advantage of the hydride heat pump in comparison to the chemical heat pump is the possibility of using different operating temperatures because these temperatures are determined by the hydriding enthalpies of the pair of hydrides chosen for the system. In the following sections some hydride systems for continuous cooling or heating effects are discussed.

4.2.1 Cooling and heating cycles with the system
$LaNi_5-H/Ti_{0.8}Zr_{0.2}CrMn-H$

Contemporary mechanical heat pumps pump heat from a temperature level of about 2 - 8 °C to a temperature level of about 40 °C by using mechanical energy which on the other hand is produced with specific efficiencies from primary energy (like coal, nuclear power stations etc.). Because of this the total efficiencies of such systems used nowadays are not very high.

The alternative to this may be hydride systems. The system $LaNi_5$-H/$Ti_{0.8}Zr_{0.2}CrMnH$ can operate on the same temperature levels as contempory mechanical heat pumps with the difference that in this case the heat on the low temperature level ($-$ 2 °C) is pumped to the medium temperature level (55 °C) by heat on a temperature level of $\geq$ 125 °C (figure 17). If we presume that one hydriding-dehydriding cycle as shown in figure 18 will last for 5 minutes altogether (ab/desorption time 2 minutes; heating up/ cooling down each 30 seconds) then for 1 kW cooling power ($_D\dot{Q}_1$) we need about 3 - 3.5 kg of $Ti_{0.8}Zr_{0.2}CrMn$-H/$LaNi_5$-H material in the four containers which have nearly the same weight again. This means, a 1 kW unit will have a total weight of 6 - 7 kg without additional items like heat exchanging medium, fan etc. The total volume will reach values between 3 and 3.5 litres. The coefficient of performance η_c (chapter 3) can be calculated by the equation

$$\eta_c = \frac{\Delta H_A - C_p \, (m_A + m_1) \cdot \Delta T_{2-1}}{\Delta H_B + C_p \, (m_B + m_2) \cdot \Delta T_{3-2}}$$

where

ΔH_A = 28 kJ/mol H_2 ($Ti_{0.8}Zr_{0.2}CrMnH_3$)

ΔH_B = 31 kJ/mol H_2 ($LaNi_5H_6$)

$_A C_p$ = 0.54 J/g grd   (hydride plus container)

$_B C_p$ = 0.42 J/g grd   ("   "   "   )

ΔT_{2-1} = 57 °C

ΔT_{3-2} = 70 °C

$(m_A + m_1)$ = 333 g

$(m_B + m_2)$ = 400 g

$\underline{\eta_c = 0.42}$

If the system is operating in the heating mode for the coefficient of performance

$$\eta_h = \frac{\Delta H_B + \Delta H_A - C_p \, (m_A + m_1) \cdot \Delta T_{2-1}}{\Delta H_B + C_p \cdot (m_B + m_2) \cdot \Delta T_{3-2}}$$

$$+ \frac{C_p \cdot (m_B + m_2) \cdot \Delta T_{3-2}}{\Delta H_B + C_p \cdot (m_B + m_2) \cdot \Delta T_{3-2}}$$

we calcualate $\eta_h = 1.42$

Both values agree closely with the claims made by J. H. Swisher /20/ and show that the total efficiency can be in the same range or can be better than the total efficiency of typical compressor heat pumps at corresponding sink temperatures.

For 1 kW heating power $(\dot{Q}_{A_2} + \dot{Q}_{A_1})/2$ one needs (cycling time 5 minutes) ~ 1.5 kg of hydride material in the four containers so that the total 1 kW unit will weigh about 3 kg without additional items like heat exchanging medium, fan etc. The total volume will be in the range of 1.5 liters.

4.2.2 Heating and cooling systems with high coefficient of performance using the example of $MNi_{4.5}Al_{0.5}$-H/$LaNi_{4.7}Al_{0.3}$-H and $MNi_{4.15}Fe_{0.85}$-H/$MNi_{4.5}Al_{0.5}$-H

As shown by Gruen et.al /10/ the coefficient of hydride heat pumps becomes higher than in the previous example, when the hydriding enthalpies of the hydrides are nearly equal and the temperature differences become smaller. A very high efficiency is guaranteed with the hydride pair $MNi_{4.5}Al_{0.5}$-H/$LaNi_{4.7}Al_{0.3}$-h (figure 19). Heat on a temperature level of 15 °C is pumped to a medium level of 55 °C by use of heat at 120 °C. To calculate the coefficient of performance η_h we use again the equation

$$\eta_h = \frac{\Delta H_B + \Delta H_A - C_p (m_A + m_1) \cdot \Delta T_{2-1}}{\Delta H_B + C_p (m_B + m_2) \cdot \Delta T_{3-2}}$$

$$+ \frac{C_p \cdot (m_B + m_2) \cdot \Delta T_{3-2}}{\Delta H_B + C_p (m_B + m_2) \cdot \Delta T_{3-2}}$$

where $\Delta H_B = 33.89$ kJ/mol H_2 /20/

$\Delta H_A = 28.03$ kJ/mol H_2 /20/

$C_p = 0.42$ J/g grd /20/

$m_A + m_1 = 370$ g

$m_A + m_2 = 370$ g

$C_{H_2} \sim 1.1$ w %

$$\Delta T_{2-1} = 40\ °$$

$$\Delta T_{3-2} = 65\ °$$

and η_h becomes:

$$\underline{\eta_h = 1.5}$$

A system with a high coefficient of performance in the cooling mode is given by the hydride pair $MNi_{4.15}Fe_{0.85}-H/MNi_{4.5}Al_{0.5}-H$ (figure 20).

The values are as follows:

$$\Delta H_A = 25.1\ kJ\ mol^{-1}\ H_2 \hspace{4cm} /20/$$

$$\Delta H_B = 28.03\ kJ\ mol^{-1}\ H_2 \hspace{3.8cm} /20/$$

$$C_{H_2} = 1.1\ w\ \%$$

$$C_p = 0.42\ J/g\ grd \hspace{4.5cm} /20/$$

$$m_A + m_1 = 370\ g$$

$$m_B + m_2 = 370\ g$$

$$\Delta T_{2-1} = 25\ °$$

$$\Delta T_{3-2} = 28\ °$$

and η_c becomes 0.66

A further advantage of this system is the relatively low higher temperature level of about 65 °C so that it is possible to use low-grade waste heat.

4.2.3 Hydride heat pump for very low sink temperatures

As already mentioned the hydride heat pumps in comparison to mechanical heat pumps may become more efficient when the sink temperature is very low. This can be shown by the example of the hydride pair $Ti_{0.9}Zr_{0.1}CrMn-H/LaNi_5-H$ which is able to pump heat from − 25 °C to + 50 °C with the aid of high temperature heat at 150 °C. The coefficient of performance is still good and η_h reaches a value of nearly 1.2 (figure 21).

whereby $\hspace{1.5cm} \Delta H_A = 22.5\ kJ\ mol^{-1}\ H_2\ (Ti_{0.9}Zr_{0.1}CrMnH_3)$

$$\Delta H_B = 31\ kJ\ mol^{-1}\ H_2 \hspace{4cm} /1/$$

$$_A C_{H_2} = 1.2 \text{ w \%}$$

$$_B C_{H_2} = 1.0 \text{ w \%}$$

$$_B C_P = 0.42 \text{ J/g grd} \qquad\qquad /1/$$

$$C_A = 0.54 \text{ J/g grd}$$

$$m_A + m_1 = 333 \text{ g}$$

$$m_B + m_2 = 400 \text{ g}$$

$$\Delta T_{2-1} = 75 \,°$$

$$\Delta T_{3-2} = 100 \,°$$

4.2.4 Hydride heat pumps for high temperature levels ($\geq$ 200 °C)

Many chemical or industrial processes produce heat on temperature levels of more than 200 °C. With different hydride pairs now it is possible to produce heat in the temperature ranges of 150 to 250 °C (TiFe$_{0.8}$Ni$_{0.2}$-H/TiVMn-H), 200 - 350 °C (TiVMn-H/Mg$_2$Ni-H) and even 250 °C - 600 °C (Mg-H/Ti-H). This offers the possibility to apply heating systems with very high efficiencies even at high temperatures into chemical or industrial processes.

Beside this the hydriding cycles can be used for heat transformation as shown in figure 23. In this case heat on a medium temperature level is used to produce heat on very high temperature levels. With the hydride pair TiH/Mg$_{95}$Ni$_5$-H it is possible to pump heat from a temperature level of about 370 - 380 °C/680 °C to a level of 760 - 780 °C. The efficiencies are comparable to those of the refrigeration cycles.

4.3 Hydride heat pumps for mobile applications

Waste heat from combustion engines in cars, trucks and buses is available on a temperature level of about 80 °C (coolant water) or between 300 °C and 700 °C (exhaust gas). Therefore, for some applications it is possible not only to use so called low temperature hydrides but also so called high temperature hydrides with high hydrogen storage capacities like Mg$_{0.95}$Ni$_{0.05}$H$_{1.9}$ which is able to store about 7 w %.

These hydrides become advantageous when the absorption-desorption time lasts longer than 5 or 10 minutes as the next example shows.

4.3.1 Preheating of engines or passenger compartments in cars and buses

Research work in the field of fuel consumption has shown that a remarkable reduction in consumption is possible, if one can dispense with cold starts in that the engine is preheated with an auxiliary heater. This effect becomes still more advantageous when the energy to heat the system comes from a heat storage system and not from fuel. The preheating times are in general 10 - 20 minutes so that the storage system have to store an energy amount for these times without having too much weight. As already mentioned the storage capacity of high temperature hydrides is very high and with the example of the hydride pair $MgH_2/Ti_{0.8}Zr_{0.2}CrMnH_3$ we can show that the total weight of such heating systems is low enough to guarantee a distinct saving in fuel.

The mode in which such a system works can be explained by figure 26. During operation, the engine heat of the exhaust gas is used ($_DQ_2$ in figure 22) to desorb hydrogen gas out of the MgH_2. This hydrogen gas is absorbed in a container with $Ti_{0.8}Zr_{0.2}CrMn$ to form $Ti_{0.8}Zr_{0.2}CrMn$-hydride when the heat $_AQ_1$ is diverted on a temperature level of about 16 °C. When the engine is stopped and it cools down to ambient temperature, for instance lower than - 20 °C one has the possibility of preheating the engine before starting by desorbing hydrogen out of $Ti_{0.8}Zr_{0.2}CrMnH_3$ with heat supply $_DQ_1$ from the ambient air and by using the heat of formation $_AQ_2$ which is caused by absorption of hydrogen in Mg. During engine operation, the hydrogen refilled again from the Mg-container into the $Ti_{0.8}Zr_{0.2}CrMn$-container and the cycle is closed.

With this hydride pair it is possible to preheat at ambient temperatures as low as - 20 °C. If it is required to preheat at lower temperatures, one can use other hydride pairs like $Mg_2NiH_4/Ti_{0.9}Zr_{0.1}CrMnH_3$ with which preheating can be accomplished at temperatures as low as - 40 °C.

The weight per kW and the operating time of such systems is determined by the

- heat of formation of the magnesium hydride

    $\Delta H = 77.4$ kJ $mol^{-1}H_2$

- H_2 storage capacity of magnesium

    $_{H_2}C_{Mg_{0.95}Ni_{0.05}} = 7$ w %

- H_2 storage capacity of the low temperature hydride

    $_{H_2}C_{Ti_{0.8}Zr_{0.2}CrMn} = 1.8$ w %

- heat capacity of the high temperature hydride + container in the temperature range which cannot be used for preheating. In

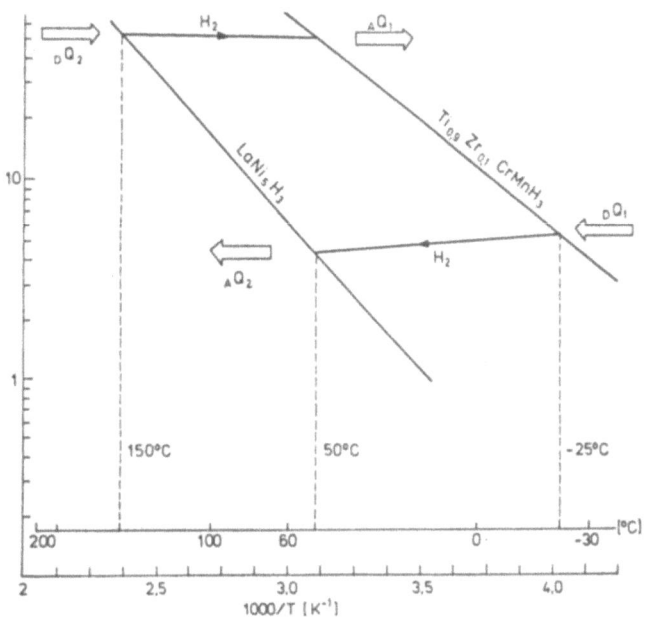

Figure 21. Plateau pressure vs 1/T for LaNi$_5$H$_3$ and Ti$_{0,9}$Zr$_{0,1}$CrMnH$_3$

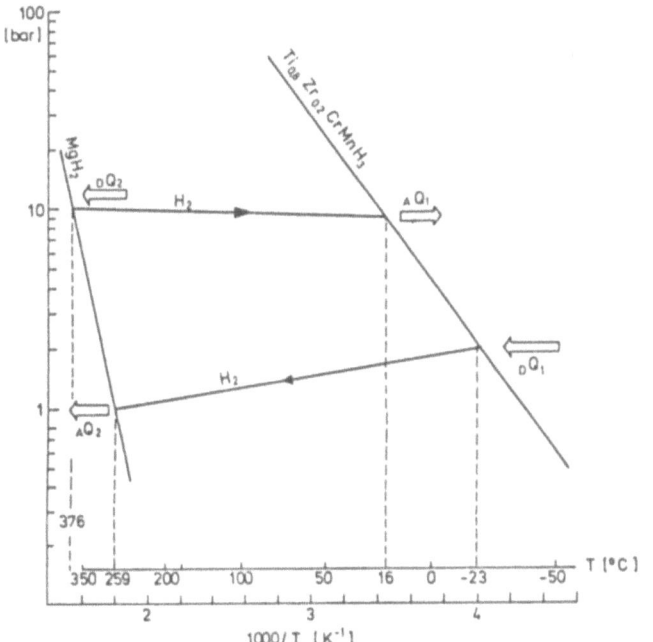

Figure 22. Plateau pressure vs 1/T for MgH$_2$ and Ti$_{0,8}$Zr$_{0,2}$CrMnH$_3$

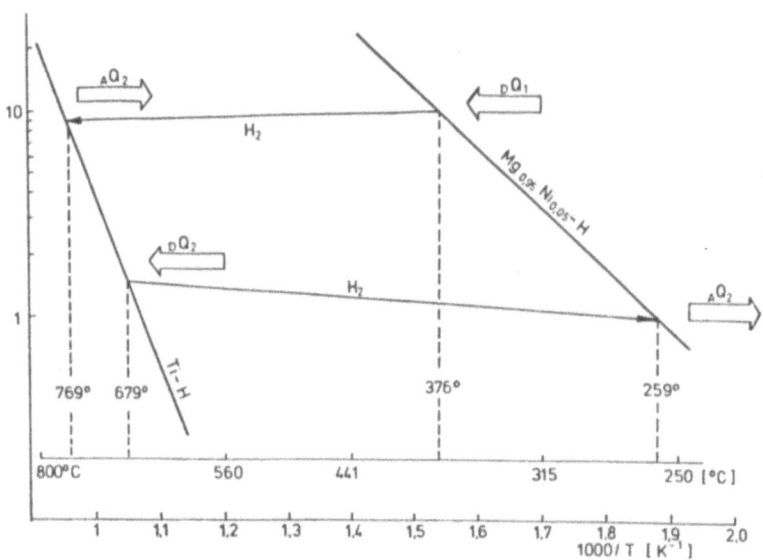

Figure 23. Plateau pressure vs 1/T for
$Mg_{0,95}Ni_{0,05}$-H and Ti-H

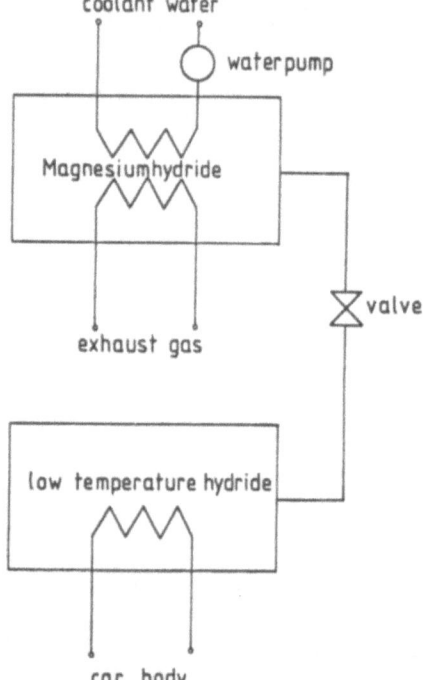

Figure 24. Principle of a hydride
auxiliary heating system

the case of coolant water < 100 °C.

$$Q_{C_p} = (1.04 \text{ J g}^{-1} \text{ grd}^{-1} \cdot m_{Mg} + 0.51 \text{ J g}^{-1} \text{ grd}^{-1}) \cdot 100 \text{ °C}$$

- weight of the container; for high temperature hydride the container amounts to 50 per cent of the total weight, and in the
case of low temperature hydrides the container amounts to 30
per cent of the total weight.

If all these factors have been taken into consideration the
usable energy content of such systems runs up to 300 kJ/kg.

From this it follows that an auxiliary heating system (as
shown in figure 23) for the engine (2.3 liters) which is able to
heat up the engine from 0 °C (we need about 6,700 kJ), weighs
22 - 23 kg. This is a value which will not influence the fuel
consumption of a car.

4.3.2 Hydride cooling systems for mobile applications

As already shown in section 5.2 hydride based cooling systems now have a power density of about 6 - 7 kg/kW. The general
cooling power of contemporary compressor cooling devices is in
the range of 5 - 10 kW. This means that equivalent hydride systems will have a weight between 30 and 70 kg with the advantage
that no extra fuel consumption for the operation of the cooling
unit is necessary. Possible hydride pairs are mentioned in
section 5.2, whereby we have to point out that a further development in the hydride field is necessary in order to find hydrides
with higher storage capacities than those of the $LaNi_5$- or
MNi_5-based alloys (1 - 1.2 w %) because then the weight of the
system can be reduced.

Hydrides with high storage capacities are for instance on
the base TiCrMn and reach values of 1.8 w %, so that the weight
of the hydride pair in comparison to hydrides on the base of
$LaNi_5$ and/or MNi_5 can be reduced by a factor 0.6 to 0.7. From
this it follows that in future power densities of 5 kg/kW might
be realized.

In comparison to actual air conditioning systems we again
get the advantage that no fuel will be consumed to operate the
system with the result that a remarkable saving of fuel is possible.

5   APPLICATIONS FOR METAL HYDROGEN SYSTEMS WITH HYDROGEN CON-
SUMPTION

Since the metals release heat during the chemical reaction
with hydrogen (and only then), special insulation as is required

by all other heat storage units is not necessary. This can lead to considerable weight and cost savings as compared with conventional heat storage units even with closed H_2-circuit.

Thus any combustion process (in industry, the household or vehicles) which consumes at least to some extent hydrogen from a hydride accumulator (employed as intermediate accumulator to hold the gas between supply and consumption) can be combined with partial or complete accumulation of the combustion waste heat. This dual function of the hydride accumulator makes possible, in practical application, the following fuel/heat coupling mechanism (fig. 24) /21/. Waste heat is generated during the combustion processes (e.g. H_2-engines or H_2-domestic heating etc.) and fed to the hydride for hydrogen liberation, it is, therefore, accumulated in the metal and not given off to the atmosphere simultaneously with the combustion. In order to ensure a continuous hydrogen yield, the waste heat from the combustion process must, of course, be always greater than the energy required for the liberation of the hydrogen from the hydride. If the heat quantity available with all transmission losses is identical to the energy necessary for the decomposition of the hydride, the combustion process is without waste heat to the outside, as a closed system, hydrogen combustion/waste heat storage, prevails in this case. Only when the metal is recharged with hydrogen (and only then), does the hydride formation result in the yielding of the previously accumulated combustion waste heat. The waste heat energy can (by changing the pressure during charging) be made available in various temperature ranges for practical use, e. g. room heating.

5.1 Stationary heat storage systems with metal hydrides

Should it become necessary, as a result of petroleum oil scarcity, to replace gasoline and/or fuel oil with a different energy medium such as hydrogen or town gas (with hydrogen content), it should be possible for reasons of energy supply to individual households, to install an infrastructure for hydrogen out of hydrogenous gas mixtures. Town gas or hydrogen could then be produced in central gas plants and supplied via pipe networks to the individual households.

As the waste heat prevailing in the household is of a relatively low temperature level, low temperature hydrides with little hydrogen binding energy are particularly suitable for hydrogen storage in the household. During the day hydrogen can be drawn from the hydride storage unit for heating purposes, storing the combustion waste heat (e. g. of the gas central heating) in the hydride. The heat quantity required for the liberation of the hydrogen can be drawn from the air in the interior of the house if desired, providing air conditioning without power consumption

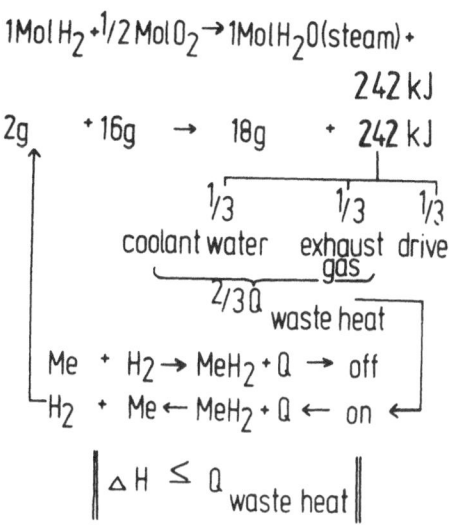

$$1\,\text{Mol}\,H_2 + {}^{1}\!/\!{2}\,\text{Mol}\,O_2 \rightarrow 1\,\text{Mol}\,H_2O\,(\text{steam}) + 242\,kJ$$

$$2g \quad + 16g \quad \rightarrow \quad 18g \quad + \quad 242\,kJ$$

$\tfrac{1}{3}$ coolant water $\tfrac{1}{3}$ exhaust gas $\tfrac{1}{3}$ drive

$\tfrac{2}{3}Q$ waste heat

$$Me + H_2 \rightarrow MeH_2 + Q \rightarrow \text{off}$$
$$H_2 + Me \leftarrow MeH_2 + Q \leftarrow \text{on} \leftarrow$$

$$\left|\; \Delta H \leq Q_{\text{waste heat}} \;\right|$$

Figure 25. Reaction process during H_2-combustion

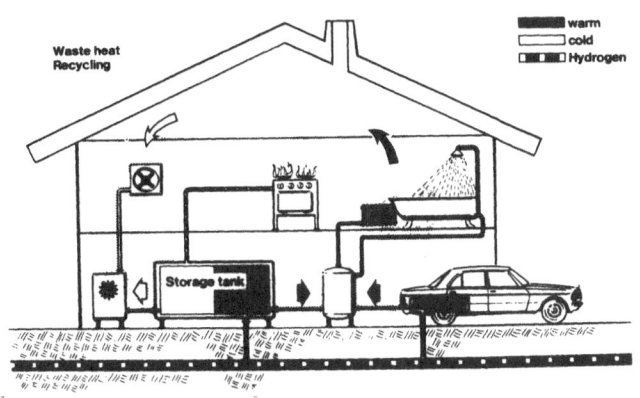

Figure 26. Hydrogen storage in metal hydrides
for house and car application

(similar to the application in motor vehicles) free of charge, or be drawn from the ambient atmosphere.

During the night the discharged alloy can be recharged with hydrogen from town gas and the heat quantity obtained can be used for heating purposes. If heating is not required (summer), the hot water obtained during hydride formation can be collected in a boiler. Both the air conditioning in the interior of houses, without consumption of primary energy, and the recovery of waste heat, e. g. from central heating systems, with the aid of the hydrogen hydride storage unit are factors which lead to a considerable reduction in the primary energy requirements and to corresponding cost savings.

In connection with a hydride storage unit in the car, it is possible to couple the waste heat in a car with the heat requirement in the household (fig. 25 and 26). Assuming there is a general hydrogen supply for households, the stationary hydride storage units can perform the following functions:

- Gas intermediate storage unit (consumption peak).
- Cooling system (air conditioning systems, deep freezer etc. without power consumption!).
- Fuel tank for motor vehicle refuelling.
- Heat storage unit for the household and motor vehicle waste heat.
- Heat pump (when heat is taken from the ambient air for H_2-liberation).

5.2 Energy infrastructure for the use of hydrogen hydride technology

Since, however, the distribution of pure hydrogen via widespread gas networks appears to be unrealistic, even as a medium-term project, emphasis in Daimler-Benz' research work is placed on the development of hydrides which are largely insensitive to gas mixtures. In this way it could be possible in future with the aid of selective hydrogen absorption to use hydrogenous gas mixtures like town gas for hydride formation instead of pure hydrogen. The selective absorption of hydrogen via hydrides could replace the conventional pressure swing absorption units used today for hydrogen production from gas mixtures. As in both cases the existing gas network can be used, a special hydrogen infrastructure for motor vehicles would not be required.

Recent work which was performed at the Institute of Gas Technology (Chicago) /22/ furnished proof that the Titanium-Nickel hydrides /17/ developed at Daimler-Benz are able to selectively extract the hydrogen from a methane/hydrogen mixture with a hydrogen content of only 15 % by volume. This makes it possible even now to stretch the natural gas supplies by adding hydrogen (from coal) and to filter out these hydrogen quantities, as required, at the final user. The pure hydrogen obtained in this way can be stored both in stationary (house) or mobile (vehicle, mainly passenger cars with gasoline/H_2-mixture operation) hydrides and subsequently used. A vehicle with hydrogen drive and an annual kilometrage of 15,000 km would then supply a heat energy quantity during refuelling from the domestic gas tap corresponding to a calorific value of approximately 500 ltrs. oil per year. For this reason, the heat storage function of the hydrides (which in this case store the fuel hydrogen as their main function) can be additionally utilized to advantage as long as a hydrogen consumer is available (vehicle, household).

As shown for domestic gas taps the technical preconditions for hydrogen extraction from water by means of electrolysis (current) or from natural gas by means for reformer units are known as well. Therefore, theoretically, every household which is connected to an electrical or gas network has the preconditions for its own hydrogen generation, even today. An energy balance for hydrogen production and vehicle refuelling from the "domestic gastap" using as an example electrolytic water decomposition is given in /23/.

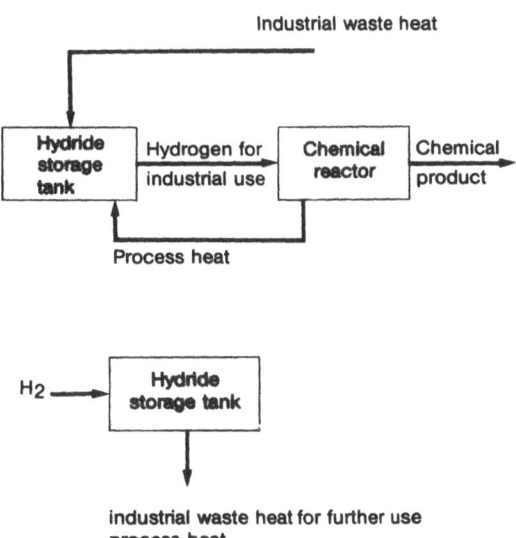

Figure 27. Industrial waste heat recovery
with the aid of metal hydrides

5.3 Heat storage in metal hydrides for the recovery of industrial process heat

As, in industry, several 100 millions m³ of hydrogen are produced, transported and used annually, it would be possible to combine all exothermic chemical hydrogen reaction processes (e. g. NH_3 sythesis) with hydride storage tanks. Because of the hydride enthalpy the accumulable heat of the hydrogen-consuming exothermic processes is 15 % in the low-grade range and up to 30 % in the high-grade range (based on the lower calorific value of a given volume of hydrogen) (fig. 27).

If insufficient heat for the decomposition of the hydride is produced in the corresponding temperature range, e. g. in highly efficient combustion processes, a fuel mixture (for example natural gas/hydrogen, oil/hydrogen, coal/hydrogen or similar) is employed and part of the combustion heat of the natural gas, oil or coal is also used for desorbing the hydrogen and is thus accumulated. Conversely, only up to 30 % of the lower calorific value of hydrogen can be accumulated in hydrides in chemical reactions having very high exothermic reaction enthalpies. The remaining heat would have to be accumulated in a different manner or be dissipated into the environment as energy loss.

For industrial waste heat storage and reuse, it is therefore recommended to combine existing industrial processes in which H_2 is consumed, with hydride accumulators, when the reuse of the heat is technically feasible and is desired for reasons of energy economy or environmental protection.

6   CONCLUDING REMARKS

The technical feasibility of heat storage by hydride systems was shown. The high heat storage capacity with respect to weight and volume and the rapid kinetics of the reaction processes offer the possibility to use such systems for different purposes in household, vehicular and industrial applications. It could be explained that in the case of very high (> 120 °C) and very low temperature levels (< 0 °C) the hydride systems are advantageous in comparison to conventional systems.

Therefore, these hydride systems (open and closed H_2-cycles) might become attractive in the future especially when the hydride development with respect to heat storage means and hydride heat pumps is further pursued.

References

/1/ F. A. Kuijpers
Philips Res. Rep., Suppl. 2 (1973)

/2/ G. D. Sandrock
Proc. 12th Intersociety Energy Conversion Engineering
Conf., Washington, D.C., 1977, Vol. 1

/3/ J. J. Reilly
Proc. Intern. Symp. Hydrides for Energy Storage,
Geilo, Norway (1977)

/4/ Y. Machida, T. Yamadaya, M. Asanuma
Proc. Intern. Symp. Hydrides for Energy Storage,
Geilo, Norway (1977)

/5/ I. Jacob, D. Shaltiel, Dr. Davidor and I. Miloslavski
Solid State Commun, 23 (1977) 669

/6/ T. Gamo, Y. Moriwaki, N. Yanagihara, T. Yamashita,
T. Iwaki
Int. Journal of Hydrogen Energy, to be published 1981

/7/ J. J. Reilly, R. H. Wiswall, Jr.
Inorganic Chem. 7, 2254 (1968)

/8/ Ger. Patent 28044453 (1973), to B. Bogdanovic.

/9/ D. M. Gruen, I. Sheft
Metal Hydride Systems for Solar Energy Conversion,
Proc. NSF-ERDA, Workshop on Solar Heating and Cooling
of Buildings, Charlottesville, Vancouver (1975)

/10/ D. M. Gruen, M. H. Mendelsohn, I. Sheft
Solar Energy Vol. 21, pp. 153 - 156, Pergamon Press
Ltd., 1978

/11/ H. H. van Mal
Philips Res. Repts. Suppl., 1976, No. 1,
Stability of Ternary Hydrides and some Applications

/12/ H. Buchner
Das Wasserstoff-Hydrid-Energie-Konzept, Chemie-Technik
9, 7. Jahrgang (1978)

/13/ J. Toepler, O. Bernauer and H. Buchner
Journal of the Less Common Metals, 74 (1980)
385 - 399

/14/ Thomas E. Duffy, David A. Rohy
United States Patent 4, 161,211, Jul. 17, 1979,
Methods of an Apparatus for Energy Storage and
Utilization

/15/ M. A. Gutjahr, H. Buchner, K. D. Beccu, H. Säufferer
A New Type of Reversible Electrode for Alkaline
Storage Batteries Based on Metal Alloy Hydrides,
8th Intern. Power Sources Conf., Sept. 1972, Brighton
/U.K., Conference Proc.

/16/ O. Bernauer and H. Buchner
Spezielle Probleme der Metallhydridspeicher,,
Mobile Stromversorgung im Felde, Conf. Proc.,
May 1980, Deutsche Gesellschaft f. Wehrtechnik e.V.

/17/ H. Buchner, M. A. Gutjahr, K. D. Beccu, H. Säufferer
Wasserstoff in intermetallischen Phasen am Beispiel
des Systems Ti-Ne-H, Z. Metallkunde 63, 417 - 500
(1972)

/18/ E. Schmidt-Ihn, H. Buchner and M. A. Gutjahr
Metallhydride für die elektrochemische Energie-
speicherung und -erzeugung, Mobile Stromversorgung
im Felde, Conf. Proc., May 1980, Deutsche Gesellschaft für Wehrtechnik e. V.

/19/ K. D. Beccu, Swiss Patent 4199/69

/20/ James H. Swisher
Hydrides Versus Competing Options for Storing
Hydrogen in Energy Systems, Journal of Less-Common
Metals, 74 (1980) 301 - 320

/21/ H. Buchner
Das Wasserstoff-Hydrid-Energie-Konzept, Chemie-Technik,
9, 7. Jahrgang (1978)

/22/ V. Cholera, D. Gidaspow
Hydrogen Separation and Production from Coal Derived
Gases Unsing Fe_xTiNi_{1-x}, 12th IECEC, Washington D.C.,
USA (1977)

/23/ H. Buchner
Progress in Energy and Combustion Science (1980),
Heft 4

HEAT PUMPS COMBINED WITH THERMAL STORAGE

H. van der Ree

TNO-Division of Technology for Society
Laan van Westenenk 501, 7334 DT   APELDOORN
The Netherlands

SUMMARY

Thermal storage can be applied at either the
high-temperature or the low-temperature side of heat pumps.
High-temperature storage is applied for shifting the
electricity load on the grid, for reducing the on/off switching
frequency of the heat pump and for improving a system's energy
balance. Low-temperature storage can improve the averaged COP
and will reduce the size of the heat pump. It is indispensable
in solar assisted heat pump systems and particularly interesting
when long-term storage is considered. For the same heating
requirements electrical heat pumps will require larger thermal
storage capacities at the evaporator side than heat activated
systems.

Heat pumps together with thermal storage can be realised
in many forms. An interesting option is based on the
intermittent sorption cycle, which can provide for thermal
storage, incorporating a heat pump or a heat transformer
effect.

1.   INTRODUCTION

Since in the early seventies energy conservation became
increasingly important the heat pump has been subject of a non
ceasing interest. The heat pump is sharing this interest with
various other energy saving systems, of which solar-energy
systems were most on the fore in the early years. In later years
this role was taken over by the heat pump. A major advantage of

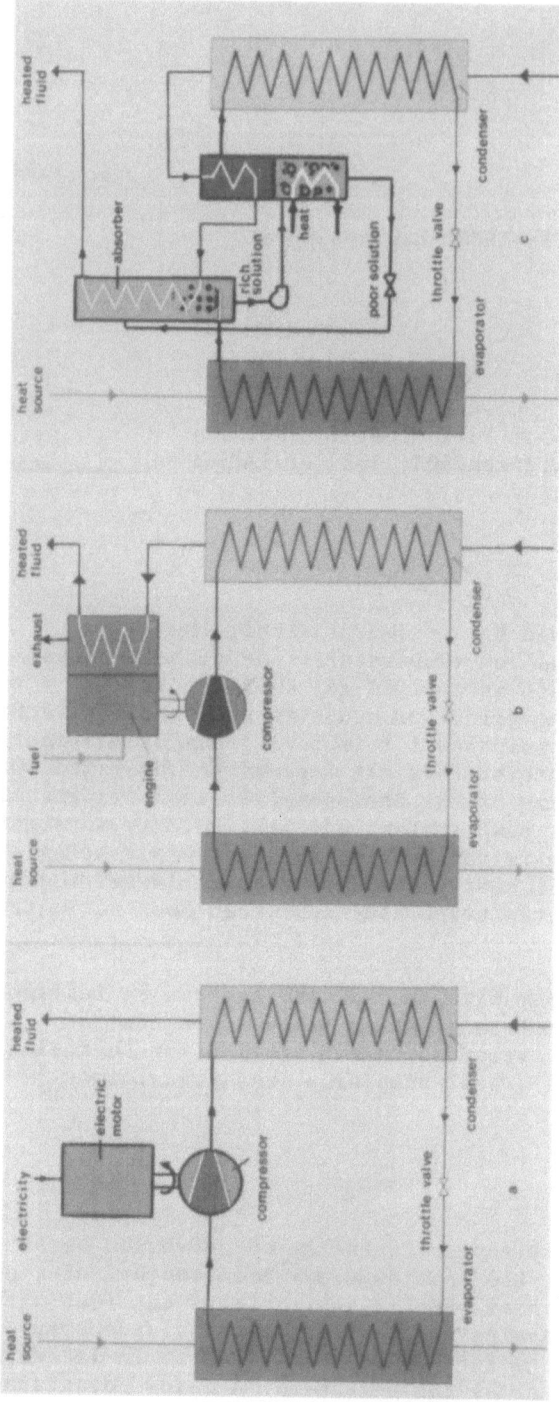

Figure 1a: Electrically driven heat pump
b: Engine-driven heat pump
c: Absorption heat pump

heat pumps is their flexibility regarding the process on which they can be based, and the many designs that can be derived. Because of these characteristics heat pumps will lend themselves to a great spectrum of applications, and it is therefore technically feasible for the heat pump to penetrate to a great extent into various market segments. Moreover, by using heat pumps, scarce fuels like oil can be replaced by other fuels or energy-generating means that can be relied upon for longer times. These aspects indicate that the use of heat pumps can contribute to achieving energy conservation and fuel diversification, these being two major goals in the energy policy of many countries.

As the heat pump field is pretty broad and has not been fully explored up till now, this subject will offer matter for an exchange of information for many years to come. At the scientific/technical level a "heat pumps" commission has been established within the International Institute of Refrigeration (IIR) in addition to the 10 commissions already existing. Also the International Energy Agency is aware of the need of exchanging information on heat pumps and a project has been proposed to establish an International heat pump centre. One of the tasks of this centre will be to collect and disseminate information on heat pumps. Finally it can be mentioned that the European Community has sponsored a number of projects on heat pumps within the framework of two successive R and D programmes, which has already given rise to an international dissemination of knowledge and experience (1).

In most heat pump applications the system is based on the use of low-grade environmental heat. This is all stored solar energy and consequently low-temperature natural thermal storage is generally inherent in heat pump systems. In spite of this it can be observed that for different reasons an additional thermal storage facility can be useful in connection with heat pumps in some cases. It should be clearly mentioned, however, that in the vast majority of heat pump systems which are applied in practice, thermal energy storage other than that included in the low-temperature heat source, has not been incorporated.

2. HEAT PUMP BASICS

The heat pump, as its name already indicates, converts low-grade heat which is freely available from many sources, to useful heat of a higher temperature. Thermodynamically the heat pump is identical to the refrigerator, its application is in a different field, however. A refrigerator provides useful cold, whereas the heat pump provides useful heat. Because of the thermal contribution of the low-grade heat source the heat pump always delivers more heat than the amount of energy required for driving.

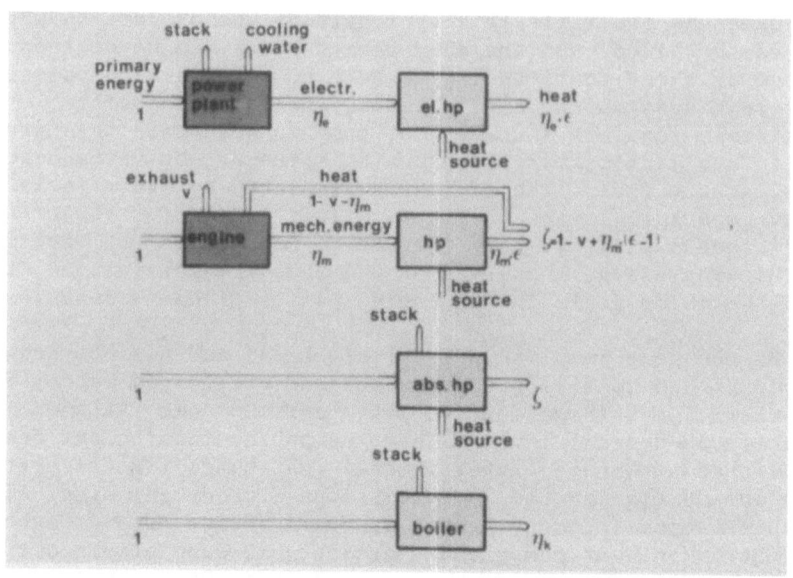

Figure 2: Primary energy to useful heat conversions for various heat pump types and a central heating boiler

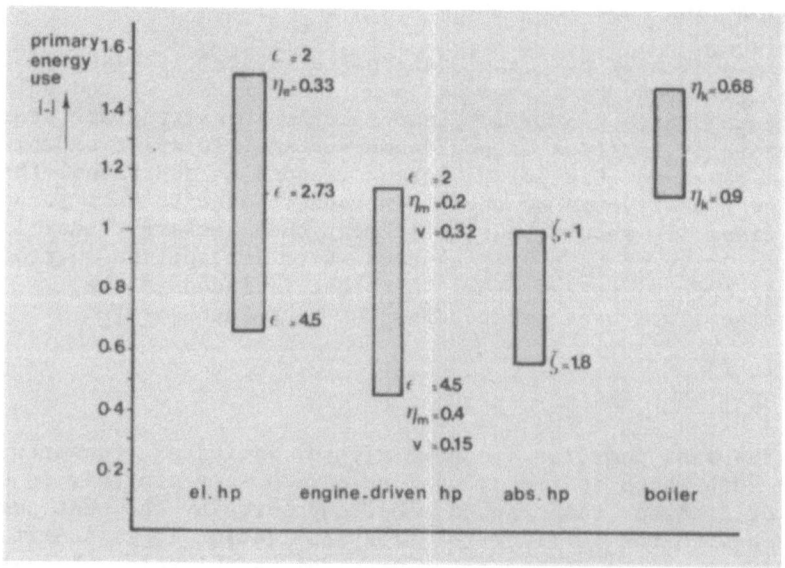

Figure 3: Ratio of primary energy use to heat production, based on gross calorific values

It is due to this characteristic feature that the heat pump can play an important role in conserving energy.

The most important heat pump types are the electrically driven compression heat pump, the engine-driven compression heat pump and the absorption heat pump (Figure 1). The components that are the same in the three diagrams are:
- the condenser, where the fluid (refrigerant), that is circulating in the heat pump cycle, is condensed at high pressure under heat release to the system to be heated;
- the evaporator, where the refrigerant is vaporised at a low pressure and heat is absorbed from the low-temperature heat source;
- the throttle valve for reducing the pressure of the liquid refrigerant on its way from the condenser to the evaporator.
The three processes are distinguished by the way in which the vapour compression between evaporator and condenser and the drive of the cycle are realised. In case a) this is done by a mechanical compressor, driven by an electric motor. In case b) the mechanical compressor is connected with a combustion engine, which has the advantage that the engine's waste heat will contribute to the system's heat production. In case c), the absorption heat pump, there is no mechanical energy component in the compression process, like in the other cycles. In a "thermal compressor" an absorbent is circulating which is able to absorb the refrigerant vapour at a low pressure. With this process heat is released, and the heat of absorption forms part of the heat pump's heat production. The liquid mixture (strong solution) is pumped from the absorber to a desorber, which is on the high-pressure side of the process. In the desorber the refrigerant is boiled off from the solution, which involves an input of higher temperature (drive) heat. The lower concentrated solution (poor solution) is fed to the absorber after pressure reduction in a throttle valve.
Looking to the drive energy that goes into the various heat pumps, it will be clear that a distinction can be made between electrically driven heat pumps and fossil-fuel- or heat-activated machines. The type of higher-grade input energy is a main item according to which heat pumps can be classified.

With heat pumps the ratio of the useful heat production to the drive energy required is expressed as "Coefficient Of Performance (ε)(COP)" and "Heat Ratio (ζ)" for mechanically/electrically driven systems and fuel/heat-activated systems respectively. This ratio is always better than unity, as indicated before. Unlike the characteristics of conventional heating systems, which are practically not influenced by the environmental conditions, the COP or ζ of heat pumps will depend on the temperatures on both the high- and low-temperature side.

Particularly the temperature difference between these two plays
an important role. The higher this difference is, the lower the
COP or ζ will be.

For the above reason the operating properties as well as
the economics of heat pump installations are largely determined
by the heat source. Suitable heat sources for heat pumps can be
divided into two main groups:
- natural sources, such as ambient air, direct solar radiation,
  surface water, ground water and the soil;
- waste heat from air ventilation, waste water or industrial
  processes.
Each of these heat sources has both advantages and disadvantages,
which have to be weighed one against an other. There is a growing
interest in using several heat sources in combination.

The energy savings to be reached with heat pumps highly
depend on the system selection, on the systems operating
conditions as well as on the energetic quality of the apparatus.
As long as these factors are not specified, the energy savings
can only be roughly indicated.
A simplified energy conversion model of the three above mentioned
heat pump types, as well as of a central heating boiler, is
shown in Figure 2. Furthermore in Figure 3 the approximate
ranges of energy consumption, all expressed in terms of primary
energy, are indicated.
The figures for the central heating boiler, which has been
introduced as a system of reference, apply to a conventional
gas-fired boiler and a high-efficiency boiler, respectively.
Improved boilers in which the flue gases are condensed to achieve
a high efficiency are now entering the market. They should be
considered therefore when energy saving by heat pumps, as compared
with boilers, is discussed. As against the high-efficiency boiler
the electrical heat pump should have an averaged COP of at least
2.73 before primary energy is actually conserved. Many electrical
heat pump systems have lower values than this, especially when
the heat source is outside air. When the comparison is made with
the conventional boiler, the threshold COP is 2.06. With the
engine-driven compression heat pump the upper end of the bar in
Figure 3 is based on a combination of low efficiencies of the
various components, which might arise for small units. The best
combination, as to the lower end of the bar, is resulting in a
particularly low primary energy use. Large diesel-engine driven
heat pumps connected with a favourable heat source could achieve
such a good performance. Also the absorption heat pump shows good
results in view of a saving of primary energy. It should be noted,
however, that the average heat ratios of absorption heat pumps do
not exceed a value of 1.3 with the few machines that are
available at present. However, higher values can certainly be

anticipated in future.
From Figure 3 it can finally be concluded that primary energy can
be saved to a larger extent and also more easily with
fuel/heat-activated heat pumps than with electrical heat pumps.

3.   ELECTRICAL HEAT PUMPS WITH THERMAL STORAGE

3.1 Introduction

In several cases it is advantageous to provide heat pump
systems with some sort of thermal storage. This can be realised
at either one of the two temperature levels, between which the
heat pump operates, in other words at the high-temperature side,
where the useful heat is produced or at the low-temperature heat
source side (Figure 4). Thermal storage at both sides can be
applied as well. An interesting option is storage at a low
temperature, where heat from a part of the heat pump cycle is
used for charging this storage, as will be discussed later.
In heat pump systems short-term as well as long-term thermal
storage can be applied. Long-term storage is generally found at
the heat source side, where the heat can be stored without losses
at a low temperature.

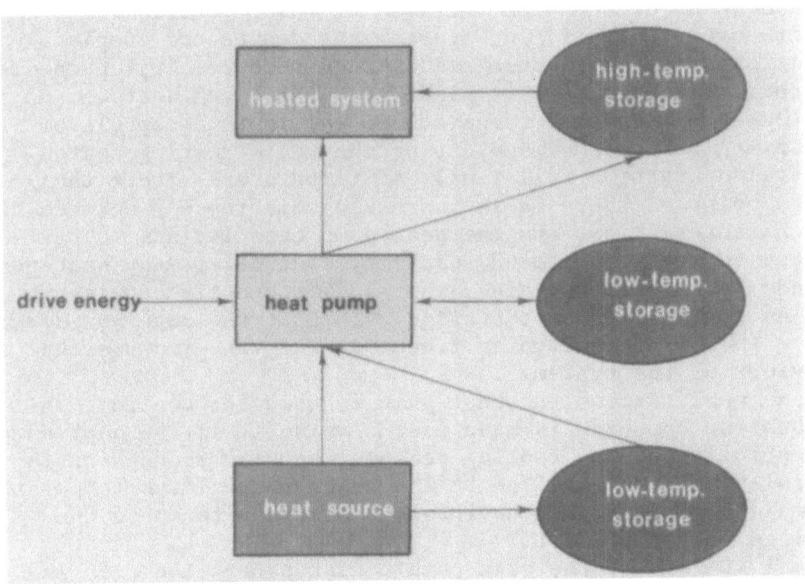

Figure 4: Thermal storage options in a heat pump system

3.2 High-temperature storage

In various heat pump applications short-term heat storage is incorporated at the heat production side. This storage is often part of the building structure and the heat distribution system. This is specifically the case with floor heating systems, where the heat distributing tubes are lying in a heavy concrete floor, which can represent a significant accumulating capacity. There are several reasons for incorporating a high-temperature thermal storage in a heat pump system:

- In various countries electricity is made available for heat pumps at reduced rates. This is done on the condition that the heat pump is off the grid during peak hours. Short-term thermal storage at the heat supply temperature is then needed to compensate for the interrupted heat production.
Sometimes it is advantageous for the electricity rates to run the heat pump only during the night. This can be realised by means of a thermal storage by which the daily heat demand can be satisfied. Concerning air-source heat pumps this mode of operation has a drawback from an energy saving point of view, since during the night the outside air temperatures are lower than the day-time-temperatures, which reduces the average COP of the heat pump.
- One of the characteristics of heat pumps is, that the heating capacity goes down at lower outside air temperatures. Air-source heat pumps are most sensitive in this respect. The space heating demand will increase at lower outside temperatures, however. In Figure 5 the demand and supply characteristics of a house and of an air-source heat pump are sketched. The heat pump is sized to the maximum heat demand at the lowest outside air temperature, and is thus capable of satisfying the heat demand during the whole heating season (monovalent system, full line). At temperatures above the design point A, there is an increasing discrepancy between the two characteristics, and the heat pump capacity has to be reduced to meet the demand. Capacity control of many heat pumps is made by on/off switching, which unfortunately reduces their COP. An appropriate heat storage volume at the supply side can reduce the on/off-switching frequency and thus improve the behaviour of the system.
- In some applications the heat pump is used for cooling, while the heating capacity is made useful as well. If the heat pump is controlled by the cooling demand, the heat production is generally not synchronized to the heat demand. High-temperature thermal storage is then obviously a solution to avoid the spoilage of useful heat.
- Thermal storage at the heat pump's heat production side is obvious where large heating capacities are required at irregular times with low or no load periods in between. This

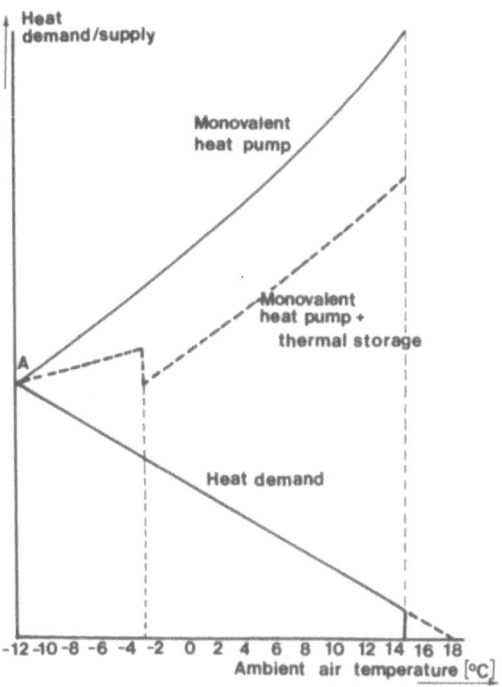

Figure 5: Typical heat production characteristics (full load)
of a monovalent air-source heat pump with and
without a low-temperature thermal storage

pattern is typical for warm tap water use and therefore tap
water heat pumps are always combined with a storage tank at
the heat supply side.

3.3 Low-temperature storage

Wether storage at the low-temperature side of heat pumps is
relevant, depends on the heat source which is available for heat
pump applications. As indicated before, sources like ground water,
the earth and to some extent surface water, are offering stored
heat of a natural kind. The temperatures of these sources are

relatively high at low ambient temperatures, and consequently there is no need for additional thermal storage. However, with ambient air as a heat source, the conditions are less favourable, resulting in a heat pump characteristic which, at most temperatures, deviates far from the demand characteristic, as discussed before (Figure 5). Apart from the drawback that in the low outside-temperature range the heat pump has a poor COP, the simple monovalent system, as indicated here, is also disadvantageous from an economic point of view, since the investment costs are high as a result of the large heat pump capacity. One of the ways to improve these points is applying a low-temperature thermal storage. If at low ambient temperatures the heat pump is provided with stored low-grade heat of a somewhat higher temperature, the heat production line will shift to a higher level in this region and a smaller heat pump can satisfy the demand. Figure 5 shows an example of a characteristic of a heat pump combined with latent heat storage (dotted line) (2). In this example the heat output and the COP are increasing when the heat extraction is switched over from the ambient air to the heat storage. Wether there is an increase or not and how large this increase will be, depends on the system's design. From Figure 5 it can be concluded that low-temperature storage will also reduce the part-load problem, which is discussed in the previous part. Latent heat storage in heat pump systems is generally based on water as a storage fluid. The water is frozen by the heat pump, setting free the heat of solidification.

Low-temperature heat storage is indispensable in solar-assisted heat pump systems to cope with the discrepancies between the fluctuating solar radiation and the heat that can be absorbed by the heat pump. In chapter 5 this will be discussed in more detail.
When both heating during winter time and cooling during summer are required, it is possible to store the cold produced by the heat pump during winter and use it for summer cooling. This type of long-term storage will also be discussed in chapter 5.
A specific advantage of heat storage at a low temperature is of course that heat losses are eliminated. If the heat storage is below ambient temperature, heat leakage will even favour the storage system. Low-temperature heat storage is therefore particularly interesting when long-term storage is considered. The fact that the thermal storage volume is not afflicted with heat losses is an important reason for connecting a heat pump with thermal storage.

In heat pump systems the low-temperature storage can be charged by heat directly from the heat source in warmer spells. An other very interesting possibility is charging by heat from subcooling the refrigerant that has just passed the condenser

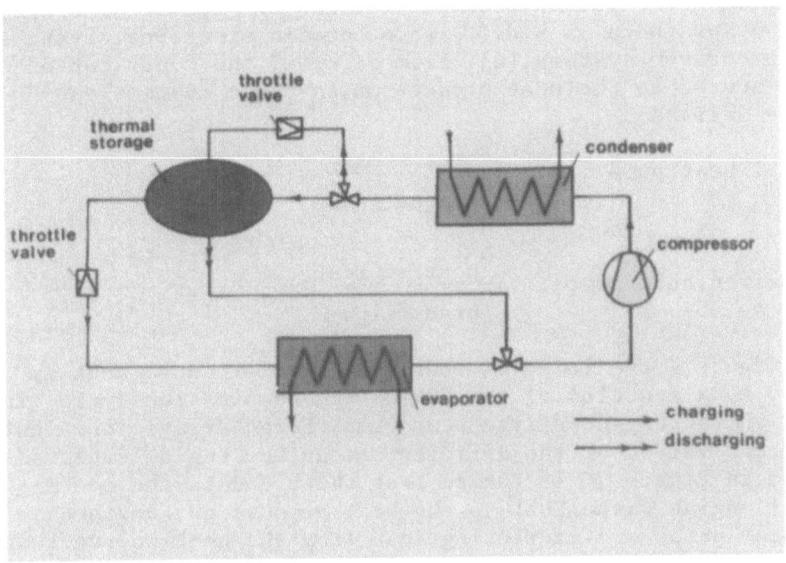

Figure 6: Heat pump with thermal storage, charged by subcooling the refrigerant

(Figure 6) (3,4,5). The condensed refrigerant releases heat to
the storage tank, when the heat pump works in the mode where the
natural heat source is supplying the low-grade heat. The liquid
refrigerant is subcooled to a temperature which will depend on
the temperature of the storage. This method of charging has the
advantage that the storage can be heated to a higher temperature
(to the condenser temperature as a maximum) than by charging
directly from the heat source. An other important advantage is
that charging by refrigerant subcooling involves no significant
extra use of drive energy, since the compressor work is practically not affected by the resulting change of the thermodynamic
cycle. To release the stored heat the liquid refrigerant is
throttled to a low pressure and then passed to the storage, where
it absorbs heat by evaporation. There are two operational modes
for this heat release. The first is that all the refrigerant
flows through the storage, which is thus representing the sole
heat source (alternative operation). An other is that the storage
supplies only the capacity deficiencies of the evaporator at low
ambient temperatures (operation in parallel). In this case much
less heat is extracted from the storage than in the former case

giving rise to a considerably smaller storage volume.

Regarding the volume required for low temperature storage for heat pumps there is a difference between electrical systems and engine-driven systems (6). From Figure 2 the contribution of the heat source to the heat production in these systems can easily be derived:

Electrical heat pump : $\dfrac{Q_{heat\ source}}{Q_{production}} = \dfrac{\varepsilon - 1}{\varepsilon}$

Engine-driven heat pump : $\dfrac{Q_{heat\ source}}{Q_{production}} = \dfrac{\eta_m \cdot (\varepsilon - 1)}{(1 - v) + \eta_m \cdot (\varepsilon - 1)}$

In Figure 7 the ratio of source heat to produced heat is presented as a function of the COP of the compression cycle. The spread with the engine-driven heat pump characteristic originates in different values of the efficiencies on the engine side, as indicated in Figure 3. It can be seen that, due to the contribution of engine waste heat to the heat production, engine-driven heat pumps are extracting less from the heat source than electrical heat pumps. Low-temperature thermal storage for comparable applications, will therefore be smaller to the same extent. The relative difference between the two heat pump types is greatest at lower COP's of the compression cycle; at increasing COP's the curves tend to approach each other. The absorption heat pump has not been included in the figure; as a heat-actuated machine it has the same characteristic as the engine-driven heat pump regarding the load on the heat source and the thermal storage, if present.

4. BIVALENT HEAT PUMP SYSTEMS

In the previous chapter low-temperature thermal storage has been brought up as a method to reduce the size of heat pumps. An other solution, which has been used far more in practice up till now, is to apply an auxiliary furnace or boiler to assist the system at low outside temperatures. In these "bivalent" systems the heat pump is comparatively small and has a capacity, which is only sufficient down to a specific temperature, the balance point. For the temperature region below this point there are two operational options:

Bivalent alternative operation. At the balance temperature the heat pump is switched off and the heat production is completely taken over by the auxiliary furnace. This mode is often favoured by the electric utilities as it contributes to curtail

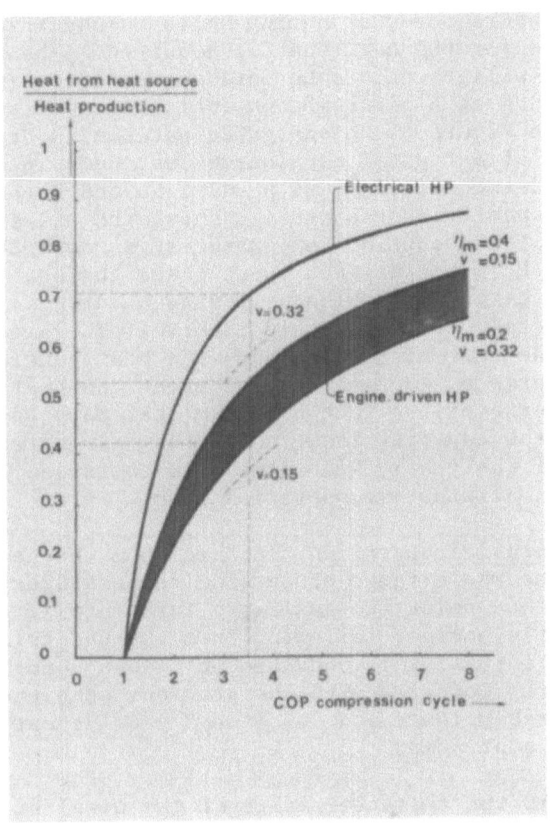

Figure 7: Relative contribution of the heat source to the
heat production of electrical heat pumps as compared with engine-driven heat pumps

the problem of the heavy load on the grid in cold spells. A
special advantage with air-source heat pumps is that the balance
temperature can be chosen such that the air cooler will never
frost. A general draw-back of bivalent systems is that a part of
the yearly heat production is not generated by the heat pump,
which will affect the energy-saving potential. In this respect
the bivalent alternative option is virtually less appealing than
other options.
At this point it is important to mention that in several West-

European countries the winters are moderate and the number of days with relatively high temperatures is rather great. So there are many hours per year during which an undersized heat pump can provide for the heat that is required. A heat pump in a bivalent mode will therefore be more energy-effective than would be concluded at a first glance from the capacity installed. As an example in Figure 8 the energetic outcome is depicted of the application of a typical air-source heat pump in a single family house under conditions, as prevail in the Netherlands (7). The heat pump capacity characteristic shows the usual pattern of lower values at lower outside temperatures, as does the COP-characteristic. P is the balance point where the heat pump capacity equals the heat demand. The bell-shaped curve reflects the heat that is needed per temperature interval. The area below this curve to the left of P represents the heat generated with the auxiliary furnace, the area to the right the heat from the heat pump. In spite of the fact that the heat pump has a capacity of only 1/3 of the capacity required at the design temperature of -12°C, 64% of the yearly heat demand is satisfied by the heat pump, even in an alternative operation mode.

Bivalent parallel operation. In this mode the heat pump is employed over the whole range of outside temperatures. To the left of the balance point the auxiliary furnace will only meet the capacity deficiencies, following from the undersizing of the heat pump. This situation is depicted in Figure 9 for the same example as before. Compared with the previous case the auxiliary heat is considerably lower and 90% of the yearly heat demand is produced by the heat pump.

In Figure 10 the characteristics of the bivalent systems, discussed here, and of a specific air-source heat pump with thermal storage at a low temperature, are plotted together. All these systems reduce the size of the heat pump which has to be installed, as indicated before. In this respect the bivalent solution is obviously more profitable than the solution based on thermal storage. The energetic merits of the three systems which are brought up in this part as examples, are presented in the table, added to Figure 10. At the lower line a coefficient is brought up, which indicates the ratio between the energy production and the primary energy involved in the various conversions. This factor includes the power plant, which is assumed to have a generating efficiency of 33%, based on the gross calorific value of the fuel. The system with thermal storage has clearly the best outcome; the bivalent parallel system as the second best, lies not far behind, however. In this case energy saving brings no sufficient arguments to justify a heat pump with low-temperature thermal storage.

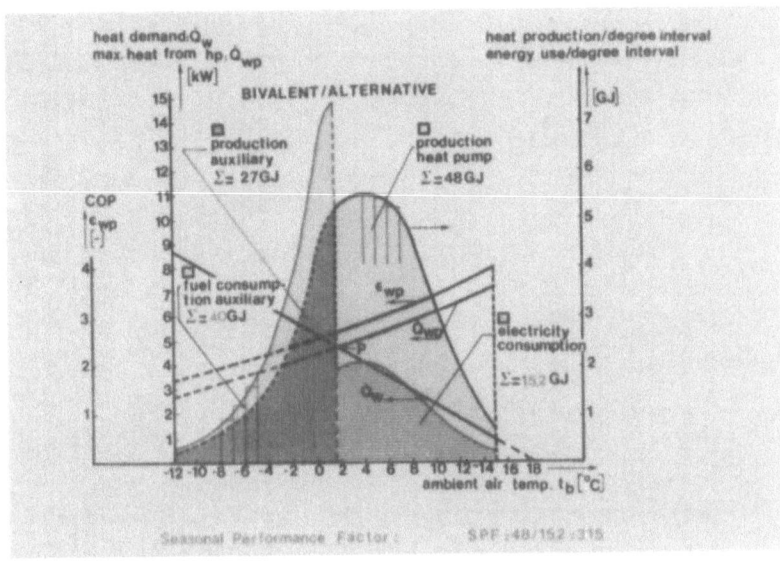

Figure 8: Example of characteristics, related to a
          bivalent/alternative electrical heat pump system

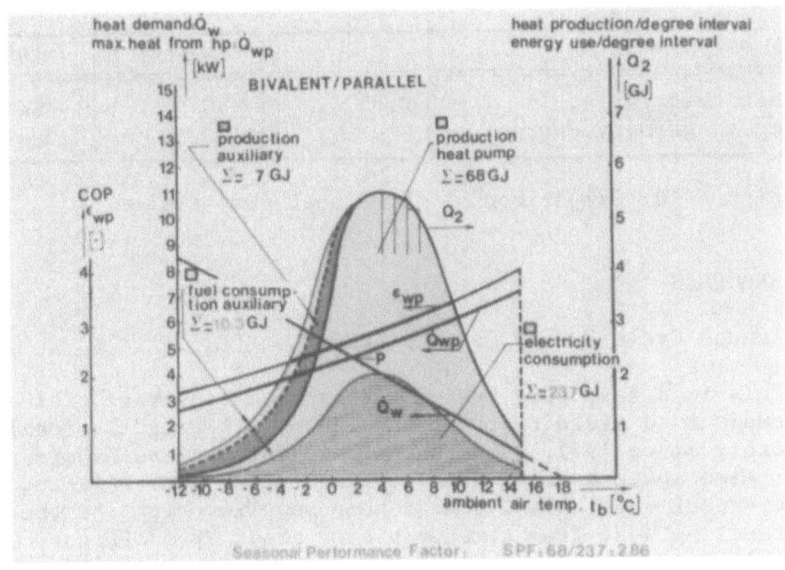

Figure 9: Example of characteristics, related to a
          bivalent/parallel electrical heat pump system

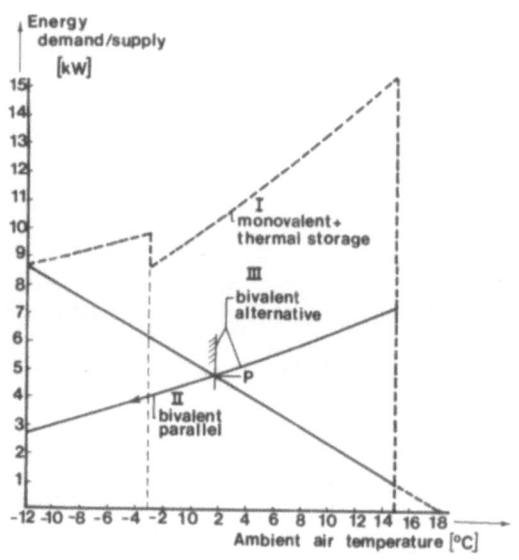

Energy Consumptions System	I	II	III
Electricity by HP	7306 kWh	6583 kWh	4222 kWh
Gas by auxiliary	-	10.3 GJ	40 GJ
System's heat ratio, related to primary energy	0.95	0.92	0.88

Figure 10: Comparison of three heat pump systems

5. EXAMPLES

5.1 Annual Cycle Energy System (ACES)

This is a heat pump system with seasonal storage, which has been subject to field tests in the USA by Oak Ridge National Laboratory since 1977. ACES is intended for premises where besides winter-time space heating also cooling during summer is required. In the experimental plant a heat pump supplies heat for the house by extracting heat from a water bin of about 70 m³ (Figure 11) (8,9,10). During winter the water is frozen; accumulated cold serves for cooling in summer. In the case that all the ice is melted, while there is still a need for cooling, the heat pump is operated during the night to freeze ice for the next day's

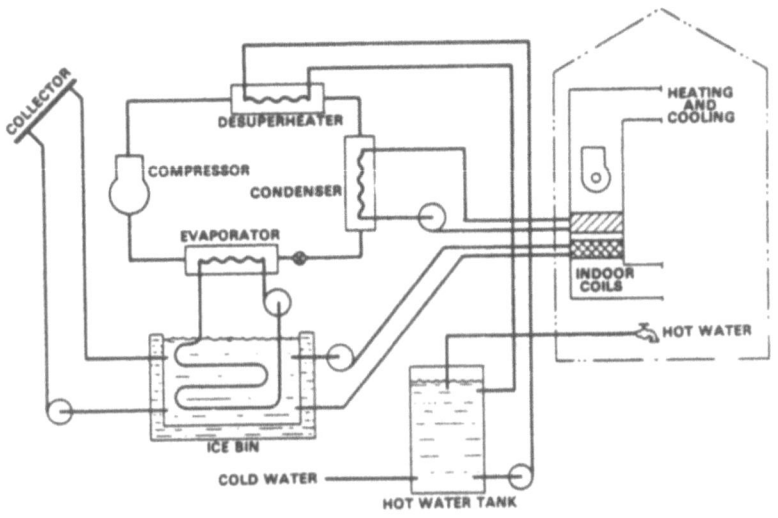

Figure 11: Simplified diagram of ACES (9)

demand. The condenser heat is then rejected through an outdoor
coil. The lower night temperatures will favour the heat pump's
COP. By this short-term storage the electricity consumption for
air conditioning is shifted to the night low-load hours, thus
reducing the summer peak load during the day. In a Northern
climate, where the heating requirement exceeds that for cooling,
the system requires a supplementary heat source. Simple, unglazed
solar panels can be used to satisfy this demand, due to the low
temperature at which the panels have to operate. Domestic hot
water is produced by desuperheating the vapour from the compressor.
With this arrangement water temperatures can be obtained exceeding
the condensing temperature of the refrigerant.
For the experimental ACES project an annual saving is reported
of about 56% on electricity against a system based on full electric resistance heat in winter and electric air conditioning in
summer. When a comparison is made with a system using an air-to-
air heat pump instead of resistance heat in the control house of
the project, a saving of 49% is found in a specific year.

5.2   Solar assisted heat pump

In countries with cloudy winters the application of solar

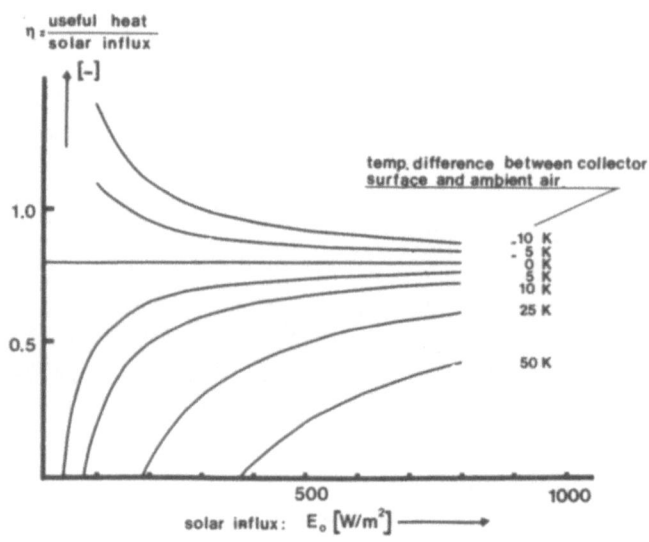

Figure 12: Efficiency characteristics of a simple solar
panel specimen

energy is hampered by the low efficiencies of flat plate solar
panels, due to the low mean radiation intensities. This situation
can be improved by introducing a heat pump in solar energy
heating systems. When a heat pump is installed between the solar
panels and the heated object, the solar heat can be captured at
lower temperatures, resulting in much better efficiencies of the
collectors (Figure 12). An important point is that the panels
can be operated at temperatures below ambient. This eliminates
the heat losses which normally occur, and consequently the construction of the panels can be simplified considerably. Based on
this an interesting development of "energy roofs" can be observed,
especially in Germany (11). In an energy roof a simple static
heat exchanging surface is incorporated, which is gaining heat
from solar radiation, but also from other sources like ambient
air and rainfall.

In solar-assisted heat pump systems thermal storage at the
low-temperature side is indispensable to cope with the highly
varying solar radiation, the nightly interruptions, and the
limited capacity of the heat pump. For reasons, mentioned in
chapter 3, also thermal storage at the high-temperature side can

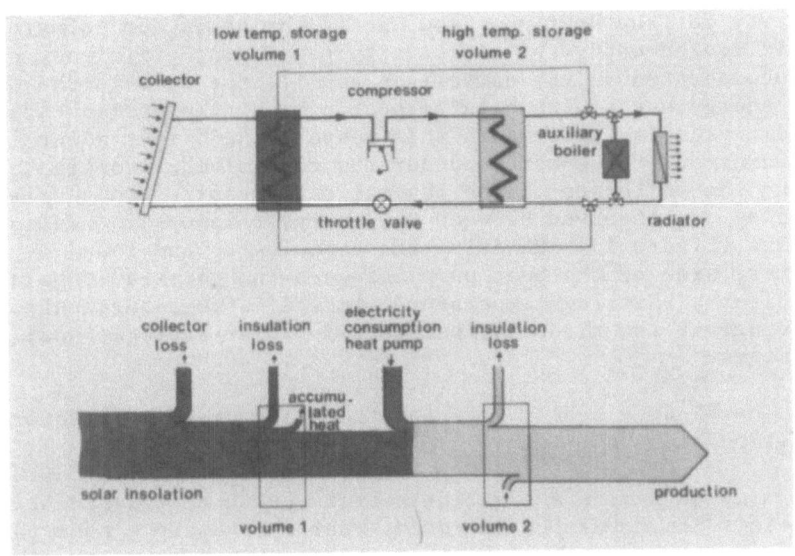

Figure 13: Solar-assisted heat pump process

be considered. In Figure 13 a simple solar-assisted electrical
heat pump system with thermal storage at the two temperature
levels is presented. It has served as a model for evaluating
the merits of a heat pump in a solar energy system for house
heating, as compared with a system using the solar energy
directly, under conditions prevailing in the Netherlands (12).
In this context it should be realized that the heat pump will
create higher collector efficiencies, as mentioned before,
giving rise to a greater influx of solar heat and thus to more
energy saving, but that, on the other hand, it will involve a
use of electric energy by the heat pump. The additional consumption of electricity is of course not found in direct solar
systems and will obviously lower the energy saving potential.

In Figure 14 the energy balance is shown for a solar system
with and without a heat pump, according to calculations with the
above-mentioned model. These figures are related to a cold, but
rather sunny February month in the Netherlands, and are the
results of computing energy flows hour by hour, using hourly
weather data as an input. It can be seen that the heat pump
brings an extra influx of solar energy of about 14% of the heat
demand, thus reducing the amount of auxiliary energy. It should

be born in mind, however, that part of this auxiliary heat is electricity for the heat pump now. So if the saving of primary energy is looked upon, the electricity generating efficiency has to be incorporated in the comparison and it appears that the primary energy use of the solar + heat pump system exceeds that of the direct solar system. In this respect the direct solar system turns out to be better under the conditions investigated. Regarding the influence of the thermal storage sizes, no dramatic differences are observed between the energy balances for various capacities (Figure 15). Shifting the capacity to the low-temperature side of the heat pump improves the solar heat gain. This indicates that low-temperature storage is more advantageous than storage at the higher temperature in solar-assisted heat pump systems.

5.3 The earth as a heat source and thermal storage volume for heat pumps

In many countries a high interest in earth heat source heat pump systems has developed in recent years. By using a heat pump seasonal storage of natural heat in the earth can be made operational in an effective manner. The thermal capacity of the earth makes, that its temperatures are attractive for heat pump applications. The subsoil heat is extracted by a heat exchanger, which is buried in the earth. The most common type is a horizontal heat exchanger which consists of a network of plastic tubes parallel to the earth's surface at a depth of 1 to 2 m. Through the tubes a fluid circulates which serves as an intermediate between the earth volume and the heat pump's evaporator. This fluid is usually a glycol/water mixture. The required earth surface area is pretty large, and amounts to 2 to 3 times the floor area to be heated in the house. The earth will recover during spring and summer through insolation, rainfall and ground water movement, depending on the underground hydrological situation.

Comparing design data of existing earth heat exchangers shows that there are appreciable differences in tube length and tube distance for about the same heat extraction .from the earth. The present design methods are obviously engaged with large margins of safety. Many companies tend to use rule-of-thumb methods (13). Because of the uncertainties in this field research has been undertaken in recent years on the transport phenomena in the earth. In the Netherlands, for instance, an extensive computer model has been developed with which the heat extraction from and the heat storage into the earth can be accurately calculated (14,15). This model is based on the application of the finite-element method. With this method a given configuration is divided into a large number of small elements. The computer program integrates, with a step-by-step method in time, the heat transfer equation among the elements. Some calculation results of this

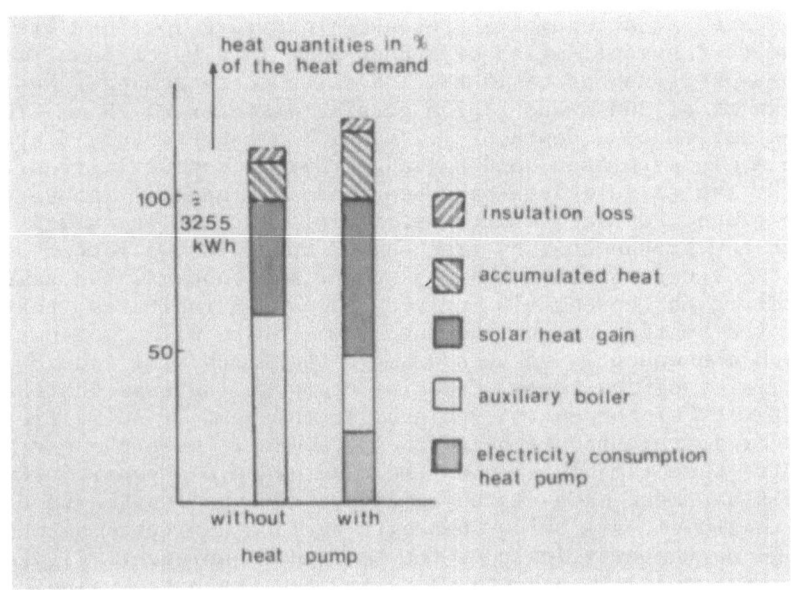

Figure 14: Example of energy balances of identical solar
systems with and without a heat pump. Low- and
high-temperature storage, both 6 m³ of water

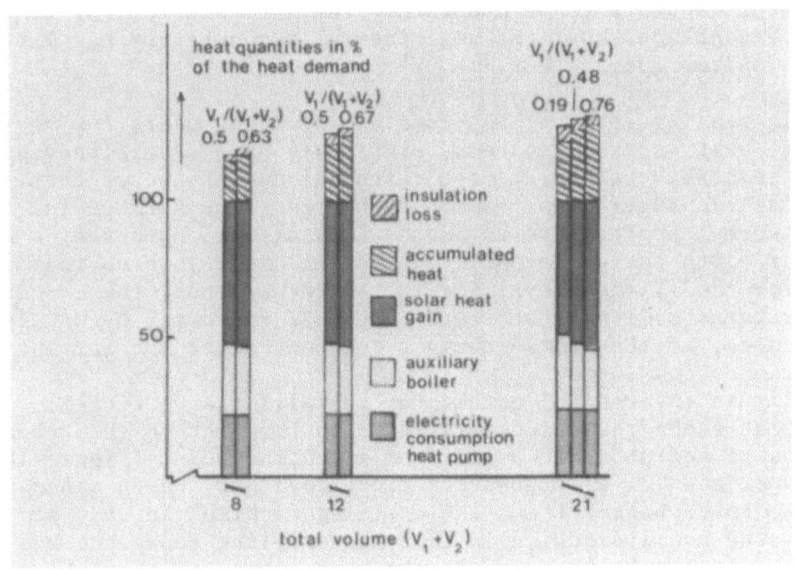

Figure 15: Influence of thermal storage capacities

program are, as an example, presented in Figure 16. They are
related to the application of a heat pump in a Dutch house with
a maximum heat demand of 10 kW. The earth heat exchanger has a
tube length of 200 m and a tube outside diameter of 25 mm. The
tube is buried at a depth of 1.2 m, with a tube-to-tube distance
of 1.6 m, in saturated sand having a thermal conductivity of
2 W/m.K. The calculations have been made for natural recovery
of the ground, as well as for artificial charging in summer
through the ground coil by free summer heat. In the latter
case it has been assumed that in summer the tube has the same
temperature as the earth's surface. Figure 16 indicates, that
during the heating season the tube temperature will drop below
0°C, and consequently in this example the earth will freeze in
the vicinity of the tubes. Freezing tends to increase the thermal
conductivity of the earth. Besides, latent heat of solidification
is set free; this has to be given up, however, when the earth is
defrosted some time later. For the same example a sensitivity
analysis has been made on the influence of tube length, tube diameter, depth of coil below the earth surface and tube-to-tube
distance on the efficiency of the heat pump equipment. Figure 17
shows the result of this analysis. In this graph the values of
the average COP of the heat pump system are expressed when
several parameters are varied in turn, keeping the others
constant. It can be seen that from the design parameters
considered, mainly the tube length has a marked influence on the
COP of the heat pump. The influence of the other design
parameters appears to be moderate. From the calculations it can
also be concluded that the soil thermal conductivity has little
effect on the average COP.

Increasing interest has come up in recent years for vertical
subsoil heat exchange systems, since they have several advantages
over the horizontal type. From calculations it appears that the
vertical exchanger requires about 40% less tube surface for the
same thermal performance as the horizontal one. Moreover, a much
smaller earth surface area is needed. It has also been found,
that for conditions prevailing in the Netherlands, the earth
temperatures will be restored fairly well in summer by natural
influences, if the heat exchanger does not stick too far into
the ground, say 10 m.
Recently an interesting method for installing a vertical
type heat exchanger was developed. With this method the tubes
are positioned under an angle from a central point (Figure 18).
The upper ends of all tubes are close together, which makes it
easy to interconnect them. The drilling machine can stay at one
point; the usual practice is that the drilling apparatus has to
be moved from hole to hole. An other advantage is that the tube
grid can be installed with a minimum of disturbance at the site,
which is obviously very important when systems are projected for
already existing premises.

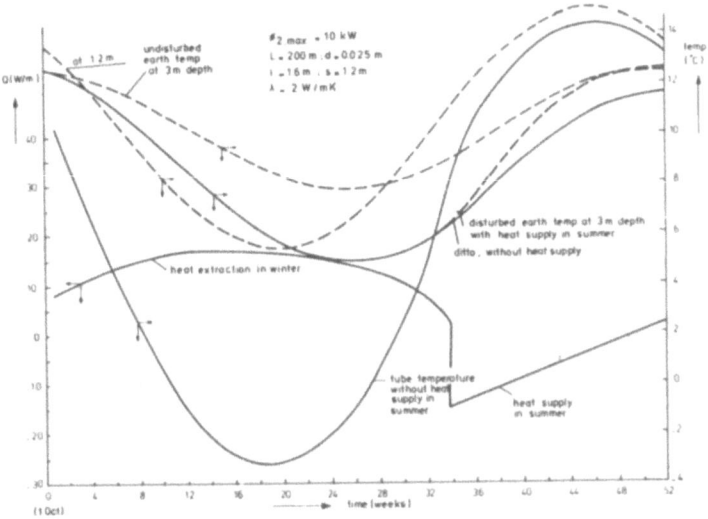

Figure 16: Computational results for the application of a
specific earth heat exchanger in the Netherlands.
Heat flow to tube and earth temperatures during
heat extraction in winter and heat supply in summer

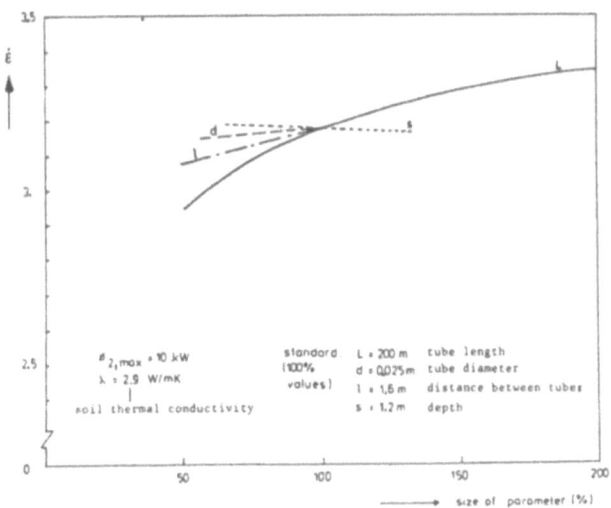

Figure 17: Influence of design parameters on average COP
of heat pump

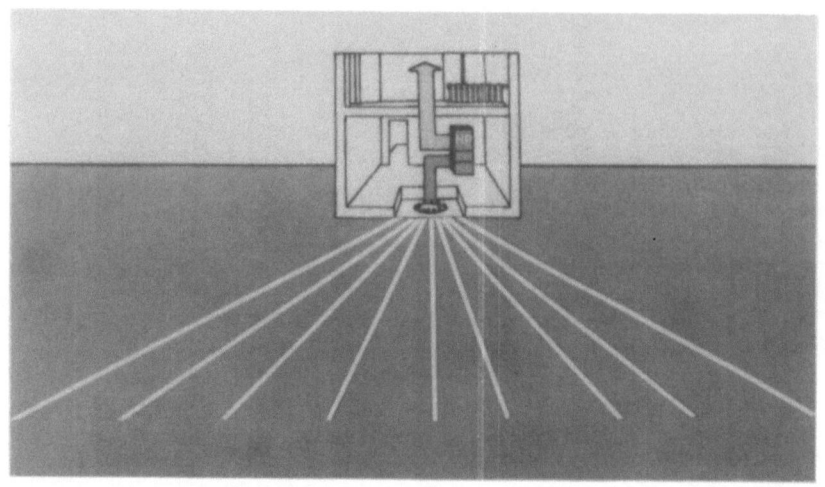

Figure 18: Vertical earth heat exchanger with tubes,
         positioned under an angle (17)

6. THE INTERMITTENT SORPTION CYCLE

In chapter 2 the continuous absorption heat pump has been
briefly described. The same principle can be applied to intermittent cycles, resulting in basically very simple machines
(Figure 19). Depending on the temperature levels of the heat
sources and sinks that are connected with the process, this
apparatus will operate as a heat pump, a thermal storage system
without a heat pump effect, or a heat transformer, by which heat
is produced at a higher temperature level than the cycle's drive
heat. In all the arrangements the cycle proceeds in two phases:
-    The desorption phase, in which the "refrigerant" is separa-
     ted from the absorbent by supplying heat to vessel I. The
     refrigerant is condensed in vessel II at a lower tempera-
     ture, during which heat is released to the system to be
     heated or to an appropriate sink.
-    The absorption phase. The liquid refrigerant in vessel II
     is evaporated and the vapour is absorbed in vessel I.
     These processes are accompanied by a heat supply and
     release, respectively.

The vessels have to be brought at the appropriate temperature levels during each phase to meet the required objective,

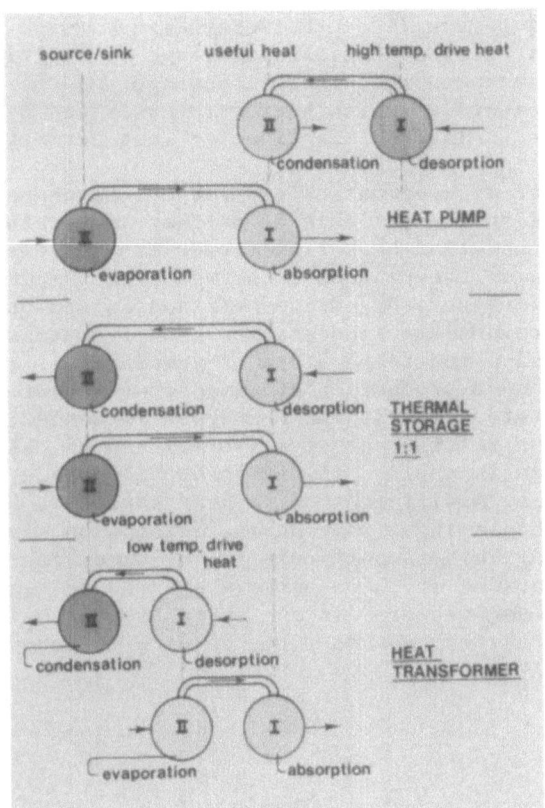

Figure 19: Various operations with intermittent sorption
process

as indicated in Figure 19. The three cases are obviously all
energy-storage systems, but the heat pump includes an extra heat
production of useful heat and the heat transformer the conversion
of low-temperature drive heat to useful heat of a still higher
temperature. The attention is called to the fact that in the heat
pump case useful heat is produced in the desorption phase as well
as in the absorption phase from refrigerant condensation and
absorption, respectively. Intermittent sorption heat pumps therefore do not give rise to an interrupted heat production.
Where in the second case the intermittent sorption cycle is used
for thermal storage only, the heat of condensation which is released during the desorption phase, is equal to the heat that is
needed for evaporation during the absorption phase later on, if
thermal and thermodynamic losses are neglected. An appropriate

thermal storage volume will be the interfacing system at the condenser/evaporator side. A clear advantage of applying the sorption process between this thermal storage and the heat user is the low temperature at which heat surplusses can be stored now, which is particularly attractive for long-term storage.

Where heat of more elevated temperatures is needed, and, on the other hand, heat is available of a lower temperature, the application of a sorption heat transformer can be considered. In the desorption phase the process is between the temperature levels of the low-temperature drive heat and an appropriate heat sink, which can remove the condensation heat from vessel II. For the absorption phase the temperature of vessel II is raised, causing a useful heat production at temperatures over the drive-heat-temperature level. The heat that is needed for the evaporation of the refrigerant in vessel II can be taken from the same source as is applied in the desorption phase, but also from other sources. A different drive-heat source is necessary, if the drive-heat playing a role during desorption, is not continuously available. This is particularly the case with solar-activated systems. One of the sources where the heat supply to vessel II can be derived from is the system itself, since part of the heat production from vessel I can be used to feed vessel II

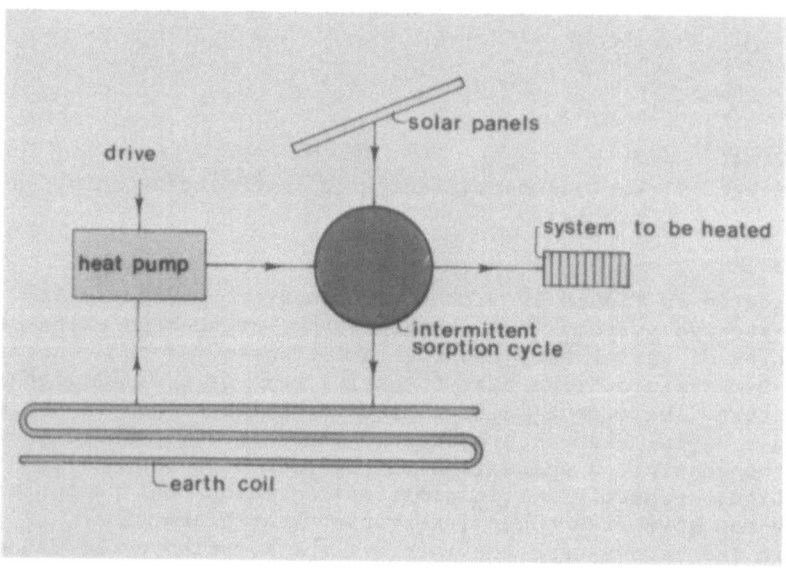

Figure 20: Intermittent sorption heat transformer on solar energy, assisted by a heat pump

(16). Of course this will lower the efficiency of the total cycle. An other possibility is supplying heat by an additional heat pump, which takes heat from the low-temperature volume, where heat is released in the desorption phase (Figure 20). The system in this mode represents an electrical heat pump with a partial sorption topping cycle. The heat pump can also be used to supply heat to vessel I in order to charge the system during the night. The electricity use can so be shifted to low-load hours.

7. REFERENCES

(1) Commission of the European Communities - "Catalogue of heat pump R, D and D projects", September 1979

(2) Kirn, H.; Hadenfeldt, A. - "Wärmepumpen", Müller, Karlsruhe (1976), ISBN3-7880-7070-6

(3) Cube, H.L. v. ; Steimle, F. - "Wärmepumpen, Grundlagen und Praxis", VDI, Düsseldorf (1978), ISBN3-18-400345-0

(4) Kirn, H. - "Deckung der Leistungsspitzen von Wärmepumpen-anlagen durch Latentspeicher", Elektrowärme International 33 (1975) A6, S. A284-A287

(5) Fluck, D. ; Kirn, H. - "Untersuchung einer monovalenten Aussenluft-Wasser-Wärmepumpe mit Latent-Speicher", KI Klima + Kälte-Ingenieur 12 (1977), S. 439-444

(6) Weissenbach, B. a.o. - "Energiespeicher in der Wärme- und Stromversorgung", Bundesministerium für Forschung und Technologie, Bonn (1978)

(7) Ree, H. v.d. - "De warmtepomp; algemene aspecten, toepassing en ontwikkelingen", Klimaatbeheersing 10 (1981), nr. 4, pg. 199-211

(8) Gilli, P.V. ; Schabkar, F. ; Halozan, H. - "Verringerung des Energieaufwandes mit Hilfe von Wärmepumpen", Verlag für die Technische Universität Graz (1978), ISBN3-7041-0007-2

(9) International Energy Agency - "Heat pump systems", Final annual report (1979/1980)

(10) Fischer, H.C. - "Seasonal ice storage for domestic heat pumps", International Journal of Refrigeration, Volume 4, nr. 3 (1981), pp. 135-138

(11) RWE Beratung - "Energie-Dach/Fassade/Zaun/Stapel", RWE informiert Nr. 166 (1980)

(12) Ree, H. v.d. - "De warmtepomp in verwarmingsinstallaties met zonne-energie", Klimaat, elementen, installaties, IG-TNO (1975), pg. 33-42

(13) Berntsson, T. a.o. - "The use of the ground as a heat source for heat pumps in urban areas", Swedish Council for Building Research, Stockholm, Sweden (1980), ISBN91-540-3357-8

(14) Nievergeld, P.G.M. - "Investigations on using the earth as a natural source for heat pumps", Final Research Report for the European Community (1978)

(15) Nievergeld, P.G.M. - "Investigations on using the earth as a heat storage medium and as a heat source for heat pumps", Nordic symposium on earth heat pump systems, Göteborg, Sweden (1979)

(16) Doroszlai, P. - "Möglichkeiten der Steuerung des Stromverbrauches von Wärmepumpen durch den thermochemischen Niedertemperatur-Speicher "PAWA" ", Elektrizitätsverwertung Jg. 53 (1978), Nr. 3, S. 27-33

(17) Commercial documentation

SOLAR HEAT STORAGE: COMPARISONS AND GOALS

Carlo Mustacchi° and Vincenzo Cena°°

°Chem. Engng Dept., Univ. of Rome, Italy
°°ADES, 23 via dei Giubbonari, 00186 Rome, Italy

ABSTRACT

A number of options for storing solar heat are presented (water tanks, pile-up of rocks, soil, solar ponds, latent heat systems). The applicability of such options to hourly, daily, seasonal storage and the relevant economic pictures are discussed.

INTRODUCTION

Most heat-intensive applications require a relatively constant heat input. On the other hand, solar radiation is a highly variable energy source. Further, this source is partially unpredictable inasmuch as the weather vagaries are intrinsically non-reproducible. The usage of heat may be occasionally out-of-phase with the source, as in the case of space heating where the night and the winter requests are opposite to the peaks of available incident radiation. Another example is that of the solar desalination plants which cannot operate with a variable heat input because the levels, pressure drops and thermal gradients in the system must be kept constant not to degrade the operational efficiency.

Solar electricity generation must also follow a time variation compatible with power usage and quite dissimilar from the availability of solar radiation.

In what follows we shall review the description of the availability of solar energy, the options for storing solar heat and the costs of these options.

SOLAR ENERGY AVAILABILITY

On a given surface of fixed orientation as a function of the latitude, the total incident radiation is in the range 1000 to 2000 kWh/m² year. However, the monthly intensity in summer may be 3 to 5 times higher than in winter. This radiation is available during the day with a mean duration 12 hours but is shortest at the winter solstice and longest at the summer solstice. Radiation is nil at sunrise and sunset and peaks at noon, with values close to 1 kW/m² for clear days.

If the atmosphere were completely transparent, the instantaneous incident radiation on a plane tilted at an angle equal to the latitude and facing South could be estimated by means of

$$H = 1353 \cos \delta \; \cos \omega \tag{1}$$

where H is the incident radiation in W/m², δ is the solar declination given by

$$\delta = 23.45° \sin(0.986(284+N)) \tag{2}$$

for the N-th day of the year, and ω is the hour angle given by

$$\omega = 15°(h-12) \tag{3}$$

at hour h.

The value estimated with (1), (2) and (3) must be curtailed by the effects of atmospheric capacity due to dusts, ozone, carbon dioxide, water vapor, air, clouds. The multiplying factor K_T for H is a stochastic variable with mean value in the range 0.4 to 0.5 and istantaneous values which may range from zero (dark overcast sky) to slightly above 1 (sunny day with white clouds on the horizon). The statistical properties of K_T are such that it exhibits a peak of frequency around the value 0.6-0.7 and a flat distribution elsewhere. More details can be found in the Appendix.

When designing or evaluating a storage system, it must be kept in mind that the values of K_T exhibit a strong measure of "memory" : the K_T of a given instant has about 80% probability of reproducing K_T of one hour before and only about 20% probability of moving to a higher or lower value. This is to say that bad weather and clear weather come in runs having a much longer duration than those which could be expected by a merely random sequence of K_T's. This again is a reason for providing in most cases a storage capacity much in excess of whatever is required by the alternating effect of day and night.

STORAGE OPTIONS

The options for storing heat can be classified by means of their operating principle into sensible, latent, chemical storage systems. Of these, only the first category is being commercially used outside of laboratories or small scale prototypes.

Sensible heat implies exploiting a medium which is raised to a high temperature whenever excess heat is available and cooled to retrieve thermal power whenever needed. Whatever medium is chosen for this purpose, it will require a containment and a heat transfer surface. The cost of these items will often be the economic limiting factor towards each specific application. The main types of sensible heat storage are water tanks, pebble beds, buried-pipe soil storage, solar ponds. In the case of latent heat storage, the few prototype applications have used wax, tar and ice with a melting-freezing cycle.

As to the chemical storage, notwithstanding the hundreds of well-known reversible chemical reactions, very few systems have been implemented in actual practice, and these make use of salt hydrates and some coordination compounds.

WATER TANKS

Steel, masonry or concrete tanks have present costs in the range $30 \div 100$ $/m^3$, the higher range belonging to containers of a few m^3 and the lower range a few thousands m^3. If the medium is water, an amplitude fluctuation ΔT enables the tank to store 4.2 ΔT MJ/m^3. Neglecting the heat losses and assuming a depreciation 10% per year, the cost of storing 1 MJ would be of the order of $1/\Delta T$ $ per year. Since the cost of fossil fuels is close to 0.3 $/kg, i.e. 0.007 $/MJ, the water tank would pay for itself only if the turnover of heat is at least $1/0.007$ $\Delta T = 14$ times a year with a $\Delta T = 10°C$.

This implies that any storage tank using water to store heat for a period longer than twenty days is an inconvenient proposition. Even if excess heat is available for free, it still pays to dispose of it and resort to fossil fuels whenever needed. This again is true under the very favorable assumptions that the heat transfer surface, the pumping and circulating systems have no cost and that losses are made neglegible by using the appropriate amount of thermal insulation.

On the other hand, there may be applications where temperature fluctuations larger than 10°C are tolerable and in these cases the range of applicability of hot water tanks is extended.

A specific example of a water storage for space heating applications is a follows. Assume that a house having a volume 500 m^3 with losses of 200 W/°C (good insulation) and one change of air per hour, must be heated in a typical Southern European location to a constant temperature of 18°C. The heat requirements would be 65.2 GJ/year. Using 20 m^2 of flat plate collectors and a cylindrical storage tank (9.5 m dia., 4.2 m depth, volume 300 m^3) the heat collected would amount to about 86.8 GJ/year, the fossil fuel makeup required would be 10.4 GJ and the water temperature would fluctuate between 40 and 60 °C.

The yearly costs obtained by depreciating the collectors and tank and adding the cost of makeup fuel would total an expense of about 1200 $/year, i.e., over twice the cost of a conventional all-fossil heating system. This solution makes use of three or four meters of earth around the tank as a storage medium. However, the heat losses to the ground have a very slow approach to steady state, so that during the first year of operation the fossil makeup fuel is three times higher than that mentioned above.

PILE-UP OF ROCKS

In this case an underground space filled with rocks is used to store heat.

The space for the rock storage can be found naturally (small creek) or it can be provided in a flat land by removing the top layer of earth and piling it in two embankments. The space between these embankments is to be filled with rocks and covered by insulation and by an impermeable layer.

This kind of storage could be combined with earth storage by digging trenches in the bottom of the storage area and piling the excavated earth between the trenches. The latter would be filled with rocks to form vein-like configurations within the earth. This version reduces the required volume of rocks and thus can be less expensive.

The top layer of the storage, sloping to the south, would serve as location for air-type solar collectors. The air from these collectors is blown in summer into the bottom of the rock-filled space (or into the bottom of the trenches) by perforated piping thus heating the whole storage space by convection. In winter hot air from the top of the storage space is blown to the buildings to be heated. The pebble diameter is related to the mean retention time for heat. A daily storage will require 2-3 cm dia. pebbles, whereas a seasonal storage may use 30-50 cm dia. blocks. To guarantee an appropriate heat exchange (5 W/m^2°C) the air velocity must be a few cm/sec through the packed bed. This entails a pressure drop of about 50 kinetic heads for each layer of bed 1 pebble high.

For estimating purposes, the containment for a pebble bed will cost about one half of equal-volume water tank, i.e. 15 to 50 $/m³, to which 5 to 10 $/m³ for the packing must be added. Since the heat cumulated per unit volume is only one third of that cumulated in water, with equal temperature fluctuations, it can be seen that the economics of pebble beds is quite comparable to that of water basins. It is economically unappropriate for seasonal storage, but convenient on a few days' turnover basis. If anything, the greater cost of a pebble bed per unit heat stored is compensated by the lower cost of air collectors and of air distribution system if the dwelling is provided with warm air space heating.

STORAGE IN THE SOIL

The type of storage appears at present one of the most promising for longterm storage and therefore a more detailed assessment is in order.

Typical properties of ground relevant to this technology are the following:

Thermal conductivity of ground K:

    dry 0.6 ÷ 1   W/m°C
    wet 1.5 ÷ 3   W/m°C

Heat capacity $c\rho$:

    dry 1.5 ÷ 2 x 10^6 J/°C m³
    wet 2   ÷ 4 x 10^6 J/°C m³

Thermal diffusivity α :

    dry 4   ÷ 5 x 10^{-7} m²/sec
    wet 7   ÷ 8 x 10^{-7} m²/sec

The depth of penetration of a heat wave in the soil is of the order of $(\alpha t)^{\frac{1}{2}}$, and with times of the order of six months (i.e. 1.55 x 10^7 sec) this penetration turns out to be 3 to 4 meters.

This implies that in order to exploit the ground as a storage medium it is sufficient to position heating or cooling pipes 3 to 4 m apart in the soil itself.

These pipes may take the shape of vertical U-tubes through which hot water is circulated in the sunny season. Whenever a heat request is to be met, cold water can be circulated in the same U-tubes and after upgrading its temperature level by means of a heat pump, the heat can be delivered to satisfy the specified requirements. Assuming a typical cost for small buried pipes (2 x 1" dia.) of 10 $/m and a tube pitch of 3 m, the cost of 1 m^3 of soil storage is of the order of 1 $/m3.

Even though the volume heat capacity of water is over twice higher than that of soil, which entails requirement of 2 m3 of soil to substitute 1 m^3 of water, the cost of the storage turns out to be at least 20 times less than water storage.

To give appropriate orders of magnitude, the heat usage of 1500 m^2 of household is about 1000 GJ per year. Assuming a peak-peak temperature fluctuation of the soil of 20°C, the volume of storage required is

$$10^{12}/2 \times 10^6 \times 20 = 25000 \text{ m}^3$$

If the shape of the storage is that with the minimum ratio surface to-volume, i.e. a hemisphere, it will be :

$$R \cong 23 \text{ m} \tag{4}$$

A more detailed evaluation shows that only 1/3 to 1/2 of this storage capacity is actually necessary, that is a reservoir with a radius of 15 m is sufficient for this type of housing.

A number of important decisions must be taken :
a) what is the relevance of heat losses
b) what is the effect of the transient periods
c) what is a good choice for mean storage temperature and its amplitude fluctuations
d) what is the effect of water table or moisture in the soil.

Points a) and c) are actually related. The istantaneous heat losses from a hemisphere are

$$Q = 2 \quad \pi \ R \ K(T - T_\infty) \tag{5}$$

where T is the warm soil temperature and T_∞ is the undisturbed ground temperature far away from the reservoir. In most European locations, T_∞ being equal to the mean annual temperature will range between 9°C (Holland) and 17°C (Sicily).

For the purpose of establishing orders of magnitude, assume T = 12°C, K = 1 W/m°C and c_ρ = 2 x 10^6 J/m3 °C. The yearly losses (one year = 3.1 x 10^7 sec) are 1.95 x 10^8 R(T - 12). On the other hand, with an amplitude fluctuation F of the ground temperature (°C), the amount of energy stored and retrieved annually is, in J/year,

$$\frac{4}{6} \ \pi \ R^3 \ F \ c \ \rho = 4.2 \times 10^6 \ R^3 \ F \tag{6}$$

The percentage losses in a year are thus

$$1.95 \times 10^8 R(T-12)/4.2 \times 10^6 R^3 F = 46(T-12)R^2 F \qquad (7)$$

If we accept an annual loss not exceeding 20%, we shall have

$$0.2 \leq 46(T - 12)/R^2 F \qquad (8)$$

from which we have

$$R \geq 15 \left((T - 12)/F\right)^{\frac{1}{2}} \qquad (9)$$

Thus, with a soil temperature between 30° and 60°C,

$$T = (60 + 30)/2 = 45°C$$
$$F = 60 - 30 = 30°C$$
$$R = 15.9 \text{ m}$$

and an acceptable soil storage must have a diameter of the order of at least 32 m.

This size of storage is sufficient for 1500 to 2000 m^2 of housing floor, i.e. 10 to 20 apartments. It is seen that this solution is not acceptable for single-family houses.

On the order hand, with the same F and T = 20°C, R must be above 7.8 m, that is the minimum storage size is about 1000 m^3 of earth which can store about 60 GJ, just about right for a single-family house.

However, with this solution the ground reservoir would fluctuate between 5° and 35°C.

Since the house would require heat at a temperature level of at least 30°C, a heat pump would be necessary to raise the temperature level of the stored heat.

As to question b), it can be shown that the thermal losses during the first t seconds of operation of the reservoir can be estimated from

$$L = 2 \pi R K (T-T_\infty)(1 + 2 R/(\pi \alpha t)^{\frac{1}{2}}) \qquad (10)$$

Referring to the previous example, with T = 20°C, T_∞ = 12°C, R=8 m, K = 1 W/m°C, α = 5×10^{-7} m^2/sec, and using t = 3.1×10^7 sec (first year of operation), one finds L = 1324 W. In fact the mean losses during the first year are seen to be about 3 times higher than the steady state losses and it will take a few years before the storage performs in its appropriate range. Question d) has been investigated in detail and the results are that a moving water table, even at speeds of a few millimeters per day, render the ground storage unacceptable. As to moisture, it has the effect of providing the ground with a higher apparent heat capacity and a slightly lower diffusivity with no major overall effect on the storage behaviour.

SOLAR PONDS

These devices are water pools into which an artificial density gradient is maintained by means of a solute to impede convective losses from the bottom towards the water surface. These pools are a few m deep and their darkened bottom surface behaves as a collecting surface for the short wavelength radiation which penetrates the water.

A high salt concentration in the bottom layers and the use of virtually pure water in the top layers stabilizes the system and allows the attainment of bottom-layer temperatures as high as 90°C. The investment cost can be very limited. It is claimed that with excavated soil lined with a plastic or rubber layer, $20 \div 30$ $/m^2$ of basin is an appropriate range. Here again all the considerations made for the case of water tanks apply; however, the striking difference is due to the fact that this particular type of storage device is at the same time a collecting device for solar radiation. With a mean cost of flat plate collectors close to 200 $/m^2$, the storage per se can be considered as having a very small or even a negative cost.

The main limitations of this device are due to the initial inventory and eventual make-up of salt. The quantities of salt stored in one m^2 of basis are of the order of 300 kg. Using the cheapest salt at mean European prices (Mg Cl_2, 0.15 $/kg) this alone would entail an additional investment of 45 $/m^2$. This, of course, except for the very limited case of installations implemented in the sites where marine salt is produced and Mg Cl_2 is a value-less discardable by-product.

Other problems, such as maintaining the appropriate transparency of water and providing the salt and freshwater make-up, add to the difficulties of this system, so that the optimistic assessment of its potential must be accepted with caution.

LATENT HEAT STORAGE

Latent Heat

In principle, a latent heat storage operates on the basis of a transition from a state A to a state B involving a large absorption or release of heat. This transition must be reversible.

The transition itself can be a change of phase such as :
- a solid-solid transformation
  (diaminopentaerythrol transforms at 69°C by absorption of 184 kJ/kg)
- a solid-liquid transformation (melting or fusion) (e.g. ice melts at 0°C by absorption of 334 kJ/kg)
- a liquid-gas transformation (vaporization) (e.g. water vaporizes at 100°C absorbing 2260 kJ/kg)
- a solid-gas transformation (sublimation)(e.g. iodine sublimates at room temperature absorbing 245 kJ/kg).

Theoretically, at one given pressure of the container, the forward and backward transition happen at the same temperature, if the compound is a simple and pure chemical compound.

Two phenomena alterate this property in real systems:
- if the compound is not pure or it is a mixture of compounds (e.g. wax, tar) the change of phase spans a range of temperatures
- if the compound is exceedingly pure the change of phase is delayed, for lack of "nucleation centers".

In actual solar practice, both with pure or mixed compounds a certain amount of solid impurities is added (dust, sand, graphite) to provide nucleation and obtain an essentially reversible system.

To all practical purposes, therefore, a system used to store latent heat is caracterized by the following data:
- initial and final transition points T_1, T_2
- latent heat of transition L
- heat capacity before and after transition

$$(C \rho)_1 \quad \text{and} \quad (C \rho)_2$$

Criteria for Selecting the Latent Medium

They fall into four categories:
a - thermodynamic
b - kinetic
c - chemical
d - economic

(a) is the requirement that T_1 fall at least slightly above the temperature at which heat must be delivered. Thus, for space heating purposes without a heat pump, T_1 must exceed 20°C. On the other hand T_2 must be at least slightly below the temperature at which the collectors operate. If T_2 is chosen to be 40°C, for instance, the collectors stop loading the system if the incident radiation falls below 150 W/cm^2; the other thermodynamic requirement is that the material should have a congruent, reproducible melting point.

The material must melt completely so that the solid and liquid phases are identical in composition. Otherwise a segregation will occur, resulting in composition changes after cycling, with a complicated, often unpredictable behaviour (zone refining).

(b) The kinetic requirement is that the material must not supercool, and this is obtained, as mentioned by providing a suitable type and amount of impurities.

(c) The material should be stable for a period of at least 20 years, should not interact with the container and should not be dangerous in case of leakage or fire.

(d) The material should be cheap and available.

Very few materials survive the screening of criteria (a) to (d).

Some of these, as examples, are shown in Table 1.

Physical Formulation

To simplify the study of propagation of the fusion front in a solid, call:

ρ the density   , kg/m^3
C_1 the specific heat of the solid phase, $J/kg°C$
K_1 the thermal conductivity   "   "   , $W/m°C$
α_1 the thermal diffusivity   "   "   , m^2/sec

The corresponding quantities for the liquid phase will have a subscript 2 (C_2, K_2, α_2). Suppose also that L is the latent heat of fusion (J/kg) and T_1 the melting point. Call X (t) the position of the surface of separation between liquid and solid. Call v_1 and v_2 the temperature of the solid and liquid phase, respectively.

The equations to be satisfied are:

$$\nabla^2 v_1 = \frac{1}{\alpha_1} \frac{\partial v_1}{\partial t} \tag{11}$$

$$\nabla^2 v_2 = \frac{1}{\alpha_2} \frac{\partial v_2}{\partial t} \tag{12}$$

i.e. two Fourier equations, one for each phase with the following boundary conditions:
- at the interface, temperature is at the melting point

$$v_1 = v_2 = T_1 \quad \text{for} \quad x = X (t) \tag{13}$$

Table 1

Material	Paraffins	$Na_2S_2O_3 \cdot 5H_2O$ Sodium tiosulphate	$Na_2HPO_4 \cdot 12H_2O$ Sodium phosphate	$Cacl_2 \cdot 6H_2O$ Calcium chloride	Tar
Freezing range (°C)	8÷70	48	35	29	50÷120
Density (kg/m^3)	780÷920	1700	1520	1500	900÷1000
Specific heat ($J/kg°C$)	2000	1500	1700	1600	2000
Latent heat (J/kg)	$1.5÷2x10^5$	$2.0x10^5$	$2.8x10^5$	$1.7x10^5$	$1.6x10^5$
Conductivity sol/liq ($W/m°C$)	0.24 0.05	0.47	0.51	0.5	0.74 0.1
Cost ($\$/kg$)	0.2	0.8	0.9	0.03	0.1

- at the interface, the difference of heat flows is due to liberation of latent heat

$$K_1 \frac{\partial v_1}{\partial x} - K_2 \frac{\partial v_2}{\partial x} = L \rho \frac{dX}{dt} \tag{14}$$

Equations (11) and (12) must be solved with conditions (13) and (14) and the initial and additional boundary conditions appropriate for the case on hand.

The simplest and most famous particular case is Neumann's solution, in which the region $x > o$ is initially liquid at constant temperature V with the surface $x = 0$ kept at zero at all times. In this case the additional boundary conditions are

$$v_1 = 0 \quad \text{for} \quad x = 0 \tag{15}$$

$$v_2 = 0 \quad \text{for} \quad x = \infty \tag{16}$$

The initial condition is

$$v_1 = v_2 = V \quad \text{for} \quad t = 0 \quad \text{and every} \quad x. \tag{17}$$

The analytical solution for this case in obtained by solving the equation

$$\frac{e^{-\lambda^2}}{\text{erf }\lambda} - \frac{K_2 \, \alpha_1^{1/2} (V-T_1) e^{-\alpha_1 \lambda^2 / \alpha_2}}{K_1 \, \alpha_2^{1/2} T_1 \text{erfc }\lambda \, (\alpha_1/\alpha_2)^{1/2}} = \frac{\lambda \, L \, \pi^{1/2}}{C_1 \, T_1} \tag{18}$$

Once λ is obtained, the solutions are:

$$v_1 = \frac{T_1}{\text{erf }\lambda} \text{ erf } \frac{x}{2(\alpha_1 t)^{1/2}} \tag{19}$$

$$v_2 = V - \frac{V - T_1}{\text{erfc }\lambda \, (\alpha_1/\alpha_2)^{1/2}} \text{ erfc } \frac{x}{2(\alpha_2 t)^{1/2}} \tag{20}$$

$$X = 2 \, \lambda \, (\alpha_1 t)^{1/2} \tag{21}$$

An important special case is that in which the liquid is initially at its melting point. Then

$$V = T_1$$

and equation (18) reduces to

$$\lambda \, e^{\lambda^2} \text{ erf }\lambda = \frac{C_1 \, T_1}{L \, \pi^{1/2}} \tag{22}$$

A few values for the function $\lambda \, e^{\lambda^2} \text{ erf }\lambda$ are given below.

λ	$\text{erf }\lambda$	$\lambda \, e^{\lambda^2} \text{ erf }\lambda$
0.0	0.	0.
0.1	0.1125	0.0114
0.2	0.2227	0.0464
0.3	0.3286	0.1079
0.4	0.4284	0.2011
0.5	0.5205	0.3342
0.6	0.6039	0.5194
0.7	0.6778	0.7745
0.8	0.7421	0.1259

With paraffin wax, for instance :

ρ = 850 kg/m^3
C_1 = 2000 J/kg°C
K_1 = 0.24 W/m°C
T_1 = (melting-surface temp.) = 60-40 = 20°C
L = 1.7 x 10^5 J/kg
α_1 = $K_1 / \rho C_1$ = 1.4 x 10^{-7}m^2/sec

Equation (22) gives

$$\lambda e^{\lambda^2} \text{erf} \lambda = 0.13$$

and from the above table λ = 0.33, erf λ = 0.36

From (21)

$$X = 2 \times 0.33 \ (1.4 \times 10^{-7} t)^{\frac{1}{2}} = 2.5 \times 10^{-4} t^{\frac{1}{2}}$$

In one night (43200 seconds) a solidification front advances
5.2 cm. A paraffin slab which must buffer a night's load of solar
heat must therefore be of a thickness of the order of 5 cm.

An important point is to be made.

Assume that latent heat is to be stored in molten paraffin wax.
Each m^2 of heat exchange surface between the hot collector water
and the paraffin my cost as much as 100 $. This m^2 will provide
the heat input to a slab having a volume 0.05 x 1 m^3 of paraffin,
worth about 8.5 $.

This shows that in most cases the problem of storing latent heat
is not the latent medium, but the cost of heat transfer surface.

Numerical Approach

With any geometry different from the flat slab and any boundary
or initial conditions more complicated than those of the preceding
paragraph, the analytical approach is difficult if at all possible.

An example will be given here of numerical treatment for one such
case.

Assume that in a reservoir where water is circulated the thermal
capacity is increased by packing in it paraffin balls having 3 cm
dia. in a suitable spherical plastic container. The range of fusion
of the paraffin is from 60° to 80°C, the specific heat 2000 J/kg°C,
the latent heat 1.7 x 10^5 J/kg, the density 850 kg/m^3, the thermal
conductivity 0.24 W/m°C, diffusivity 1.4 x 10^{-7} m^2/sec. The paraffin
is initially molten (at 80°C) and water is circulated at 40°C. It

is required to estimate the time for the complete solidification of the wax balls.

The Fourier equation in spherical geometry is written

$$\frac{\partial^2 T}{\partial r^2} + \frac{2}{r} \frac{\partial T}{\partial r} = \frac{1}{K} \frac{\partial H}{\partial t} \tag{23}$$

where T = temperature   °C
        r = radius   m
        K = thermal conductivity W/m°C
        H = enthalpy J/m^3

The enthalpy in the range 40-60°C will be given by

$$H = c \rho \ (T - 40) = 1.7 \ (T - 40) \tag{24}$$

If a reference state with zero enthalpy is chosen for the solid at 40°, in the range 60 ÷80°C the enthalpy is

$$H = 2 \times 0.85(60-40) + 0.85 \times 170(T-60)/(80-60) =$$
$$= 34 + 7.23 \ (T-60) \tag{25}$$

Divide the ball into 5 concentric spherical shells and locate point 1 to 6 at the boundary between each shell and the next. Equation (23) written around the k-th point will be:

$$\frac{T_{k+1} + T_{k-1} - 2 \, T_k}{s^2} + \frac{2}{(k-1)s} \frac{T_{k+1} - T_k}{s} = \frac{1}{K} \frac{H'_k - H_k}{\Delta \tau} \tag{26}$$

where H'_k is the enthalpy of shell k after $\Delta \tau$ seconds have elapsed. In our case, since s = 0.003 m and K = 0.24 W/m°C, solving for H' we obtain:

$$H'_k = H_k + \frac{0.24}{(0.003)^2} \ (T_{k+1} + T_{k-1} - 2T_k + \frac{2}{(k-1)} \ (T_{k+1} - T_k)) \tag{27}$$

Stability of integration requires that the coefficients of all the temperature of the second member of (27) be non-negative, i.e.

$$\frac{\partial H_k}{\partial T_k} + 26670 \ \Delta \tau \ (-2 - \frac{2}{k-1}) \geq 0 \tag{28}$$

from which we obtain

$$\Delta\tau \leq \cfrac{\cfrac{\partial H_k}{\partial T_k}}{53340(1+\cfrac{1}{k-1})} \qquad (29)$$

The minimum value of $\dfrac{\partial H_k}{\partial T_k}$ is $C_{1}\rho = 1.7 \times 10^6$

and with $k = 2$ it will be

$$\Delta\tau \leq 15.9 \text{ seconds} \qquad (30)$$

Let us therefore assume time steps of 15 sec. Equation (27) becomes

$$H'_k = H_k + 4 \times 10^5 \ (T_{k+1} + T_{k-1} - 2\ T_k + \frac{2}{(k-1)}(T_{k+1} - T_k)) \qquad (31)$$

Starting from time zero at each time equal to 15,30,45... seconds we shall find the new enthalpies by means of (31) and the new temperatures by means (24) and (25).

Nodes 1 and 6 require equations different from (31). Node 6 is constantly kept at 40°C; node 1 is treated by means of its heat balance equation:

$$K(T_2 - T_1)\ 4\pi\ (s/2)^2\ \Delta\tau /s = \pi\ (s/2)^3 (H'_1 - H_1) \times 4/3 \qquad (32)$$

$$H'_1 = H_1 + 2.40 \times 10^6 (T_2 - T_1) \qquad (33)$$

At time zero, the situation is :

T_1	T_2	T_3	T_4	T_5	T_6
80	80	80	80	80	40

H_1	H_2	H_3	H_4	H_5	H_6
178.6	178.6	178.6	178.6	178.6	0

At time 15 sec, T_6 remains at 40°C, and applying (31) to T_2 through T_5 and (33) to T_1:

H_1	H_2	H_3	H_4	H_5	H_6
178.6	178.6	178.6	178.6	154.6	0

T_1	T_2	T_3	T_4	T_5	T_6
80	80	80	80	76.68	40

The following Table 2 has been calculated for 90 sec, at the end
of which it is seen that even the outermost shell between point 6
and 5 has not reached complete solidification, but has lowered
its temperature from 80° to 65.58°C. By proceeding in the calculation
the time can be found at which all the points, including point 1,
fall below 60°C, and the spheres have totally solidified.

Table 2

time	T_1	T_2	T_3	T_4	T_5	T_6
0	80	80	80	80	80	40
15	80	80	80	80	76.68	40
30	80	80	80	79.69	73.82	40
45	80	80	79.97	79.17	71.34	40
60	80	80	79.81	78.49	69.17	40
75	80	79.97	79.73	77.71	67.27	40
90	80	79.96	79.64	76.85	65.58	40
time	H_1	H_2	H_3	H_4	H_5	H_6
0	178.6	178.6	178.6	178.6	178.6	0
15	178.6	178.6	178.6	178.6	154.6	0
30	178.6	178.6	178.6	176.4	133.9	0
45	178.6	178.6	178.3	172.6	116.0	0
60	178.6	178.56	177.7	167.7	100.3	0
75	178.6	178.4	177.6	162.0	86.5	0
90	178.6	178.3	176.1	155.9	74.3	0

OPTIMAL STORAGE SIZE

A convenient storage capacity will be tailored to buffer the solar
input for a few hours, a few days or for a season. Let us review
the second instance, i.e. the case in which the storage size is
conceived to compensate for a run of overcast days. Let us assume
to observe the system on the first day of overcast sky after a
continuous run of sunny days. Calling p the probability of having
a sunny day in the given month,

$$P = p^{n_1}(1 - p)^{n_2} \tag{34}$$

is the probability of being in the situation that the past run of
sunny days was n_1 days long and the "future" run of overcast days
will be n_2 days long. If such is the case, the thermal content of
the storage is

$$\min (n_1(\eta GA-L), V) \text{ provided that } \eta GA > L \qquad (35)$$

where η is the mean collection efficiency
- G the incident radiation for a clear day, $J/m^2 day$
- A the collecting surface, m^2
- L the daily thermal load, J/day
- V the storage capacity, J

This amount of heat will be drained during the future n_2 overcast days. If the stored heat is not sufficient to support load L, a conventional make-up Q will have to be provided:

$$Q(n_1, n_2)= \max (n_2 L - \min (n_1(\eta GA-L),V) , 0) \qquad (36)$$

A choice of the optimal storage capacity will be based upon the expected average make-up fuel consumption Q':

$$Q' = \sum_{n_1=1,30} \quad \sum_{n_2=1,30} \quad \frac{Q(n_1,n_2) \ p^{n_1}(1-p)^{n_2}}{n_1 + n_2} \qquad (37)$$

obtained by summation of all the possible events times their probability.

The optimal storage capacity will be such as to minimize the objective function

$$S = C_1 V/d + C_2 A/d + C_3 Q' \qquad (38)$$

where C_1 is the storage cost per J capacity
- C_2 the cost of 1 m^2 of collecting surface
- C_3 the cost of 1 J of make-up fuel
- d the system lifetime in days

Typical values of the constants are:

C_1	$1\div 5 \times 10^{-6}$	$\$/J$
C_2	200	$\$/m^2$
C_3	1.5×10^{-8}	$\$/J$
d	4000	days
p	0.4-0.5	(winter months in Europe).

The minimization is to be carried out for each month in a year and a decision should not be based on the results for a single specific month. For example, in the months of december and january, product GA is always smaller than L. Then Q = Q' = L and the optimal storage size is zero since there is no heat to be buffered.

Carrying out this computation with the above values of the constants,

a peculiar result will be found, namely, that on the basis of
pure economic compromise it is never convenient to buffer solar
heat for more than one day. This result is invariant even if the
unit storage tank cost decreases or the fossil fuel price increases
by one order of magnitude. Therefore it can be assumed that the
relevant conclusion has a validity embracing at least the next
twenty years.

As to the "hourly" storage, the global heat balance will be

$$cM \frac{d\,T}{d\,t} = \max \left(A \left(\tau\alpha\, W \sin \frac{\pi\,t}{D} - U(T-T_e) \right) , 0 \right) + F - P \qquad (39)$$

where   c   is the specific heat of the storage material, J/kg°C
       M   its mass,   kg
       T   the storage temperature,   °C
       W   the peak incident radiation on the collectors, W/m^2
       D   the duration of daylight,   sec
       t   the time elapsed since sunrise,   sec
       T_e the mean ambient temperature,   °C
       F   the thermal flowrate of make-up heat,   W
       P   the heat load,   W
       $\tau\alpha$ transmittance-absorptance product of collectors
       U   heat loss factor of collectors $W/m^2°C$

This differential equation is to be integrated with a number of
bounds and initial conditions. In particular, once established a
minimum temperature T_M for the heating fluid (30°C for floor coils)
it will be

$$F = 0 \quad \text{if} \quad T > T_M \qquad\qquad\qquad\qquad (40)$$

stating that no fuel make-up is required if the storage can supply
the load without falling below T_M. Further,

$$\frac{d\,T}{d\,t} = 0 \quad \text{if} \quad T = T_M \qquad\qquad\qquad\qquad (41)$$

states that no heat can be drained from storage if its   temperature
falls below T_M, and

$$T_{t=0} = T_{t=86400} \qquad\qquad\qquad\qquad (42)$$

states that the daily behaviour is cyclic.

After integration of (39) with conditions (40),(41),(42), the
consumption of make-up fuel F will be obtained and here again a
minimization will be carried out of the sum of the depreciation of
collectors (proportional to A), the depreciation of storage
(proportional to M) and the cost due to the daily make-up fuel
consumption (proportional to F integrated for the whole day).

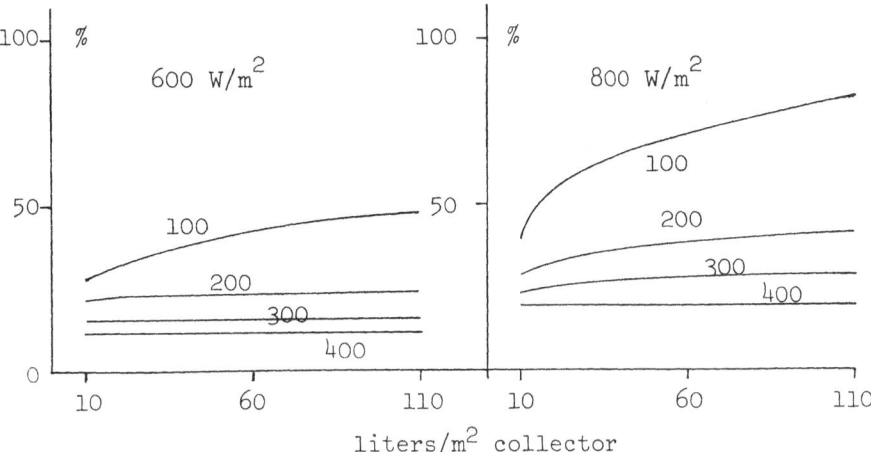

Fig. 1. Fraction of heat load covered by a solar system as a function
of storage volume (the parameter is the ratio load/col-
lector area)

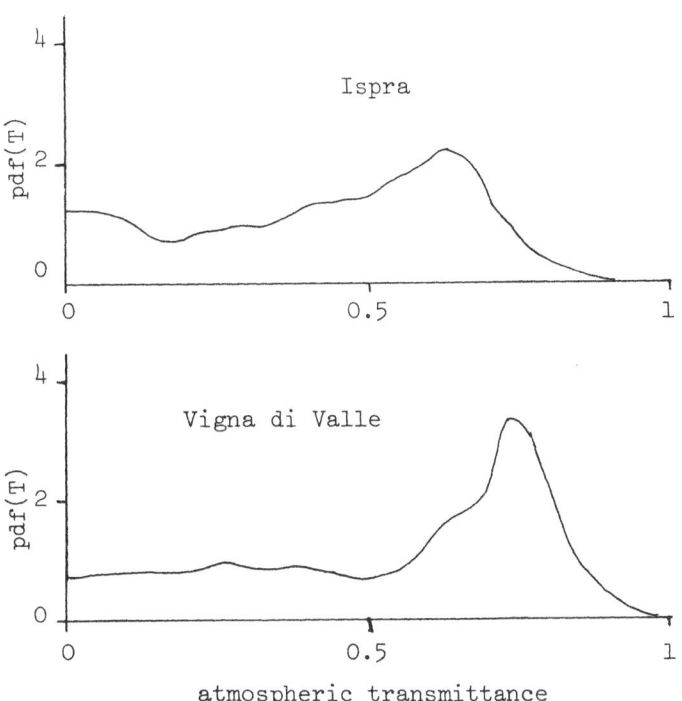

atmospheric transmittance

Fig. 2. Probability density function of atmospheric transmittance
at Ispra and Vigna di Valle (Italy)

An example of this procedure is that of fig.1, relevant to the typical case of $\tau\alpha = 0.8$, $U = 7$ W/m^2°C, $T_M = 30$°C, $T_e = 0$°C. Ordinates are percentage of thermal load obtained from the solar source. Parameter of the curves is the ratio thermal load to collector area (W/m^2). Using the cost data mentioned above, it is found that the optimal storage capacity for hot water tanks is in the range 10-30 liters/m^2 of collector. Only for installations with large collector areas (100-1000 m^2) and high peak radiation (800 W/m^2) the optimum storage capacity may attain the range the 70-100 liters/m^2 of collectors so often claimed in the technical literature.

GENERAL BIBLIOGRAPHY

Aranovitch, E., M.Ledet, C.Roumengous and D.Van Asselt. Description et performances d'un système de chauffage solaire fonctionnant à basse température. Journées Intern. d'Etude sur le Chauffage Solaire dans le Bâtiment, Univ. de Liège, 12-14 Sept. 1977

Biswas D.R. Thermal energy storage using sodium sulphate decahydrate and water. Solar Energy 19 (1977) 99

Buchberg H. and J.R.roulet. Simulation and optimization of solar collection and storage for house heating. Solar Energy 12 (1968) 31

Cabelli A. Storage tanks - a numerical experiment. Solar Energy 19 (1977) 45

Carslaw H.S. and J.C.Jaeger. Conduction of Heat in Solids, Clarendon Press, Oxford (1959)

Cavalleri G. and G.Foligno. Proposal for the production and seasonal storage of hot water to heat a city. Solar Energy 19 (1977) 677

Chauhan R.S. and V.Kadambi. Performance of a collector-cum-storage type of solar heater. Solar Energy 18 (1976) 413

Close D.J. and R.V.Dunkle. Use of adsorbent beds for energy storage in drying or heating systems. Solar Energy 19 (1977) 233

Davison R.R., W.B.Harris and J.H.Martin. Storage of sunlight underground: the Solaterre system. Chem. Technol. 5 (1975) 736

Dijkmans A.P.G., L.S.Fischer and C.W.J.Van Koppen. The longterm storage of solar heat in the ground. Journées Intern. d'Etude sur le Chauffage Solaire dans le Bâtiment, Univ. Liège, 12-14 Sept. 1977

Drew M.S. and R.B.G.Selvage. Sizing procedure and economic optimizatic methodology for seasonal storage solar systems. Solar Energy 25 (1980) 79

Eckert E.R.G. The ground used as energy source, energy sink or for energy storage. Energy 1 (1976) 315

Garg H.P. Year-round preformance studies on a built-in storage type solar water heater at Jodhpur, India. Solar Energy 17 (1975) 167

Givoni B. Underground longterm storage of solar energy - an overview. Solar Energy 19 (1977) 617

Hahne E. Heat storage. Proc. UNESCO Workshop on Engineering aspects of Solar Energy Utilization, Stuttgart 19-28 April 1977

Heine D. and A.Abhat. Investigation of physical and chemical properties of phase change materials for space heating/cooling applications. Proc. Intl. Solar Energy Congress, New Delhi, 1 (1978)

Hill J.E., G.E.Kelly and B.A.Peavy. A method of testing for rating thermal storage devices based on thermal performance. Solar Energy 18 (1976) 421

Klein S.A., W.A.Beckman and J.A.Duffie. A design procedure for solar heating systems. Solar Energy 18 (1976) 113

Lorsch H.G., K.W.Kauffman and J.C.Denton. Thermal energy storage for solar heating and off-peak air conditioning. Energy Conversion 15 (1975) 1

Lunde P.J. Prediction of the performance of solar heating systems over a range of storage capacities. Solar Energy 23 (1979) 115

Meyer C.F. Status report on heat storage wells. Water Resources Bull., 12 237

Mustacchi C., V.Cena and M.Rocchi. Longterm storage of solar heat. Energy and Buildings 3 (1981) 77

Nicholls R.L. Optimal proportioning of an insulated earth cylinder for storage of solar heat. Solar Energy 19 (1977) 711

Pearson R.W., J.A.Duffie and J.W.Mitchell. Comparison of measured and predicted rock-bed storage performance. Solar Energy 24 (1980) 199

Rabl A. and C.E.Nielsen. Solar ponds for space heating. Solar Energy 17 (1975) 1

Shamsundar N. and E.M.Sparrow. Storage of thermal energy by solid-liquid phase change - temperature drop and heat flux. Trans. ASME-C (1974) 541

Shelton J. Underground storage of heat in solar heating systems. Solae Energy 17 (1975) 137

Tanaka T., T.Tani, S.Sawata, K.Sakuta, T.Horigome. Fundamental studies on heat storage of solar heat. Solar Energy 19 (1977) 415

Telkes M. Solar energy storage. ASHRAE J. (1974) 38

Wenthworth W.E. and E.Chen. Simple thermal decomposition reactions for storage of solar thermal energy. Solar Energy 18 (1976) 205

Yoneda N. and S.Takanashi. Eutectic mixtures for solar heat storage. Solar Energy 21 (1978) 61

APPENDIX - THE STOCHASTIC SIMULATION OF HOURLY GLOBAL RADIATION
SEQUENCES

Introduction

The thermal time constants of solar collectors range from a few
minutes to a few hours, at most. Yet, for lack of suitable data on
solar radiation transients, evaluations of these systems are often
carried out on the basis of mean values of insolation.

Since the response of the collectors to the wide fluctuations of
insolation about the means can entail large variations in
performance, a need is often felt for an appropriate model to
generate solar radiation noise and thus simulate realistic operating
conditions. Thus, a typical flat plate collector operated at 60°C
with ambient temperature 10°C with a $\tau\alpha = 0.82$ and an overall
heat loss coefficient 7 W/m²°C exposed to 800 W/m² for an hour
and to a heavily overcast sky during the following hour collects
806 kJ/m² whereas the same collector exposed for two hours at 400
W/m² actually lose 317 kJ/m² unless the control system switches
it off, in which case the net collection is zero.

A number of possible models have been tested and evaluated; their
structure is hereafter described and their limitations discussed.

Detrendizing Radiation Sequences

An insight is gained from a Fourier transform of the radiation
sequences /1, 2 /:

$$F(s) = \int_{-\infty}^{+\infty} W(t) \exp(-ist)\, dt \qquad (A.1)$$

which provides a description of the power spectrum, i.e., of the
contribution of the different frequencies to the radiation.
This spectrum exhibits a number of peaks due primarily to obvious
periodic circumstances, such as the regular succession of the
seasons (period 3.15×10^7 sec) or of night and day (period
8.64×10^4 sec). To avoid the burden of carrying along these low-
frequency components, a procedure has to be chosen for cancelling
from the data the effects of such deterministic trend.

Three main actions have been experimented with:

1 - subtraction from the radiation value of some means of the data,
    such as the average value of radiation at the same hour for
    each day of the same month, or season.
2 - Subtraction or division of the data by radiation values

obtained from a physical model which accounts for the extra-atmospheric radiation and for the absorption and scattering effects of the atmosphere.

3 - Division of the data by the value of the extra-atmospheric radiation on a similarly oriented plane.

The first detrendization technique, though very satisfactory, is the least usable, since it does not lend to generalizations and requires an extended collection of hourly data at the location. The second involves a knowledge of local instantaneous values of turbidity, humidity, cloud type and frequency. Experiments carried out with the Dogniaux model/3/, once acquired such data, were in very fair agreement with observations (relative standard deviation lower than 10%).

The third technique, although grossly neglecting the local climatic and meteorological effects, was finally adopted because it is the only one which can be practically used in the absence of data.

An apparent transmittance T of the atmosphere was thus obtained by dividing observed hourly radiation values by the values of the corresponding extra-atmospheric radiation/4/. The T's thus obtained are not, of course, fully detrendized inasmuch as their mean is not zero. The probability density function (pdf) of two among the dozen European locations studied by us is shown in Fig.2 (respectively Ispra, Italy, and Vigna di Valle, Italy). These curves are all very skewed, with modal values between 0.5 and 0.8 in different locations and mean between 0.4 and 0.6 (one-half of the extra-atmospheric radiation is a fair rule-of-thumb in most places).

Peaking around the modal value is mostly due to values of T in the summer. However, no greatly significant deviations are observed in the different seasons. The stochastic models will therefore by judged on the basis of their capability of reproducing the main statistical features of the T's. The latter are not at all uniquely described by the probability density functions, but will have to account for the joint probabilities in time sequences.

ARMA Models

A class of models which has been widely used or suggested for the description of time series /5/,/10/, is the ARMA (Auto-Regressive Moving Average). If V_t is a random number associated with time t model ARMA (m,n) is defined by

$$T_t = \sum_{i=1,m} a_i \, T_{t-1} + \sum_{j=1,n} b_j V_{t-j} + V_t \qquad (A.2)$$

i.e. it predicts the transmittance at time t by means of a linear combination of the m previous values of the T's and the n previous values of the random variable V. The number of terms m and n is chosen on the basis of a satisfactory residual V, autocorrelation function and power spectrum of the T's /1/.

In particular, the autocorrelation function of T is calculated from

$$\rho(k) = \lim_{t \to \infty} \sum_{t} T_t \cdot T_{t+k} \Big/ \sum_{t} T_t^2 \qquad (A.3)$$

The speed of decay for the ordinates indicates that a substantial amount of memory is retained for T over times of the order of a day or so.

For the cases of models ARMA (m,0), coefficients a_1 can be estimated by solving the Yule-Walker equations:

$$
\begin{Vmatrix}
1 & \rho(1) & \rho(2) & \rho(3) & \cdots & \rho(m-1) \\
\rho(1) & 1 & \rho(1) & \rho(2) & \cdots & \rho(m-2) \\
\rho(2) & \rho(1) & 1 & \rho(1) & \cdots & \rho(m-3) \\
\cdot\cdot & \cdot\cdot & & & & \\
\rho(m-1) & \rho(m-2) & \cdot\cdot & \cdot\cdot & \cdots & 1
\end{Vmatrix}
\times
\begin{Vmatrix}
a_1 \\ a_2 \\ a_3 \\ \cdot\cdot \\ a_m
\end{Vmatrix}
=
\begin{Vmatrix}
\rho(1) \\ \rho(2) \\ \rho(3) \\ \cdot\cdot \\ \rho(m)
\end{Vmatrix}
\quad (A.4)
$$

Using, as an example, a model ARMA (6,0), the following values are obtained:

Ispra

$a_1=0.752$ $a_2=0.060$ $a_3=0.0005$ $a_4=0.044$

$a_5=0.010$ $a_6=0.005$

Vigna di Valle

$a_1=0.586$ $a_2=0.078$ $a_3=0.037$ $a_4=0.022$

$a_5=0.033$ $a_6=0.018$

The low value of all the constants beyond a_1 shows that a model ARMA (1,0) or ARMA (2,0) is sufficient, if an-regressive description is accepted.

This result is similar to that found by /5/ and /10/ to describe, respectively daily and hourly radiation sequences.

The models turn out to be:

Ispra

$$T = 0.759 \ T_{t-1} + 0.096 \ T_{t-2} + V$$

where V has a variance 0.133.

Vigna di Valle

$$T = 0.602 \ T_{t-1} + 0.127 \ T_{t-2} + V$$

where V has a variance 0.106.

These expressions were used to compute, for a sample of about 8000 hourly measurements, the values of residuals V. Hystograms of such residuals indicate that V is not a normal deviate. The distribution is bimodal and skewed and chi-square tests with any reasonable confidence level reject the possibility of generating V with a Gaussian algorithm. Further, although the autocorrelation of residuals V showed these to be uncorrelated, they will retain a high degree of cross-correlation to the T's, thus invalidating the model.

Factor Analysis

A more firmy based statistical procedure to construct solar sequences is based upon the principal component method of factor analysis (FA) /6/.

In essence, assuming that an ARMA (m,0) expression holds, a search must be made on the redundancy of information carried by the n values of T in the previous times. New variables (the principal components) are constructed so that they are mutually orthogonal, i.e. statistically independent from each other. There are obtained as linear combination of the T's as follows:

- The eigenvalues are sought of the matrix which appears in the left-hand side of (A.4). For istance, using m = 3 for location A the eigenvalues are 2.6, 0.3 and 0.1 and for location B 2.3, 0.5, 0.2. The sum of the eigenvalues equals m and the sum of the first p eigenvalues measures the fractional variance of T accounted for by using p principal components. Using only two of such components will thus account for 29/30 of the variance of T at A and 28/30 at B.

- The value of each principal component is found by scalar multiplication of vector $(T_{t-1}, T_{t-2}, \ldots)$ by each eigenvector of the matrix.

- A linear regression is found by least squares, linking T_t and the principal components.
  This regression constitutes the "deterministic" part of the forecast.

- Residuals V are computed and a check is made on whether these are still autocorrelated.
  If they are, the procedure is iterated, increasing the number of principal components. This procedure yields a stable relation, which requires fewer observations than the ARMA models to converge to a statistically significant expression.
  Two eigenvalues (or principal components) were found to be sufficient in a dozen locations to account for over 90% of the variance of T. It can be shown that the use of two components is equivalent to linearly relating T_t to the two variables T_{t-1} and $(T_{t-1} - T_{t-2})$. In other words, the hourly transmittance "remembers" its value and its speed of variation during the previous hour.
  The simulations conducted with this type of model exhibit better statistical properties, as compared with the ARMA models but both models suffer from T's with a normal probability distribution function. This difficulty is unavoidable, with any ARMA and FA model. Since the pdf's of Fig. 2 are very far from being normal, other approaches must be sought.

Markov Transition Matrix

The ARMA and FA models can be visualized as linear forecasts based on the past values of T, on which a random disturbance V is applied, the latter being independent from the value of T or of time. The next step in sophistication is to assume that the probability distribution of the V's is a function of the local value of T. This is physically obvious, since if the sky is heavily overcast, random disturbance will tend to increase its clarity whereas if the sky is very clear it will be more likely to decrease its clarity in the next hour. Thus, a dependence of the pdf of V on T has to appear in the model.

The simplest representation is that of a Markov process, where a transition function is given, p (a,b) such that if T_t= a, the probability of having T_{t+1}= b is p (a,b). The transition probability is such that it depends only on the present state (a) and not upon the more distant past.

The simplest form of function p (a,b) is that of a transition matrix, with elements p_{ab} giving the probability of changing from state a to state b in one hour. Such matrices were constructed for the twelve sites examined by building up the two-dimensional hystogram obtained from the transitions of several thousand contiguous couples of hourly observations of T. A sufficient graininess of these matrices turned out to require 25 x 25 matrices, i.e. the range of T from 0 to 1 was split into intervals of width 0.04.

Simulations are conducted by selecting row "a", corresponding to T_t, and generating a random number R with a rectangular distribution in the range (0,1). Elements p_{a1}, p_{a2}, p_{a3}..p_{ab} are added until their sum is greater than R. State "b" is then assigned to T_{t+1}. The simulations thus conducted led to a very good reconstruction of the distributions of Fig. 2 and of all the main statistically important features of radiation sequences. In fact, a speedy check on the quality of the transition matrix can exploit the fact that by raising this matrix to a high power (typically 200), a new matrix is obtained with all its rows equal. The elements of the rows reproduce the ordinates of the probability density function of T.

This model was found so consistently satisfactory that a generalized computer procedure was written to analize hourly data, build the transition matrix and simulate the hourly insolation. Corrections for collector tilt use the Liu and Jordan /11/, Orgill and Hollands /7/ or the LASL /8/ methods and results exhibit discrepancies of 2-3% at most.

Transmittance Transition Tensor

A step further in the direction of using a combined ARMA (n,0) and a transition matrix approach is that of generalizing the latter description by constructing a transition tensor of order n. Such a tensor has n-1 dimensions corresponding to the values of T in n-1 successive times. The elements along the n-th dimension yield a value for the probability of having a given transition at the next time step. Thus, a model TTT (3) would use a third-order tensor which has elements p_{ijk} measuring the probability that if $T_t = j$ at present and $T_{t-1} = i$ one hour go, then, the probability of having $T_{t+1} = k$ one hour from now is given by p_{ijk}.

This approach was experimented with a third-order tensor obtained by analyzing several thousand triplets of transmittances at one-hour interval. The simulations derived from this model were but a slight improvement as compared with the Markov approach. The statistical significance of the tensor element decreased for lack of a sufficiently extended sample. It was therefore felt that such sophistication was not justified.

Gaussian Mapping

To overcome the difficulty of unavoidably generating a normally distributed T with any ARMA model, the following method (GM) can be used. Once obtained g(T), the probability density function of

T, find the value of a new statistical variable Y, such that

$$(2\pi)^{-1/2}\int_{-\infty}^{Y} \exp(-x^2/2)dx = \int_{0}^{T} g(T) \, dt \tag{A.5}$$

The new variable Y is normal with means 0 and unit variance and has a one-to-one correspondence with T. The next step is the obtainment of an appropriate model ARMA (m,n) for Y. This model can generate a hourly sequence of Y (mapped transmittance). Each Y is then anti-mapped by finding the corresponding T from (A.5).

This procedure was found to be fully satisfactory but for a slight loss of memory: the Y's are somewhat less autocorrelated than the T's because of the filtering action of the truncation errors of (A.5). The times required for computer mapping, simulation and anti-mapping are two or three times higher than with the Markov approach. Yet, we feel that the effort may be justified in view of the wide acceptance of the ARMA type of models.

REFERENCES

/1/ G.E.P. Box and G.M. Jenkins, Time Series Analysis – Forecasting and Control. Holden-Day, San Francisco (1970).
/2/ A. Papoulis, Probability, Random Variables and Stochastic Processes, McGraw-Hill, New York (1965).
/3/ R. Dogniaux and M. Lemoine, Programme de calcul des éclairements solaires énergétiques et lumineux de surfaces orientées et inclinées. Institut Météorologique de Belgique, Série C no.14 (1976)
/4/ K.Y. Kondratyev and M.P. Fedorova, Radiation regime of inclined surfaces. Proc.UNESCO-WMC Symposium and Solar Energy, WMO no. 477, Geneva (1977)
/5/ T.N. Goh and K.J. Tan, Stochastic modelling and forecasting of solar radiation data. Solar Energy 19 755 (1977).
/6/ H.H. Harman, Factor analysis, in Mathematical Methods for Digital Computers (edited by A. Ralston and H.S. Wilf). Wiley, New York (1960).
/7/ J.F. Orgill and K.G.T. Hollands, Correlation equation for hourly diffuse radiation on a horizontal surface. Solar Energy 19 357 (1977).
/8/ International Energy Agency, Solar Heating and Cooling Program Task 1, Provisional Report (edited by O. Jorgensen) (2nd draft may 1978)
/9/ W.R. Petrie and M. McClintock, Determining typical weather for use in solar energy simulations. Solar Energy 21 55 (1978).
/10/ B.J. Brinkworth, Autocorrelation and Stochastic Modelling of Insolation Sequences, Solar Energy, 19 343 (1977)
/11/ B.Y.H. Liu and R.C. Jordan, The interrelationship and characteristic distribution of direct, diffuse and total radiation, Solar Energy, 4 1-19 (1960).

OVERVIEW OF THE THERMAL STORAGE WORK WITHIN THE ENERGY R AND D PROGRAMME OF THE EUROPEAN COMMUNITY

P.ZEGERS

COMMISSION OF THE EUROPEAN COMMUNITIES
DG XII , ENERGY R AND D

SUMMARY

This overview is divided in three parts. First the CEC Energy
R and D programme is described in which research on thermal
storage is carried out. Then a general state of the art
is given for heat storage R and D giving the advantages and
disadvantages of the best known heat storage systems, listing
the most important applications, and making some rough estimates
on their economic feasibility. Finally the results and conclusions of CEC thermal storage research are discussed. This work
includes latent heat storage with paraffins and inorganic salts,
studies on chemical heat storage and long term heat storage with
hot water storage tanks, pebble beds, solar ponds and aquifers.

INTRODUCTION

        If one tries to get an overview of heat storage systems
one is overwhelmed by the large number of possible technical
solutions and the variety of storage systems. It is generally
much less clear for which applications they may be used.
Also the cost of these storage systems is often not known
and rarely is the economic feasibility considered of a storage
system together with its application. Before discussing
the state of the art of thermal storage work executed by the
CEC I will therefore devote the first half of my lecture to
a general overview. I will discuss storage applications, advantages and disadvantages of the best known storage systems and
make some rough estimates on the economic feasibility of
complete systems.
I will begin with a short general description of the CEC
Energy R and D Programme.

THE ENERGY R AND D PROGRAMME OF THE C.E.C.

        The European Communities are extensively active in
energy R and D. They spent about 300 Mio ECU */year for
"in house" and contract research in different fields such
as coal and nuclear energy, solar energy, geothermal energy,
fusion and energy conservation.

        Research on thermal storage is executed in the
framework of the so-called "indirect action" Energy R and D
Programme of the European Community. By decision of the
Council of Ministers a first Energy R and D Programme was
carried out running from 1975 to 1979 with funds totalling
59 Mio ECU.

        A second Energy R and D Programme (1979-1983) was
approved in 1979 with a budget of 105 Mio ECU .
These programmes are sub-divided in five sub-sectors : Solar,
Geothermal, Energy Conservation, Hydrogen and Energy Modelling.

* 1 ECU = 1 $ US

Table 1

	1975—1979	1979—1983
Energy conservation	11.4 MEUA	27 MEUA
Solar energy	17.5	46
Production and use of hydrogen	13.2	8
Geothermal energy	13.0	18
Energy systems analysis	3.9	6
TOTAL	59.0	105

Heat storage research is carried out in the Energy Conservation and the Solar Energy programme.
Since 1975 some 55 heat storage research contracts have been concluded with a total cost amounting to about 12 Mio ECU (for a list of contracts see Annex I).

CHARACTERIZATION OF HEAT STORAGE APPLICATIONS

The domestic sector is by far the most important area for heat storage. The different systems used for house heating and for the use of storage for domestic and industrial applications will be discussed below.

Domestic heating

Solar heating is characterized by the fact that solar heat is most abundantly available when it is not needed. Solar heating systems require therefore back up of conventional systems. Cheap seasonal heat storage may be a way to make solar heating more attractive provided that the cost of solar components can be reduced. The required temperature for solar heat storage lies between 30° and 70°C.

Heat pumps offer several possibilities for heat storage applications.

- Air is most often used as a heat source for heat pumps. This has the disadvantage of a somewhat lower efficiency which brakes down completely at low outdoor temperatures. These heat pumps therefore require back up by conventional heating systems. The use of soil as a heat source may avoid this draw back as the temperature at a few meters below the surface is constant at 5—10°C throughout the season. Heat may be extracted from soil with a system of horizontal or vertical tubes with a circulating

fluid.In the summer the heat in the soil is restored. The required soil temperatures lie between 0° and 20°C.
- Daily heat storage of 60-80°C in combination with electrical heat pumps will allow operation of the heat pumps during the night and use of the heat during the day. Heat storage could thus contribute to load levelling of the electricity production if electrical heat pumps will find a large scale introduction. Such a small heat storage system has the additional advantage that the heat pump is running continuous- ly for a few hours. In this operation mode the efficiency is considerably higher (up to 10%) than during usual operation where the heat pumps is switched off and on to keep stable room temperatures.

Industrial_waste_heat_and_heat_from_urban_waste_incineration : waste heat from industry and heat from urban waste incineration is normally available free of cost and is produced throughout the year. With district heating networks this heat can be used for domestic heating during the winter. Heat delivered in the summer and partly in autumm and spring however would still have to be discharged. The availability of cheap seasonal heat storage would enable profitable use of this heat. It would increase the heating capacity and avoid thermal pollution during the summer.

Co-generation_of_heat_and_power : with a power plant in back pressure operation the efficiency of the electricity production reduces from 40% to 30%, but 60% of the heat input of the plant which first was lost,is now available at 120-150°C as useful heat for domestic heating.
- In such a heat-power plant, heat can be most efficiently produced during the night when electricity demand is low. To provide heat during the day, daily heat storage is necessary. Such a daily heat storage system may also be useful when the power plant is repaired and heat production has to be assured. Also here a district heating network is needed. The required heat storage temperatures lie around 120°C.
- Heat produced by heat-power coupling is cheap. The conti- nuation of heat production by cogeneration during the summer and the storage of this heat in cheap seasonal heat storage systems for use in the winter,may be an interesting option. In this way also the problem of thermal pollution is avoided.

Heat_storage_and_power_production: with heat storage at temperatures between 250 and 600°C load levelling could be achieved for different types of power production (fossil fired, nuclear, solar). From the fact that these systems would replace gasturbines one may deduce a rough estimate for the allowed cost.

Heat storage in industry

Large amounts of waste heat are produced in industry. This
waste heat is most efficiently used in the factory, provided
that applications can be identified. Economical feasibility
for storage systems in industry however is very difficult
to achieve. This is partly due to the short payback times of
2-3 years required in this sector. It is also the experience
in the EC thermal storage programme that economically viable
applications for thermal storage in industry are not easily
found. A study to explore heat storage opportunities in industry
is presently being executed.

CHARACTERIZATION OF THE BEST KNOWN STORAGE SYSTEMS

Thermal heat storage is possible in a large variety of ways and
I will give here some general characteristics for the most
important systems.

Sensible heat storage in insulated hot water containers

The cost of such storage systems is of the order of 10-15 $
for a heat storage capacity equivalent to one liter oil (LOE).
This type of heat storage is only useful for short and medium
term heat storage as seasonal heat storage would require
prices between 2 and 4 $ per liter oil equivalent (LOE).

Advantages	Disadvantages
. Cheap storage medium	. Large volume required
. Heat exchangers often not required	. Heat discharge at fluctua-
. Technology available.	ting temperatures
	(unless stratified heat
	storage is used).

Latent heat storage
These systems deliver heat at constant temperatures, have
higher energy densities and are commercially available in the USA
at a cost of about 400 $ per liter oil equivalent (LOE). Also
experiments executed in the Commission's programme gave
cost values of this order.

Advantages	Disadvantages
. The volume required at 80°C is about half the volume of hot water storage with an equal storage capacity	. Need for heat exchangers which reduces the overall efficiency

- Heat discharge at constant temperature

- Heat conductivity of the solid phase is often bad
- Problems with supercooling and segregation
- Expensive

Due to the high cost these systems are only interesting for short term thermal heat storage.

Chemical heat storage

Chemical heat storage is brought about by a reversible chemical reaction with absorption of heat in one direction and the discharge of heat in the reverse direction

$$A + B \rightleftharpoons C + D$$

Many different reactions offer a large variety of storage temperatures.

Advantages

- Large energy density up to 10 times higher than hot water storage at 80°C (e.g. Na_2S/H_2O)

- Heat discharge at constant temperatures

- Storage of the components at room temperatures for indefinite time

- Possibility for transport of stored heat.

Disadvantages

- High cost due to a complex installation . A Na_2S/H_2O system which is close to commercialization in Sweden is believed to be of the order of 50 $ per liter oil equivalent. But from the complexity of the installation it seems more likely that the cost will be of the same order as latent heat storage systems.

- Few systems are available

- The cooling and heating of the storage components between the reaction temperatures and room temperatures impose thermal losses and decrease the storage efficiency

- Much R and D is needed to obtain information on kinetics, stability corrosion, heat capacity, thermal conductivity and on the cost of such a system.

Of the large number of chemical storage systems, chemical heat pumps seem to be the most interesting option. If the cost of the system could be reduced considerably, chemical heat storage could become interesting for seasonal storage. But it seems unlikely that prices of 2-4 $/LOE will be obtained.

Aquifer heat storage

Seasonal storage of heat up to 150°C is believed to be possible in natural aquifers of about 20-30 m high, confined on the bottom and on the top by clay layers at depths ranging from 50 to 500 m. Heat is introduced and extracted via drilled holes with hot water.
Feasibility studies in Sweden and Denmark calculated an investment cost of 2 $/LOE.

Advantages

. One of the few possibilities for seasonal heat storage which may be economically viable.

Disadvantages

. Suitable sites for aquifer heat storage may be rare (e.g. thickness, permeability , use of aquifers for other purposes, such as drinking water, availability of cheap heat and heat users).
. Many technical problems still have to be solved
. Very little practical experience. An experiment in Auburn (USA) gave heat storage of 2-3 months (two monghs storage one month production) with a 70% heat recovery efficiency.

Soil and ground water heat storage

This type of heat storage up to 20°C serves as a heat source for heat pumps. The heat can be extracted from the soil with a system of tubes with a circulating fluid. The systems with horizontal (1-2 m deep) and vertical tubes (down to 10-15 m) require a surface of two times and half the heated surface respectively. The cost of this system lies between 2 and 4 $ per LOE and is close to economical feasibility.

Cost of heat source and heat storage systems

The investment cost of 2-4 $/LOE for which seasonal heat storage becomes economically feasible was based on the assumption that the heat to be stored is free of charge and that for seasonal storage the allowed investment cost for a storage capacity of one liter oil equivalent equals the cost of a liter oil saved per year, multiplied by the lifetime of the system which was estimated to be 10-20 years.

When the number of storage cycles increases the storage system
is economically feasible for higher investment costs. For
n storage cycles per year with a storage capacity of one LOE,
n liters of oil will be saved per year and the allowed investment cost is n times larger than for seasonal storage. For example
weekly and daily storage during the cold period of the year
gives allowed investment costs of about 50 $/LOE and 300 $/LOE
respectively. From the table one can see which systems are
suitable for daily and weekly storage. Only aquifer storage
and possibly chemical storage may turn out to be useful for
seasonal heat storage.

Table 2 : Cost of different storage systems for
a heat storage capacity equivalent to
one liter oil

Storage system	Cost/LOE	Storage system	cost/LOE
Hot water sto- rage (steel tank)	10-15 $	Aquifer heat storage	2-4 $
Latent heat storage	400 $	Pebble bed storage	20-40 $
Chemical heat storage	50-400 $	Soil and ground water storage	2-4 $

As the most important contribution to energy saving will be
brought about by seasonal heat storage, we will elaborate
somewhat on these calculations which are a strong simplification
of the real situation.
In reality several other factors interfere : heat losses and
running costs will decrease the allowed investment cost, the
installation of a storage system may save investment cost on the
installation of a gas boiler thus increasing the allowed investment cost for the storage system. Finally the assumption that
heat is without cost is only valid for a few applications.
Industrial waste heat, heat from urban waste incineration plants
and solar energy may be assumed to be without cost. Heat produced
with co-generation is roughly half the price of heat produced
with a boiler. Then in all these cases, additional investments
have to be made to bring the heat to the user. In the case of
solar energy the investment cost for this additional equipment
amounts to about $ 10.000 for a one family house. The district
heating network, required to bring waste heat or heat from power
plants to the user costs per one family house roughly $ 5.000.

Industrial waste heat and urban waste incineration seem therefore to be most attractive for seasonal heat storage. Co-generation and solar energy are less attractive.

Conclusions on the present state of the art

- Most-interesting heat storage applications lie in the domestic heating sector
- Seasonal heat storage has the biggest potential for energy saving but economic feasibility is most difficult to achieve
- Aquifers and to a less extent chemical heat storage are candidates for seasonal heat storage. The first solution requires still much experimental work but has the outlook of economic feasibility. Chemical seasonal heat storage is attractive due to the possibility of indefinite storage at room temperature. It is technically feasible but the cost has to be decreased by at least one order of magnitude. Much R and D is required to obtain cost reductions
- Seasonal heat storage applications for industrial waste heat and urban waste incineration are more attractive than for co-generation and solar energy
- Applications for daily storage may be found with heat pumps and co-generation. Practically all these short term storage systems are economically feasible
- Possibilities for economically viable storage applications in industry are limited. An exploratory study to this extent is being executed.

COMMISSION'S R AND D WORK ON HEAT STORAGE

The work executed in the EC programme has four focal points :

- Latent heat storage
- Chemical heat storage
- Long term heat storage
- Soil and groundwater storage for heat pumps

Latent heat storage

The attention being given to latent heat storage is based on two potential advantages : high energy density and heat discharge in a narrow temperature range.
Several studies assessed a large number of chemical compounds on their suitability for latent heat storage. For the selection of the phase change materials the following criteria have been used :
- Temperature range - 25 to + 150°C
- High storage capacity i.e. . high latent heat of transition
(130 kJ/kg)
. high specific heat
- Long term stability under temperature cycling conditions

- Abundant and cheap (e.g. by-products of chemical industry)
- Good thermal conductivity. No super-cooling
- Non flammable. Non toxic . Non corrosive.

A few of the selected compounds were : Glauber salt, sodium acetate, sodium orthophospate, KF_4H_2O, $LiClO_3.3H_2O$ and paraffin.

Inorganic salts for storage below 100°C : As for the inorganic salts two problems arise :

. supercooling
. congruent melting

Supercooling could be solved in most salts by creating cold spots in the salt solution. In $NaSO_4.10 H_2O$ (Glauber) special additives (e.g. burax) solved the problem of supercooling.

Congruent melting occurs in all inorganic salt hydrates. This means that on melting two phases appear, a fluid phase and a new solid phase, generally the anhydrous salt. Without special care, this phenomenon results in segregation and consequently a gradual decrease of heat storage capacity. Research in this area has led to two different solutions :

- micro encapsulation in small volumes (10-100 micron)
- adding extra water and stirring.

Several systems using these techniques have been built and tested. They were marginally better than stratified hot water storage and required half the volume for an equivalent heat storage capacity.

Paraffin : the second type of material which was selected for more detailed studies was paraffin. The following problems were encountered :

- Not a sharp melting point, but a melting range with a width varying from 4 to 20°C
- Enthalpy change of 30%, in a solid-solid transition, which results in a lower heat density at melting temperatures
- Bad heat conductivity of paraffin (this can be improved by adding charcoal)
- Losses of paraffin up to 25% in 50 days (1 cycle day) take place due to precipitation in the low velocity parts of the system.

The use of paraffin is not very promising due to the losses and the solid-solid transition. Also the cost of a latent heat storage system is very high. The volume of a paraffin system is about half the size of an equivalent hot water store.

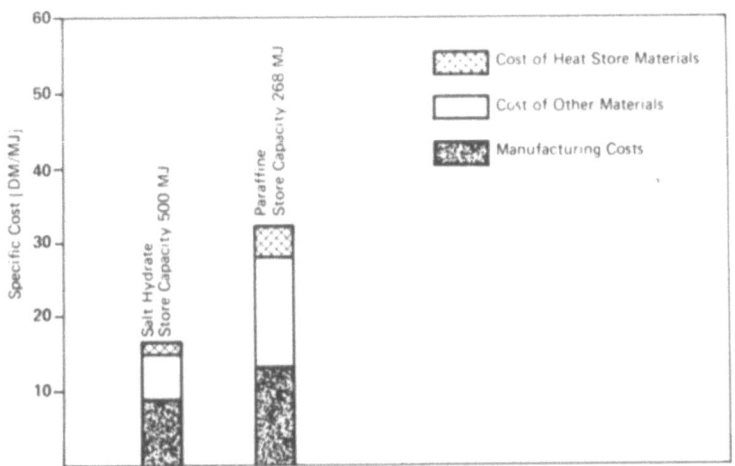

Fig. 1   Cost of latent heat storage

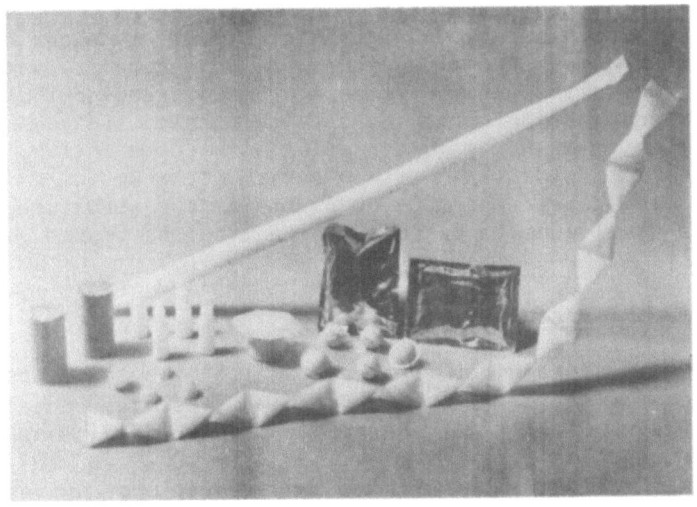

Fig. 2 Encapsulation methods for latent heat storage

Encapsulation _ :latent heat storage systems are expensive
in particular those with a circulating heat storage medium
and heat exchangers. Systems with encapsulated heat storage
media and air or water as heat transport sectors may be considerably cheaper. For this reason encapsulation of both paraffins
and inorganic salts has been investigated ; in particular
the compatibility of encapsulation materials with the heat
storage media. For $NaSO_4.10H_2O$ polythene performed best
as an encapsulation material. With 440 cycles of 8 hours
between 20 and 60° C no damage was found. Polyesters which also
performed well did not resist water.
Encapsulation of paraffin wax has also been investigated.
Encapsulation in the form of 30 mm thick cushions in double
foil of aluminium and polyethylene withstood 400 cycles
from 18 to 85°C but failed a 60 hours test at 95°C. A better
result was obtained with polyamide tubes, but here water,
as a heat transfer fluid, had to be replaced by paraffin
oil because of the sensitivity of this type of oil to water.
The size of the cushions or the diameter of the tube had
to be kept smaller than 30 mm in order for the store to have
acceptable charging and discharging times (approx. 1 hour).

Cost_ : For some of the heat storage systems the cost has been
calculated . The cost of a system consisting of tubes serving
as heat pipes with fins, was 500 8 /LOE when paraffins
were used and 250 8/LOE with inorganic salts.

High temperature latent heat storage with inorganic salts :
With industrial heat storage in mind we started R and D
on high temperature heat storage materials. These experiments
dealt with thermal stability, aging and compatability with
container materials. In the temperature range 200-450°C the
following combinations were found to be compatabile :
mild steel with Na NO_3 ; boiler steel with KCL - $ZnCL_2$,
$NaNO_3$ and $MgCl_2$ - NaCl and NiCu30Fe with KCL - $ZnCl_2$.
In the temperature range 700-900°C none of the structure
materials were found to be compatabile with the tested
salts.

Chemical heat storage

Chemical heat storage has the advantage of high energy density
and storage of heat for indefinite time. The disadvantages
are a complex installation and high cost. The Commission's R and
D was executed in two areas :
- exploratory studies mainly on chemical heat pump systems
- absorption

Exploratory studies : in the field of chemical heat storage
more than 30 systems have been investigated . In particular

chemical heat pumps are interesting for heat storage. This
basically because low grade heat, which may be absorbed
by evaporation of a chemical compound e.g. NH_3, is added
to the reaction heat. In this way heat storage efficiencies
higher then 100% may be obtained. The energy densities of
these systems can be high. A H_2O /Na_2 S heat storage system
which is marketed in Sweden has at 70°C an energy density
which is ten times higher than hot water storage.

Absorption   of water by different absorbents such as silicagel,
sorbead, activated alumina, molecular sieves gives absorption
heats corresponding to energy densities which are 1 to 3 times
higher than hot water storage. All examined absorbents show
an aging effect. The best absorbent in this respect was
sorbead ⁓ R which showed a reduction of the absorbing capacity
of 10% after 400 cycles.

Seasonal heat storage

The major part of heat storage applications require cheap
long term heat storage. The development of such systems
is therefore of upmost importance. The cost below which seasonal
heat storage is expected to be economically feasible is
between 2 and 4 $ for a heat storage capacity equivalent
to one liter oil. In this programme a variety of systems has
been and are being investigated such as hot water storage,
pebble bed stores, aquifers, solar ponds, soil and ground
water storage.

Hot water storage

A 6000 m3 plastic water tank partly dug into the ground and
insulated on top will be tested. The heat extraction is brought
about by an air heat exchanger.

Solar pond

A solar pond of 12 m3 will be built and tested and the results
will be used to design a pond of 450 m3.
These ponds are 1⁓2 meter deep and due to the fact that salts
are dissolved in the water, there is a density and temperature
gradient, with the highest densities and temperatures at the
bottom. Heat is stored in the water and in the ground below
the pond. The cost of these systems is 13 $ /m2, not taking
into account the cost of the ground.

Pebble bed stores

A 4 m3 pebbel bed was tested with air as heat transport
medium. For seasonal heat storage the size of the pebbles
was found not to be critical. Pebbles of 50 mm are required
for short term heat storage.
A 250 m3 doubled lined pebble bed with water as heat transport
medium was constructed. The heat loss coefficient for this
storage system was 0,5 W m3° C-1. The installation was working
to full satisfaction and the construction cost was 20 $ /LOE
when pumping of ground water is not required . With pumping
of ground water the cost doubles.

Aquifers and soil storage at 70°C

- Feasibility studies have been made for a large variety
  of shallow aquifers with the following basic characteristics :
  hot water injection in a shallow aquifer confined at the
  bottom by a clay layer at a depth of about 15 m, artificial thermal insulation at the top, sideways confined
  by a double wall of betonite by which the system is hydraulically insulated from water currents in the aquifer.
  Different types of storage aquifer systems have been investigated varying the following parameters : size ranging from
  5000 m3 to 175.000 m3, soil preparation (natural aquifer,
  removal of soil and replacement by material with the right
  proporties e.g. permeability), different types of walls.
  The price estimate for the aquifers varied between 7 and
  40 $/LOE. The thermal and hydraulic insulation makes this
  system too expensive for seasonal, heat storage, although
  some options come close the economic feasibility.

- A project which is rather similar to the project discussed
  above is a soil storage system of 40 m diameter and 20 m deep,
  thermally insulated on top and hydraulically insulated
  with a double concrete wall. The heat exchange system in the
  soil will consist of a continuous flexible hose with
  vertical loops with which heat is discharged in or extracted
  from the soil. This storage system will be used for a group
  of 100 houses. During the test period the heat load will
  be simulated. Heat storage will be possible up to 70°C .The
  project is now being built.

- Recently a project started on seasonal heat storage by hot
  water injection via drilled holes in a confined aquifer of
  500 m deep. Heat will be available from a waste incineration
  plant. The heat will be used for domestic heating via a
  district heating system. The storage temperature will be above
  100°C.

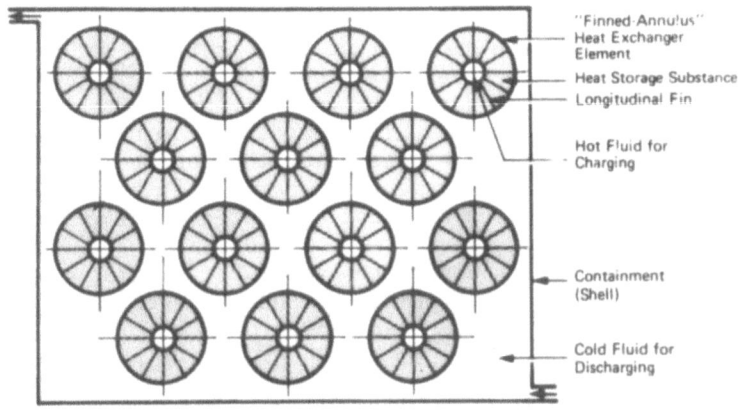

Fig. 3 Latent heat storage with a finned heat pipe

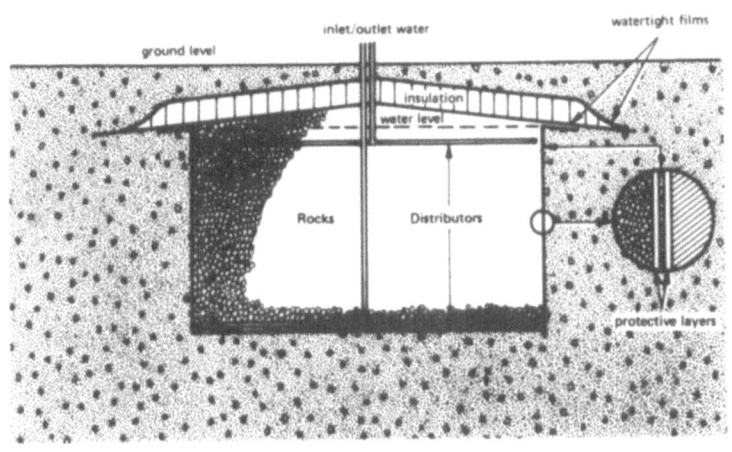

Fig. 4 250 m3 pebble bed heat store

Soil and ground water storage up to 20°C for heat pumps

Most of the projects in this area deal with extraction of heat
for heat pumps from soil with horizontal or vertical tubes
with a circulating fluid (e.g. glycol-water). For horizontal
and vertical tubes the required soil surface is twice and half
the surface which has to be heated, respectively. The
smaller soil surface required for the vertical heat exchanger
makes it a promising candidate, but it has the draw back that
soil temperatures in particular deep below the surface
(e.g. 10-15 m) often have to be restored artificially during
the summer. Allthough no accurate cost estimates are available,
one may foresee investment costs of 2-4 $ /LOE.
In one project groundwater pumped from a natural aquifer is
used as a heat source for heat pumps. The heat is restored
during the summer with help of solar collectors.

ACKNOWLEDGEMENT

I want to express my gratitude to Mr.Steemers for making
available the results and conclusions of heat storage R and D
executed in the C.E.C. programme on solar applications for
dwellings.

EC ENERGY R & D PROGRAMME - LIST OF CONTRACTS

ORGANISATION   TITLE

1. Latent heat storage

Technisch Physische Dienst TNO - TH The Netherlands	Development of a thermal storage system based on encapsulated PC-materials
Technical University of Denmark	Heat storage in a solar heating system using salt hydrates
Technical University of Denmark	Investigation of heat storages with salt hydrate as storage medium based on the extra water principle
Technical University of Denmark	Heat storage units using a salt hydrate as storage medium based on the extra water principle
Rhône-Poulenc, France	Report on perfecting low temperature thermal storage products
Société Nationale Elf/Aquitaine, France	Heat storage using latent heat of fusion of a substance
Institut für Kerntechnik und Energieumwandlung, Germany	Latent heat thermal energy storage Determination of properties of storage media and development of a new heat transfer system
Brown, Boverie & Cie Germany	Latentwärmespeicherung für Sonnenenergiesysteme
Battelle, Germany	Kompakte Latentwärmespeicher bei Benutzung von Zustandsänderungen im Temperaturbereich zwischen 320 und 340° K
Univeristy College Cardiff, UK	The design , construction and testing of a phase charge storage device.

Katholieke Universiteit Leuven, Belgium	Short term storage with paraffine of solar heat from air cooled solar collectors
Technisch Physische Dienst TNO- TH The Netherlands	Heat storage systems based on organic phase change materials with improved heat conductivity
Lab.de Marcoussis Marcoussis, France	Study of thermal storage through melting of organic materials
Lab.de Marcoussis + Université de Nantes	Heat storage based on the melting of paraffine and coupled with a heat pump, for application to heating in housing
Cranfield Bedford, UK	Thermal energy storage systems
IKE Stuttgart,Germany	Investigation of medium and high temperature phase change materials for thermal energy storage

2. Chemical heat storage

Philips GmbH Aachen, Germany	Investigation and development of systems for latent and chemical heat storage of thermal energy in the temperature range between - 25°C and 150°C
Faculté Polytechnique de Mons, Belgium	Storage as chemical bond energy of compounds undergoing thermal decomposition
Faculté Polytechnique de Mons, Belgium	Optimization of heat storage systems using silicagels
Technisch Physische Dienst TNO - TH The Netherlands	Development of a thermal storage system based on the heat of absorp- tion in hydroscopic materials
Technisch Physische Dienst TNO - TH The Netherlands	Development of a reaction vessel of a thermal storage system using the heat of adhesion
Faculté Polytechnique de Mons, Belgium	Storage de l'énergie solaire dans des réactions chimiques reversibles

| Messerschmitt–Bölkow
Blohm GmbH | Chemical heat storage with salts
between 300 and 650° K |

3.Long term storage

Centre d'Etudes Nucléaires,France	Study of a heat storage system for dwellings
Centre d'Etudes Nucléaires, France	Stockage hétérogène eau/pierres de 250 m3
Technical University of Denmark	Seasonal heat storage in underground hot water
University of Sussex England	A salt gradient solar pond for solar
L'Università degli Studi della Calabria	Long term solar heat storage by a water tank for the heating of a solar building
Delft Soil Mechanics Laboratory The Netherlands	The use of soil as storage medium for seasonal storage of solar energy
Delft soil Mechanics Laboratory The Netherlands	Field test to investigate the performance of an undeep prototype seasonal heat storage system with a heat capacity for 100 houses
Université de Montpellier II, France	Etude du stockage de chaleur, dans les sols non saturés
Fiat research Centre Italy	Seasonal storage of solar energy through heating of subsoil with cylindrical elements buried in the ground
CEA – Commissariat à L'énergie Atomique	Prediction and measurements of the behaviour of a confined aquifer at a depth of 500 m used as an induced geothermal reservoir at a tempera– ture above 100 degrees C.
MBB München, Germany	Development of an aquifer heat storage system

4. Soil heat storage serving as heat source for heat pumps

EHC Ltd 2820 Gentofte Denmark	Ways and means of increasing the COP of a heat pump
TNO Apeldoorn, Netherlands	Investigation about using soil as a natural heat source for heat pumps
Laborelec St.Genesius Rhode,Belgium	Investigation in using soil naturally and artificially reheated by solar energy as a heat source for heat pumps
European heatpump Consultors Ltd	Experimental investigations on using the earth as a storage medium and as a heat source for heat pumps
Organisation for applied Scientific Research	Use of soil as a heat source and heat storage medium for heat pump experimental investigation of a complete heat pump system in one family dwelling
Organisation for applied Scientific Research	Vertical pipe heat exchangers for heat extraction from and heat storage in the earth with heat pump systems
Dornier System GmbH	Domestic heating with earth-heat-pipes as a heat source for a heat pump
Laborelec Company	Room heating with a combined soil & energy roof heat pump
Roots Energy Limited	The technical and economic investigation and evaluation of heat pumps using ground/water heat source
A.R.M.I.N.E.S. France	The heliothermal doublet of seasonal heat storage up to 20°C with groundwater for heat pumps

5.Miscellaneous

Institute of Applied Physics TNO - TH, Delft The Netherlands	Recommendations for uniform reporting of research work in thermal storage of solar energy
Stichting Bouwcentrum en Ratiobouw, The Netherlands	Development of solar energy houses based on an integrated collector storage system combined with warm-air heating
The University of Leeds UK	Development and optimization of Trombe's solar wall
Bertin & Cie, France	Numerical simulation of solar heating of buildings
Bertin & Cie, France	Development of a pebble bed heat storage
Cranfield Institute of Technology	Investigation of the potential for energy cascading combined with thermal energy storage in industry
Belgonucleaire	Optimization of a heat storage module with a high energy density

UNITED STATES DEPARTMENT OF ENERGY
THERMAL ENERGY STORAGE PROGRAM*

J.H. Swisher
U.S. Department of Energy
Washington, D.C. 20585 U.S.A.

W.A. Frier
Pacific Northwest Laboratory
Richland, Washington 99352 U.S.A.

ABSTRACT

The objective of the U.S. Department of Energy Thermal Energy Storage Program is to develop devices, processes, and subsystems that permit domestic energy resources to be supplied at the time and locations where they can be used. The program emphasis is in five principal areas: (1) daily storage for active or passive solar and conventional heating and cooling, (2) seasonal storage for building heating and cooling, (3) thermal energy transport, (4) thermal power storage, and (5) chemical heat pumps. The program has a budget of $15.35 million for 1981.

1. INTRODUCTION

Thermal energy storage is a generic title spanning the class of technologies in which a storage device accepts and returns energy in the form of heat. The process is analogous to other more familiar methods for storing energy including electrochemical batteries for electric energy, flywheels for mechanical energy, and the damming of water for generation of hydroelectric power. A device for thermal storage may become "hot" and serve as an energy <u>source</u> at the time of use. Thermal storage also

---

*Work supported by the United States Department of Energy, Office of Conservation and Renewable Energy under Contract No. EM-78-I-01-5217 and Contract No. DE-AC-06-RLO-1830.

includes the accumulation of "cold," in which case the device becomes a thermal energy sink at the time of use (typically in air conditioning and refrigeration).

The various thermal energy storage technologies may be divided into three categories: sensible heat, heat effects during change of phase, and thermochemical reactions.

Sensible Heat: Energy storage as sensible heat is accomplished by raising or lowering the temperature of the storage medium. This form of storage is historically the most familiar, with tanks of hot water and beds of hot rocks used in a variety of applications. Energy storage density depends on the heat capacity of the medium and a maximum acceptable temperature swing. Typical density values range from 23.3 to 233 kJ/kg (10 to 100 Btu per pound). Major disadvantages of this approach are that stored heat is generally returned at an ever-decreasing temperature level and storage densities are relatively low. Advantages include low materials cost and simplicity. Most current commercial thermal storage systems employ sensible heat devices.

Phase Change: Energy storage involving a phase change in the storage medium makes use of latent heat effects of melting, vaporization, or other changes in the physical state of the medium. (The storage of winter ice for use during the summer is a historical example of this technique.) Energy storage densities in a phase-change system are generally greater than those for sensible heat storage, resulting in a potentially lower overall system cost, and stored heat is returned at a nearly constant temperature level. Phase-change storage technologies are currently less advanced than those for sensible heat. It is expected that continued development will lower the cost and improve the performance of phase-change materials for energy storage.

Thermochemical Reactions: Reversible thermochemical reactions can be used to store and transport energy. The energy is absorbed in an endothermic process at higher temperatures; the release of the energy occurs when the reaction is reversed, which occurs generally at lower temperatures. The reaction rate is enhanced normally by the use of a catalyst. The products can be cooled to room temperature and stored or shipped with no heat loss. This lack of thermal loss over long periods of time along with the relatively high energy storage per pound of material makes the reversible reactions promising for energy storage and transport technologies. The applications include storage, transport, and heating and cooling with a chemical heat pump. The chemical heat pump can also be used to upgrade waste heat.

The chemical system employed for thermochemical storage should ideally exhibit rapid, reversible kinetics involving a large heat effect. The chemical changes in the storage medium must be totally controllable, either through easy physical separation of reactants and products through the use of a catalyst, or through temperature control. An advantage of the thermochemical approach is that it allows storage at ambient temperature with essentially no time-dependent heat losses. In addition, the energy storage densities are potentially an order of magnitude higher than that attainable with either the sensible heat or phase change methods. The controllable nature of the energy release allows the matching of energy availability (temperature level) to the requirements of the end user or, under some circumstances, "pumping" thermal energy to temperatures above the source level (at the expense of a portion of the stored energy).

2. THE DOE THERMAL ENERGY STORAGE RESEARCH AND DEVELOP-MENT PROGRAM

The Federal Nonnuclear Energy Research and Development Act of 1974 (Public Law 93-577) directed the Energy Research and Development Administration to advance energy storage technologies in the areas (among others) of batteries for electric vehicles, energy storage systems for electric utility load-leveling applications, and thermal energy storage systems for use in residential and commercial buildings. These responsibilities were transferred to the Department of Energy (DOE) in 1977.

The Solar Heating and Cooling Demonstration Act of 1974 (Public Law 93-409) specifically required federal R&D efforts in technology for thermal energy storage systems.

The mission of the DOE Thermal Energy Storage Program is to develop devices, processes, and subsystems which permit domestic energy resources to be supplied at the time and locations where they can be used. The payoff is mainly in reduced oil imports through expanded production and use of coal, nuclear, and solar energy sources.

Work within the DOE Physical and Chemical Energy Storage Division consists of applied research, exploratory development, and technology base activities in several project areas. Emphasis is placed on achieving significant improvements in state-of-the-art technology through efforts which the private sector is not expected to undertake.

The DOE Office of Energy Systems Research has followed a policy of decentralization whereby the day-to-day managerial activities are delegated to DOE Field Offices. The organizational structure is depicted in Figure 1. The Headquarters staff is responsible for the broad technical management which is required for the integration of complex research and development programs. In addition, Headquarters staff interfaces with several federal agencies, private industry, Congress, and their counterparts in foreign countries.

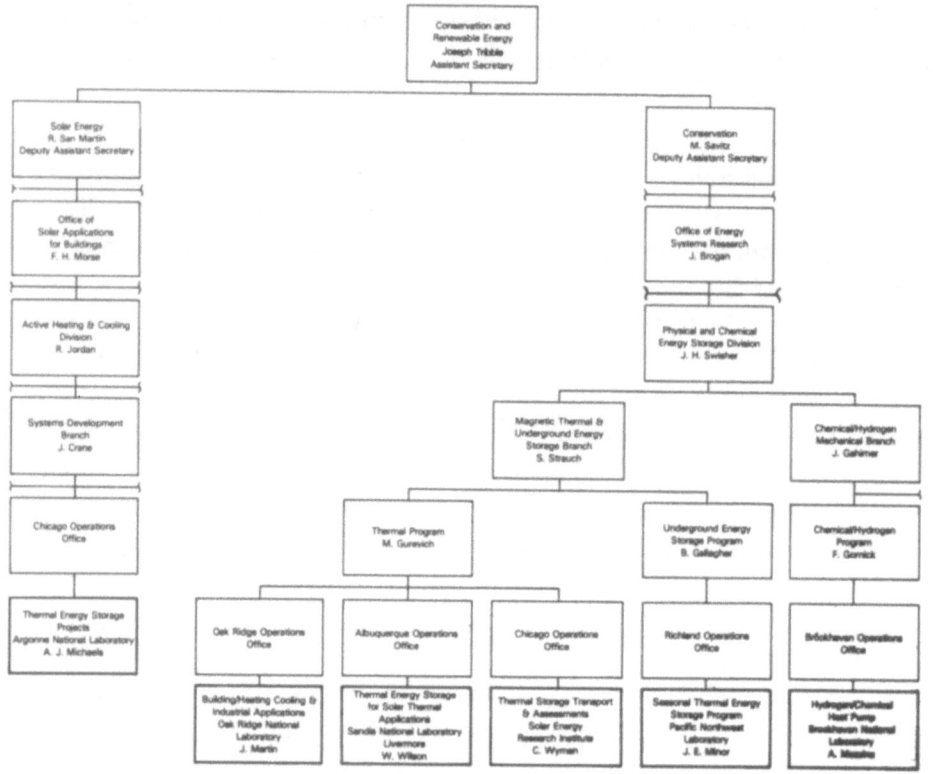

Figure 1. Organizational Structure for Thermal Energy Storage Programs

Another DOE organization, the Office of Solar Applications for Buildings, Systems Development Branch, has been given the mission to develop, construct, test, and analyze integrated solar systems that include thermal energy storage subsystems for the purpose of verifying proof-of-concept operation and system effectiveness. The objective is to ensure that these advanced solar systems are recognized, accepted, and carried forward into the marketplace through normal industry/market activities.

The DOE Office of Energy Systems Research participates in the International Energy Agency (IEA) Program of Research and Development on Energy Conservation through Energy Storage. The objective of Annex I of the IEA program is to exchange R&D information among participating countries. Each country pursues its area of research in long-term storage technologies and then provides data to the other participants. Those countries supporting Annex I include the Commission of European Communities (Belgium, Denmark, Germany, Sweden, and Switzerland) and the United States.

The DOE Office of Solar Applications for Buildings participates in the International Energy Agency Solar Heating and Cooling Program, Annex VII, Solar Central Heating Plants with Annual Storage. A portion of this program involves thermal energy storage technology.

3. THERMAL ENERGY STORAGE APPLICATION AREAS

3.1 Building Heating and Cooling

Solar heating and cooling systems need thermal energy storage to meet building demands during periods of darkness, inclement weather, or intermittent cloud cover. Rock beds and water tanks are mature, well understood methods for storing thermal energy. Latent heat (phase change) and thermochemical (chemical heat pump) technologies require further development. Daily storage technologies include not only active and passive solar systems, but also customer-side-of-the-meter storage of heat and cold derived from off-peak electricity to accomplish utility load leveling and thereby reduce oil and natural gas consumption. The principal application for chemical heat pumps is in active solar systems, while phase-change materials are being developed for active and passive solar systems and for customer-side-of-the-meter storage.

Customer-Side-of-the-Meter Storage: DOE has developed a commercialization readiness assessment and implementation strategy for distributed thermal energy storage in the residential sector.[1] This assessment confirmed that customer-side-of-the-meter thermal energy storage for residential space heating and hot water heating can capture a market large enough to yield an annual oil savings of approximately 80 million barrels

(0.3 quads) by the year 2000. That assessment also concluded that thermal energy storage for residential space and hot water heating is technically and economically ready for commercialization. The primary barriers to the immediate commercialization of thermal energy storage heating systems were found to be economic or institutional in nature. These barriers are listed below in decreasing order of importance.

- Provision of electric power rates which encourage thermal energy storage,
- Size of required rate difference between off- and on-peak,
- Availability of market estimates,
- Manufacturing capability,
- Equipment certification, and
- Financing.

Only two areas of environmental concern were identified--the impact of shifting electric power generation from peaking units to baseload units and the impact of mining activities to provide a domestic source of refractory brick.

The Electric Power Research Institute reported in its March 1981 journal that 56 utilities are conducting 74 thermal energy storage projects. It is also reported that preliminary results indicate that heat storage can be cost effective if there is a rather modest differential between regular electricity rates and off-peak rates.[2]

The Long Island Lighting Company is under subcontract to Oak Ridge National Laboratory (ORNL) and is also co-funded by the New York State Energy Research and Development Administration (NYSERDA) to demonstrate, in a residential house on Long Island, New York, the use of "cool storage." In load leveling, the energy storage component is charged at night during off-peak hours. By using cool storage, the size of the air conditioner can be reduced by half. The thermal storage medium is a phase-change material (PCM), Glauber's Salt, packaged in a sausage-like configuration called "chubs" (Figure 2). The chubs have a storage capacity of 105 to 117 kJ/kg (45 to 50 Btu/lb) and a transition temperature of $13^{\circ}C$ ($55^{\circ}F$).[3] Operational data, reliability, efficiency, and economic data will be gathered during the 1981 summer season.

For storage heating, North Carolina State University, under subcontract to ORNL, is to analyze and evaluate ceramic refractories for thermal energy storage in electrically charged heaters. This effort

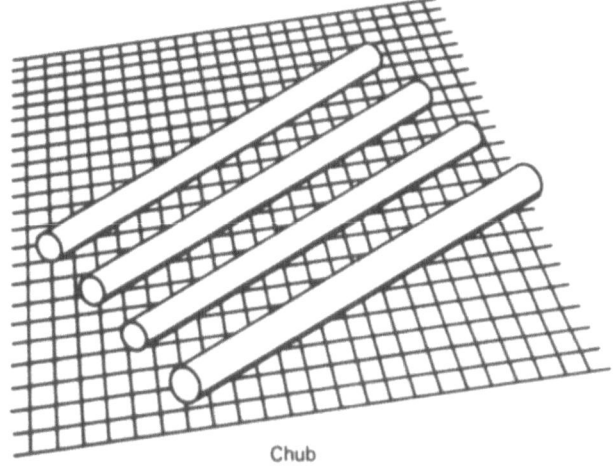

Chub

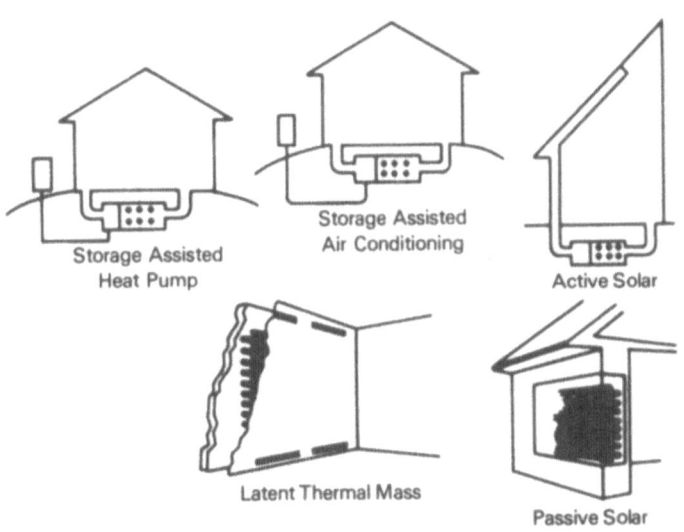

Storage Assisted
Heat Pump

Storage Assisted
Air Conditioning

Active Solar

Latent Thermal Mass

Passive Solar

81-156

Figure 2.   Chub Applications

consists of thermal testing of ceramic refractories made at the University using olivine from North Carolina and English refractories made from Norwegian olivine. Data have been collected and are being analyzed to gain insight into the development and propagation of isotherms through the brick core.[4]

R.J. Schoenhals, Purdue University, under subcontract to ORNL, installed a thermal energy storage testing facility to develop standard performance testing procedures in the United States. To develop the standard testing procedures, a German standard (DIN 44572) was used for electrically heated thermal energy storage units and the American Society of Heating, Refrigeration, and Air Conditioning Engineers (ASHRAE) Standard 94-77 was used for other types of thermal energy storage units. The test facility is used to compare testing of room-size thermal energy storage (TES) units with olivine bricks from North Carolina and with conventional bricks for ORNL. Test results should provide an evaluation of the possibility of using domestic sources of olivine for storage heaters.

Phase-Change Materials: Dow Chemical Company developed and is marketing a chemically modified form of $CaCl_2-6H_2O$, a phase-change material for thermal energy storage, called "Thermol 81" [T_m = 27.2°C (81°F)] which is congruently melting and has minumum supercooling. PSI, Inc., of Fenton, Missouri, is fabricating parts called "Energy Storage Rods" which are 8.9-cm (3.5-in.)-diameter, 1.8-m (6-ft)-high, carbon-black-coated, high-density polyethylene (HDPE) cylinders filled with 59.4 kg (27 lb) of Thermol 81 and having a storage capacity of 2595 kJ (2460 Btu) at $29.80 per rod.[5] Figure 3 illustrates a storage unit with Thermol 81.

Calmac Corporation of Englewood, New Jersey, is developing thermal energy storage systems using bulk containment of a latent heat storage media in 0.9-m (3-ft) or 1.2-m (4-ft)-diameter, 1.2-m (4-ft)-high plastic cylinders, with internal spirals of close-spaced plastic tubing for heat exchange.[6] A small pump operates intermittently when the salt is in a liquid state to prevent stratification. The Calmac "Heat Bank" unit is illustrated in Figure 4. The Heat Bank 115 contains a hydrated sodium sulfite salt [$Na_2S_2O_3-5H_2O$; T_m = 46.1°C (115°F)] and an Ice Bank Unit where water serves as the latent heat storage medium. Units with a 115.6°C (240°F) salt ($MgCl_2-6H_2O$) and a 7.2°C (45°F) salt are also available. The Calmac Heat Bank units have been cycled extensively with no

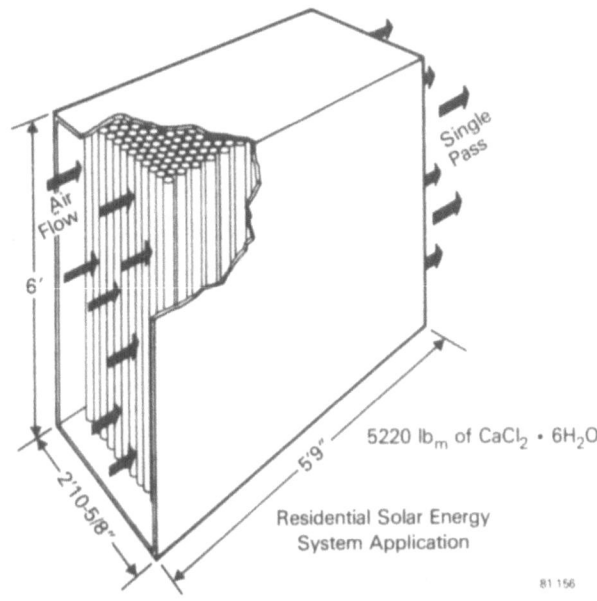

Figure 3. Full-Scale Storage Unit: Calcium Chloride Hexahydrate Encapsulated in HDPE Cylinders

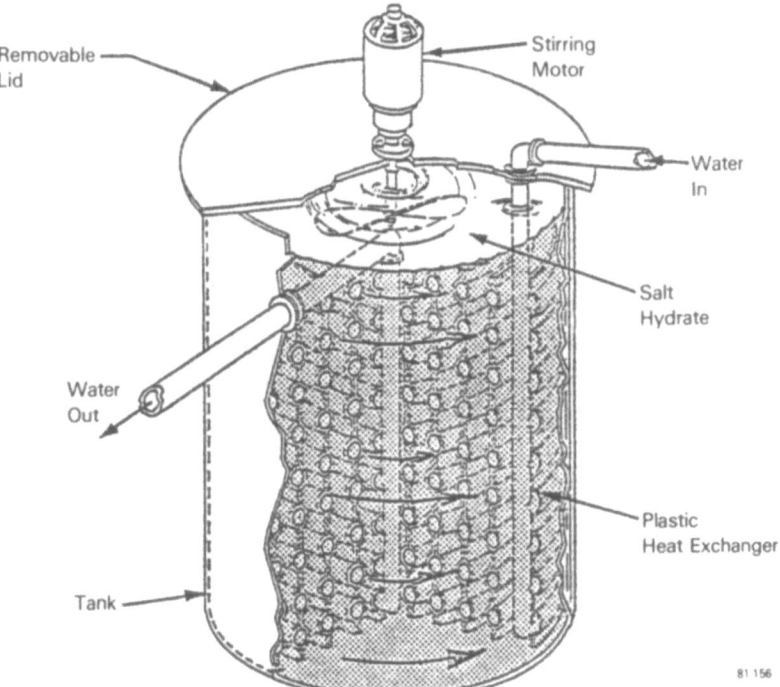

Figure 4. CALMAC "Heat Bank" Latent Heat Storage Unit

loss in storage capacity. The Heat Bank 115 unit has a storage capacity of 400.9 MJ (380,000 Btu) for a Δt of 19°C (66°F). The primary application is for commercial/industrial buildings. A price of $3000 was quoted for a single unit equivalent to $7.48/MJ.

In a generic study of phase-change materials, ORNL mathematical modelers have monitored temperatures at five points in a phase-change component. A 10.2- x 15.2- x 8.9-cm (4- x 6- x 3.5-in.) box was constructed of 2.54-cm (1-in.) thick plexiglass with one face being a 2.54-cm (1-in.) aluminum plate. N-Octadecane paraffin wax was placed inside the cell and vented to allow for thermal expansion. Experimental data were studied and the following concluded.[7]

- The effective thermal conductivity of the solid is 3.5 times the literature value.

- Natural convection or other similar effects enhance the liquid PCM behavior significantly; nonetheless, the system can still be simulated by a pure conduction model in which the liquid thermal conductivity is replaced by a suitably chosen parameter.

Research Triangle Institute of Research Triangle Park, North Carolina, a subcontractor to ORNL, is to evaluate heat pump/storage combinations for residential space cooling or heating. The method is based on computer simulations of the heat pump and a bulk phase-change storage unit of Research Triangle Institute design. A mathematical model of convection in the melted phase-change material has been programmed at ORNL using N-Octadecane and results are being compared to data obtained for a natural convection system.

Crawl Space Heat Pump: A variety of experiments have been performed to determine the feasibility of conditioning the source of air of an air-to-air heat pump using ground heat or cool stored in the earth under a house to improve the performance of the heat pump and to produce higher coefficients of performance (COP) and net energy savings.[8] Experimental results verify that outside air can be preheated by the stored heat in the earth consistently throughout the winter (Figure 5). Plans are to obtain and compare data from three similar houses. One house is to be heated with a conventional air-to-air heat pump, a second house is to be heated with an open cycle crawl space heat pump, and a third house is to be heated with a total recycle crawl space heat pump.

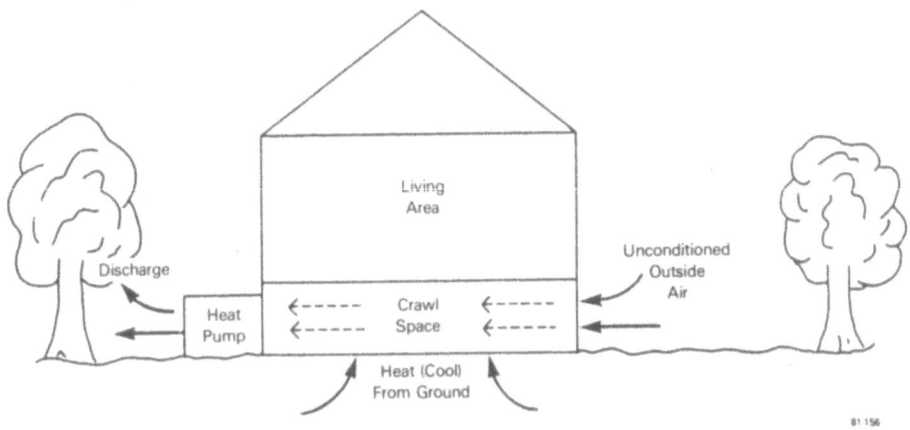

Figure 5. Crawl Space Heat Pump in a Single-Pass Mode

Stratified Tank Storage: The University of New Mexico, under subcontract to ORNL, is to determine cost and energy savings from operation of a thermal storage system incorporating labyrinth-stratified water storage tanks. Data are to be collected from the Mechanical Engineering Building at the University to determine economic and energy savings for 1981 and 1982.

The University of Tennessee, an ORNL subcontractor, is to determine the cost and energy savings from the operation of an Oliver Springs, Tennessee, elementary school building that uses membrane-stratified thermal energy storage tanks. Installation of the data collection system and instrumentation is complete. Energy produced for and used from the heating and cooling water storage is to be compared with estimated characteristics without storage during two seasons in 1981.

Direct Contact Heat Exchanger: The Solar Energy Research Institute (SERI) is studying direct contact heat exchange as an attractive method to reduce the cost of heat exchange for latent heat thermal energy storage systems (Figure 6). Specific areas of study with the direct contact heat exchange are the characteristics of start-up/shut-down cycling and the operational problems of fluid distribution and carryover. Comparisons of

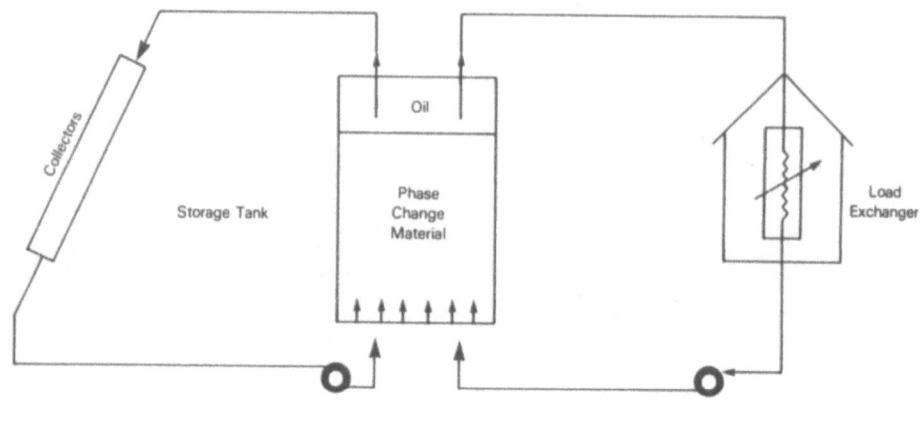

Figure 6. Direct Contact: An Effective Means for Heat Transfer

theoretical and measured drop size for a single-drop heat transfer column were made at SERI with a heptane/water system.[9] At low flow rates, the equation of Kagen, et al.,[9] best describes the experimental data (Figure 7). The velocity at which the transition to jetting (the point at which the fluid exits as a jet instead of individual drops) occurs is, however, lower than predicted (Figure 8).

Chemical Heat Pumps: Rocket Research Company, Redmond, Washington, a Brookhaven National Laboratory (BNL) subcontractor, is developing an industrial chemical heat pump. The heat pump uses a sulfuric acid/water system. The coefficient of performance can range from 46 to 5 for a given waste heat source inlet temperature of 99°C (210°F) and waste outlet temperatures that range from 104°C to 160°C (220°F to 320°F).[10] A 158.3-MJ/hr (150,000-Btu/hr) verification test unit is currently being fabricated. The unit is to be operated in both an industrial heat pumping mode and a heating and cooling with storage mode.

Southern California Gas Company (SOCAL) was recently subcontracted by BNL to develop a prototype metal-hydride heat pump. This type of heat pump is being investigated because it represents a technology which

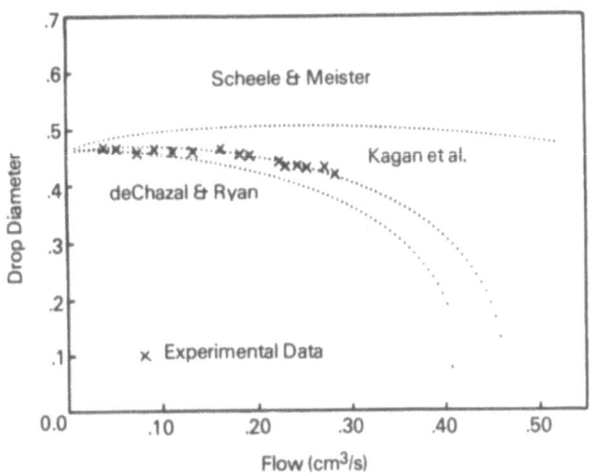

**Figure 7. Drop Diameter vs Flow Rate for the System
Heptane/Water (nozzle diameter = 0.13 cm)**

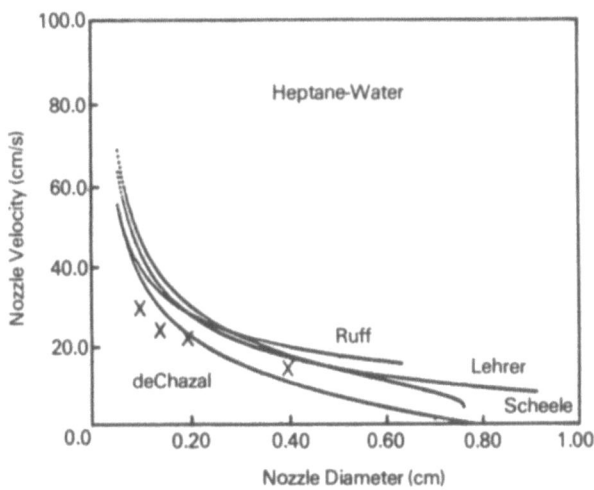

**Figure 8. Jetting Velocity vs. Nozzle Diameter
for the System Heptane/Water**

could conserve large amounts of energy and has a rapid cycle time of importance in air conditioning. Preliminary analysis indicates a potential heating COP of 1.55 is possible.

Industrial Applications: Franklin Research Center, Philadelphia, Pennsylvania, has been subcontracted to create an inventory of existing thermal storage installations for either hot or cold storage or both in commercial and industrial use in the United States and Canada. The inventory is to be published by the American Society of Heating, Refrigerating, and Air Conditioning Engineers (ASHRAE) to disseminate information to engineers and users. A breakdown by storage medium, storage system, and country is given in Table 1.[11]

Table 1. Storage Media and Systems by Country

Storage Type	United States	Canada	Total
Heat Storage			
Pressurized Water (>100°C)	27	15	42
Unpressurized Water (<100°C)	16	5	21
Sand	29	0	29
Concrete, Brick	8	1	9
Cold Storage			
Water	10	10	20
Ice	20	0	20
Combined Heat and Cold Storage			
Water	25	3	28
Water/Ice*	6	0	6
Total	141	34	175

*Water Is Used for Heat Storage and Ice Is Used for Cold Storage.

Rocket Research Company, an ORNL subcontractor, is currently developing technology for the recovery of waste heat from exhaust ducts of an aluminum pot line at the Intalco Aluminum Corporation at Bellingham, Washington. The program focuses on heat exchanger development and corrosion testing in the waste heat stream at 125°C (260°F) in the exhaust duct. The program did include the design and construction of an 8-block area with a 3-MW_t peak demand and the design of a commercial district heating system. However, with the current administration's philosophy of increasing the supply of energy and encouraging industry to sponsor demonstration programs, the design and construction of an 8-block area and the design of a commercial district heating system is no longer in the DOE program. The district heating service area had been optimized to include 3200 small residential, 115 large residential, 600 commercial, and 90 institutional structures. The baseline storage system incorporated two insulated steel tanks of approximately 3400 m³ (900,000 gallons) each representing energy storage of 200 MW_th.[12] Economic analyses indicated that the total system cost would be slightly in excess of $60 million ($US).

A pulp and paper industry study recommended the use of a variable-pressure steam accumulator to allow substitution of wood waste ("hog fuel") for fossil fuel in generating process steam.[13] When a search was made a few years ago for an industrial partner to participate in a site-specific demonstration of the technology, it was discovered that at least one U.S. company was already using the concept for steam demand smoothing with excellent results, but had not disclosed the technique to the industry at large. The results of the DOE-funded study[13] and the experience from existing energy storage activities in the American, Canadian, and Scandinavian pulp and paper industries will be documented by Howard Edde, Inc., of Redmond, Washington. Seminars are planned to disseminate this information to the industry at a Western Conference at the University of Washington (Seattle) and in Atlanta, Georgia, in September 1981.

3.2 Seasonal Thermal Energy Storage

The Seasonal Thermal Energy Storage Program at Pacific Northwest Laboratory (PNL) was composed of a demonstration task and technology support tasks. The demonstration task was intended to stimulate the interest of industry by demonstrating the feasibility of utilizing an aquifer for thermal energy storage. Because DOE is emphasizing long-term R&D work, this task will not be continued. Demonstration contracts were underway at three sites:

Site	Prime Contractor	Energy Source	Storage Temper- ature	User Application
Bethel, Alaska	TRW, Inc.	Diesel-Electric Utility	<95°C	District Heating
State Univ. of New York, Stony Brook Campus	Dames & Moore	Winter Chill	4°C	District Cooling
St. Paul, Minnesota	Univ. of Minnesota	University Heating Plant	150°C	District Heating

The objectives of the demonstration task were to demonstrate the economic storage and retrieval of energy on a seasonal basis, using heat or cold available from waste sources or other sources during a surplus period to reduce peak period demand, to reduce electric utilities peaking problems, and to contribute to the establishment of favorable economics for district heating and cooling systems for commercialization of the technology. Aquifers, ponds, lakes, and soil have potential for seasonal storage. The initial thrust of the seasonal thermal energy storage program was toward utilization of groundwater systems (aquifers) for thermal energy storage. The concept involves withdrawing water from the aquifer passing it through a heat exchanger where it is heated using an existing energy source, and reinjecting the water into an aquifer for storage (Figure 9). The program had the further objective of evaluating other methods for seasonal storage, both from existing literature and by following current work in other countries.

The objective of the technology task was designed to conduct research and development studies required to provide a sound technical base for the demonstration of seasonal thermal energy storage concepts through:

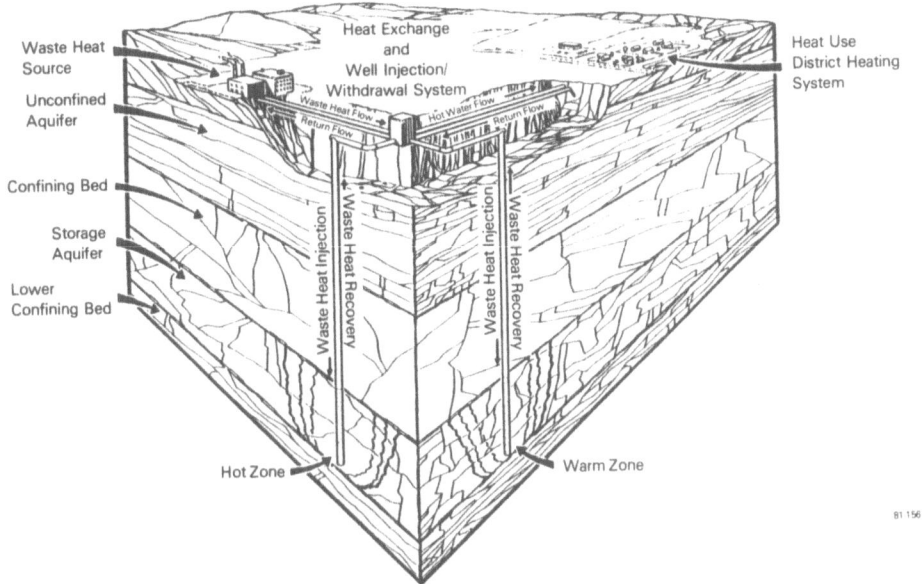

Figure 9. Seasonal Thermal Energy Storage in Aquifers

- Development of field test facilities to obtain advanced data;
- Performance of analytical, laboratory, and field research activities;
- Provision of a center for knowledge and information for seasonal thermal energy storage; and
- Technology transfer.

The above was to be accomplished through eight subtasks:[14]

- Legal/Institutional,
- Economic Assessment,
- Environmental Asessment,
- Field Test Facility,
- Compendia of Existing Technology,
- Physiochemical Analysis,
- Numerical Simulation, and
- Nonaquifer Seasonal Thermal Energy Storage.

Legal/Institutional: A study was undertaken to catalog and analyze the legal, regulatory, and institutional issues at the federal, state, and local government levels that are likely to affect implementation of seasonal thermal energy storage concepts. A technical document containing the results of this study was published in October 1980.[15]

Economic Assessment: Development of an aquifer thermal energy storage economic evaluation computer code, AQUASTOR, was completed. AQUASTOR combines demand, aquifer, and supply characteristics with climate, cost functions, and economic parameters into one systematic model. It provides the flexibility to evaluate individually or collectively the impact of different economic and technical parameters, assumptions, and uncertainties on the cost of providing heat (chill) from an aquifer storage system. AQUASTOR was used to determine the cost assessments indicated in Table 2.

Table 2. Cost of Aquifer Thermal Energy Storage Heat Storage ($1979/MBtu)[14]

1 MWt		10 MWt		100 MWt	
150°C Source Temperature					
150-ft Wells					
$50/ft	$100/ft	$50/ft	$100/ft	$50/ft	$100/ft
8.37	8.63	1.62	1.65	0.97	0.98
1000-ft Wells					
10.28	12.03	2.31	2.31	1.78	1.91
90°C Source Temperature					
150-ft Wells					
9.23	9.50	2.66	2.71	2.04	2.09
1000-ft Wells					
11.89	13.65	4.81	5.16	4.29	4.62

Note: $50/ft and $100/ft Indicate Well Drilling Cost, Casing Included.

81-156

- 482 -

A "Guideline for Conceptual Design and Evaluation of Aquifer Thermal Energy Storage" was completed.[16] The guidelines will assist in the preliminary design effort and in evaluation of the technical and economic feasibility of the aquifer thermal energy storage technology.

Environmental Assessment: Major emphasis was placed on preparation of environmental documentation requirements (federal and state) for aquifer thermal energy storage projects for compliance with the National Environmental Policy Act (NEPA). A "Findings of No Significant Impact" for the Aquifer Thermal Energy Storage Program was announced by DOE.

Field Test Facility: Field studies were conducted to assess technical problems associated with development of aquifer thermal energy storage systems. Auburn University, a subcontractor to PNL, operates a field test facility at Mobile, Alabama. Objectives of the field test facility are to operate a doublet well system (a storage system that uses a two-well injection and withdrawal system) for thermal energy storage in a confined aquifer at injection temperatures of 55°, 90°, and $125^\circ C$ (131°, 194°, and $257^\circ F$). The initial injection cycle included a total injection of $25,700$ m^3 ($6,800,000$ gallons) at an average temperature of $59^\circ C$ ($138^\circ F$). Initial evaluation of the field data collected indicates that the test should provide a reliable data set for analysis of the aquifer response at $59^\circ C$ ($138^\circ F$). No field problems were encountered with plugging using the true doublet configuration. A partially penetrating aquifer pump test was also run to ascertain the ratio of horizontal to vertical permeability. Analysis of the test shows the ratio to be 7:1.

Compendia of Existing Technology: The collection, summarization, and transmittal of seasonal thermal energy storage related information was undertaken. A working library of more than 1925 seasonal thermal energy storage related documents was assembled. A seasonal energy storage quarterly newsletter is also published and distributed by Lawrence Berkeley Laboratory, a PNL subcontractor, to representatives in 23 countries. Those countries are Belgium, Canada, China, Denmark, England, Finland, France, Hungary, Iceland, India, Israel, Italy, The Netherlands, Japan, Mexico, Poland, Portugal, Sweden, Switzerland, Turkey, the United States of America, the Union of Soviet Socialist Republics, and the Federal Republic of Germany. A seasonal thermal energy storage technology information system data base was initiated at the

Lawrence Livermore National Laboratory. The data base contains four resources: (1) administrative, (2) bibliographic, (3) computational, and (4) communications.

Physicochemical Analysis: The objectives of the physicochemical analysis were (1) to conduct laboratory tests on the stability of physico-chemical properties of aquifer materials subjected to seasonal thermal energy conditions, (2) to develop laboratory and field facilities dedicated to the support of seasonal thermal energy storage projects, and (3) to develop geochemical modeling capabilities tailored to seasonal thermal energy storage operational and environmental concerns.

Laboratory permeability/creep compaction tests were completed on samples of Massillion sandstone and Ottawa sand. At increased temperatures (25° to 150°C) and pressure [150 bars (2175 psi) confining pressure, and 60 bars (870 psi) confining pressure], a decrease in permeability was observed which agrees with published literature.

Numerical Simulation: The objective of this task was to develop numerical simulation codes to assess aquifer thermal energy storage concepts and application of these codes in support of field tests. A report[17] was published that reviews aquifer state-of-the-art modeling efforts including (1) physical, chemical, and energy mechanisms; (2) mathematical treatment of transport mechanisms; (3) simulation techniques for permeable media; (4) techniques for analyzing geochemistry; and (5) site-specific considerations.

Lawrence Berkeley Laboratory, under subcontract to PNL, is continuing work on numerical modeling with the Conduction, Convection, and Consolidated (CCC) Model. Specific tasks are (1) to perform generic simulations and study the relationships between parameter groups describing the Aquifer Thermal Energy Storage System, (2) to conduct site-specific simulations and provide input for design and operational scenario at field test facilities, and (3) to complete preliminary investigations leading to nonisothermal modeling of unconfined aquifers overlain by unsaturated porous media.

Nonaquifer Seasonal Thermal Energy Storage: Nonaquifer thermal energy storage literature was surveyed and evaluated; selection of potential nonaquifer seasonal thermal energy storage concepts was completed. Based on the preliminary evaluation, the most promising nonaquifer seasonal thermal energy storage concepts worthy of detailed evaluation today are[14]

- Large engineered insulated-pond thermal energy storage with <95°C (sensible seasonal thermal energy storage heating),

- Wet soil thermal energy storage with <95°C water (sensible seasonal thermal energy storage heating),

- Natural lakes and ponds, solar ponds, rocks, and water in mines, caverns, and large tanks (sensible seasonal thermal energy storage heating),

- Ice and compacted snow thermal energy storage (latent seasonal thermal energy storage cooling), and

- Sulphuric acid/water heat pumps at 66° to 200°C (150.5° to 392°F) (thermochemical seasonal thermal energy storage heating and cooling).

3.3 Process Heat and Heat Transport

SERI is currently investigating thermochemical energy storage and transport processes, chemical and physical rates involved, required equipment, and resultant first and second law thermodynamic efficiencies. The projected applications include solar thermal electric and industrial process heat applications to more diverse transport applications with sources such as nuclear reactors and coal gasifiers (Figure 10).

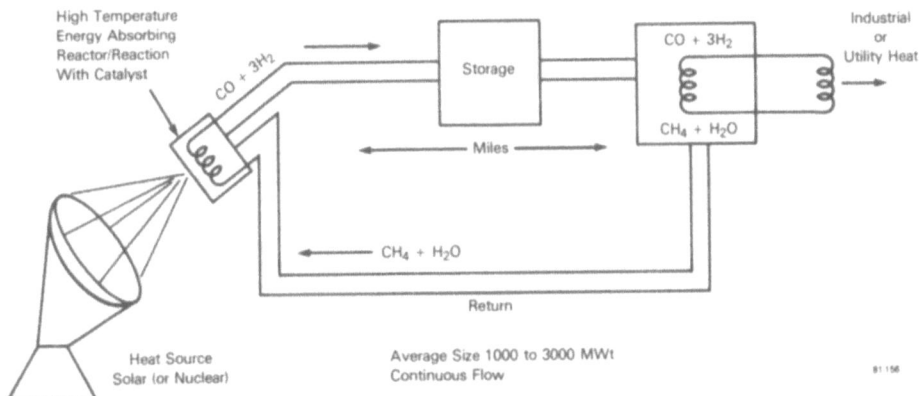

Figure 10. Heat Storage and Transmission by Reversible Chemical Reactions

Mass and energy balances, heat and mass flows, along with primary processing performance requirements and operating conditions are used to determine the equipment necessary for the process. Total installed plant investment costs are determined based upon equipment costs.

A parallel laboratory support task is structured to answer questions for both catalytic and noncatalytic reaction systems. Side reactions are analyzed because impurities tend to accumulate in a recycle process necessitating purge and purification facilities. Laboratory reaction data and literature data will provide the basis for reactor design and optimization.

Some of the conclusions reached during the feasibility study in FY 1980 are noted below.[9]

- Thermochemical energy transport systems have efficiencies equal to or better than direct thermal energy transport systems at distances greater than 32.2 kilometers (20 miles).

- The cost of chemicals and pipelines for thermochemical energy transport can be substantially less than the cost of materials and hot pipelines for direct thermal energy transport.

- The cost of delivered energy for thermochemical energy transport is inadequately defined at present.

- Thermochemical energy storage has a lower efficiency than sensible heat storage in molten salt for diurnal storage. The media cost for direct thermal storage can be significantly greater than for thermochemical energy storage (i.e., 40 to 60 times).

Tables 3 and 4 summarize the basis for the above conclusions.

The value of thermal storage is calculated for identified solar thermal applications and used for thermal storage cost goals. A ranking methodology is employed to define the most promising thermal storage concepts for specified solar thermal applications.

Stearns Rogers Services, Inc., a SERI subcontractor, is to provide calculated cost data and ranking for thermal storage concepts. The solar thermal storage concepts are for water/steam, organic fluid, and air/Brayton solar thermal systems. Applications of the storage concepts include (1) storage for electric power generation, (2) process heat, and (3) total energy systems. Power-related costs, energy-related costs, and first-year variable costs, including operations and maintenance and other annually recurring costs, are determined for the "value of thermal storage" in solar thermal power applications. A report on this activity is to be published June 1981.

Table 3. Preliminary Transport Study Results

Reaction	Endothermic Reactor Temperature (K)	Delivery Temperature Isothermic/Cascade (K)	Efficiency* (%)	Effective Media Cost (¢/kBtu)
$SO_3 = 1/2O_2 + SO_2$	1150	750/1100	89	21.50
$CH_4 + CO_2 = 2CO + 2H_2$	1100	800/1000	70	4.89
$CH_4 + H_2O = CO + 3H_2$	1100	800/1000	66	2.45
$2NH_3 = N_2 + 3H_2$	1023	723/750	52	13.20

*Efficiency = Heat Out/Heat In + Power In.

Note: Draw Salt Costs 147.8 ¢/kBtu.

81-156

Table 4. Preliminary Storage Study Results

Reaction	Endothermic Reactor Temperature (K)	Delivery Temperature (K)	Efficiency* (%)	Effective Media Costs (¢/kBtu)	Energy Density (Btu/lb)
$Ca(OH)_2 = CaO + H_2O$	866	650	69	3.71	634
$CH_4 + CO_2 = 2CO + 2H_2$	1000	800	69	4.96	1773
$CH_4 + H_2O = CO + 3H_2$	1000	800	65	2.49	2608
$BaO_2 = BaO + 1/2 O_2$	1023	873	62	114.90	421

*Efficiency = Heat Out/Heat In + Power In.

Note: Draw Salt Costs 145.8 ¢/kBtu.
      Draw Salt Energy Density 159 Btu/lb Salt.

81-156

3.4 Solar Thermal Power

Sandia National Laboratories Livermore (SNLL) manages the Thermal Energy Storage for Solar Thermal Applications (TESSTA) for DOE. The DOE Central Solar Technologies and Physical and Chemical Energy Storage Divisions have established a technology development program in direct support of solar thermal power applications. Solar thermal applications of interest to this program include concentrating troughs, dishes, and central receivers with working fluids at various operating conditions. The overall objectives of the program are[18]

- To develop second-generation storage subsystems offering cost/performance improvements over the first-generation storage subsystems currently being developed for solar thermal power applications, and

- To develop first-generation storage subsystems for those solar thermal applications that currently have no storage subsystems under development.

As shown in Figure 11, the program has been divided into seven major elements:

- General activities,
- Storage for water/steam cooled collectors/receivers,
- Storage for molten salt cooled sensible heat collectors/receivers,
- Storage for liquid metal cooled sensible heat collectors/receivers,
- Storage for gas cooled sensible heat collectors/receivers,
- Storage for organic or silicone fluid cooled sensible heat collectors/receivers, and
- Dish-mounted latent heat buffer storage.

The development status for these tasks is described in Table 5.

General Activities: These activities include a technology base to support storage subsystem development for future solar thermal power application.

Storage for Water/Steam Cooled Collectors/Receivers: Advanced storage concept development for water/steam cooled receivers and for superheated steam receivers is being conducted by Combustion Engineering and by Babcock and Wilcox, respectively.

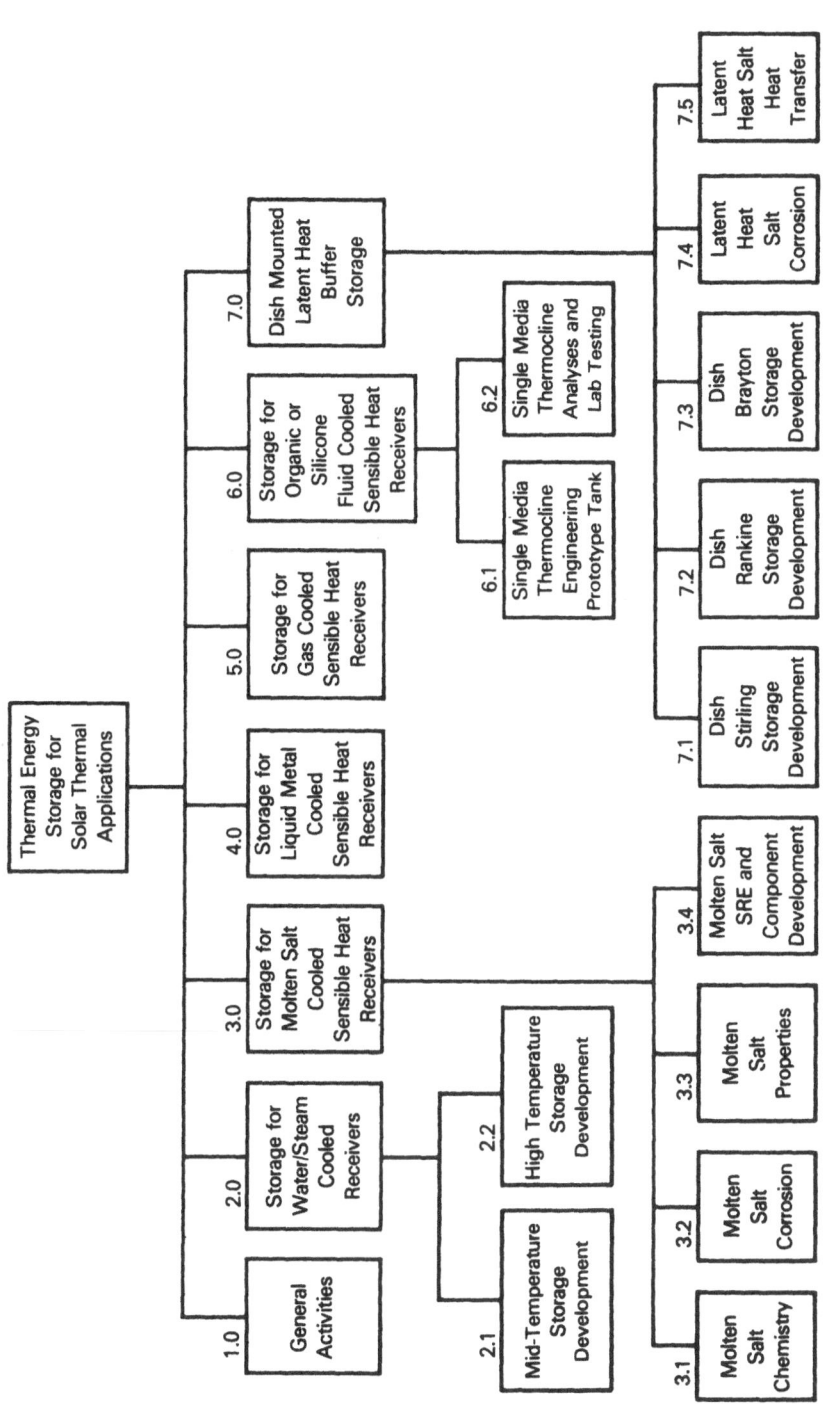

Figure 11. SNLL Task Outline

Table 5. Storage Technology Development Status for Solar Thermal Systems[18]

Name	Solar Interface	Baseline Storage Technology	TESSTA Task Element	TESSTA Development Activity Status
Barstow*	Central Receiver— Superheated Steam	Organic/Rock Thermocline	2.0	Second-Generation Conceptual Design Complete
Utility Repowering	Central Receiver— Superheated Steam	None	2.0	Assessment of Barstow Second-Generation Design Complete
	Liquid Metal	Liquid Metal 2-Tank	4.0	No Activity
	Molten Salt	Molten Salt 2-Tank	3.0	Subsystem Development Underway— Materials Studies and SRE
IPH Retrofit	Central Receiver— Saturated Steam	None	2.0	Latent Heat Storage Conceptual Design Complete
	Air	None	5.0	No Activity
	Organic	None	6.0	MSTF Testing Underway
Cogeneration	Central Receiver— Superheated Steam	TBD	2.0	No Activity— Awaiting Requirements Definition
	Air	Slag Thermocline	5.0	
	Molten Salt	Molten Salt 2-Tank	3.0	
IEA (1)*	Central Receiver— Liquid Metal	Liquid Metal 2-Tank	4.0	No Activity
Shenandoah*	Dish—Silicone	Silicone Thermocline	6.0	No Activity
Small Community System Experiment (SCSE)	Dish—Organic	None	7.0	Buffer Storage Requirements Definition and Conceptual Design Complete
Military Power Application Experiment (EE-2)	Dish—Air	None	7.0	Buffer Storage Requirements Definition and Conceptual Design Underway
Coolidge*	Trough—Organic	Organic Thermocline	6.0	MSTF Testing Underway
IEA(2)*	Trough—Organic	Organic Thermocline	6.0	MSTF Testing Underway
Modular Industrial Solar Retrofit (MISR)	Trough—Organic	TBD	6.0	No Activity— Awaiting Requirements Definition

Note: Applications With an Asterisk (*) Are Either Operational or in the Construction Phase. For These Cases Storage Development Is Directed Toward a Retrofit.

The preferred storage concept recommended by the Combustion Engineering study is a latent heat storage module design (Figure 12). The module consists of a 4- x 4- x 12-m (13- x 13- x 40-ft) rectangular, externally insulated carbon steel tank in which five tube assemblies are placed. The storage media is an 18.5 $NaNO_3$–81.5 NaOH (mole %) salt eutectic with a melting point of 256°C (493°F).

The storage module is charged by condensing 288°C (550°F) saturated steam into saturated liquid. On discharge, the storage module produces 232°C (450°F) saturated steam from saturated liquid. The storage capacity of a module is 19.0 MW_th.

The preferred storage concept recommended by the Babcock and Wilcox study is a sensible heat storage design. The design uses rectangular bin-type steel tanks which contain three separate and independent compartments (Figure 13). Each tank is 28 m (92 ft) wide and 82.3 m (270 ft) long with the vertical sides being 5.1 m (16.8 ft) high. The storage media is sand.

A tank compartment is charged by moving sand with an Archimedes lift from a cold tank compartment to a charging heat exchanger mounted at the top of a hot tank compartment. The sand travels through the heat exchanger over the water- and steam-filled tubes which heat the sand from 218°C (425°F) to 332°C (630°F). Discharging storage to produce steam at a temperature of 299°C (570°F) is accomplished in a similar manner. The storage capacity of a four-tank configuration is 2796 MW_th.

The Babcock and Wilcox study results indicate that the sand-bed concept offers marginal improvement in cost and performance over the first-generation oil/rock thermocline design.. An assessment suggested that the concept may be more appropriate for use with higher temperature systems, such as liquid-metal or gas-cooled receivers.

The first-generation storage task supported testing for the thermal storage tank for the 10-MW_e central receiver pilot plant near Barstow, California. The storage tank will contain Caloria HT 43 heat transfer oil and rock/sand storage media, with the thermocline principle applied to store the hot and cold media in the same tank. The system operates over the temperature range of 218° to 304°C (425 to 580°F) and is sized to deliver 7 MW_e over a 4-hour period. The tank has an inside diameter and total height of about 18.3 m (60 ft) and 14.6 m (48 ft), respectively.

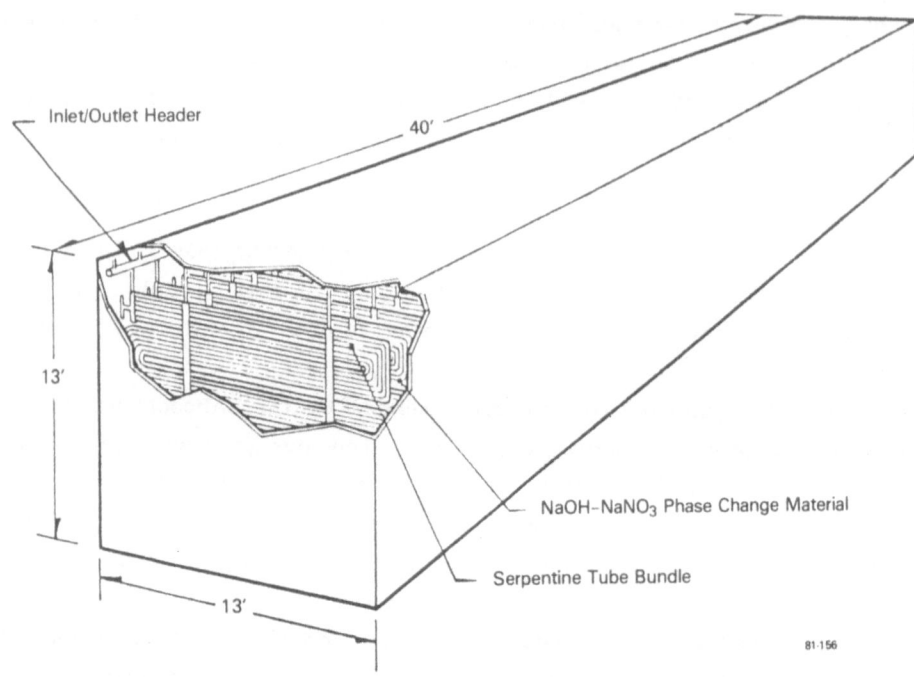

Figure 12. Latent Heat Thermal Storage Module

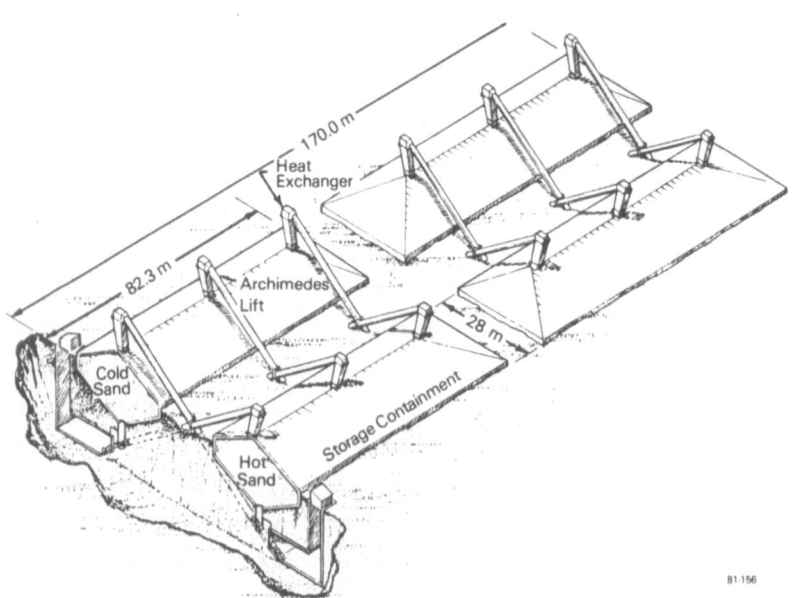

Figure 13. Moving-Bed Thermal Energy Storage

- 492 -

Storage for Molten Salt Cooled Sensible Heat Collectors/Receivers: The goal of the molten salt technology element is to provide technical information and data on molten nitrate salt to future designers and engineers of near-term central receiver applications.

The first-generation storage subsystem used a molten nitrate salt ($NaNO_3/KNO_3$ mixture) storage media with externally insulated hot and cold tanks. Containment materials lifetime and storage media lifetime/maintenance were issues identified in the development of the first-generation systems.

A molten salt physical properties (viscosity, surface tension, and density) study of the molten equimolar mixture $NaNO_3$-KNO_3 over the temperature range of 300° to 600°C (572 to 1112°F) in argon and oxygen is complete.[19] The report summarizes (1) what new information and data are available, (2) how and when the information and data will impact near-term solar thermal power applications, and (3) the experimental work being performed outside SNLL.

Corrosion tests with three alloys [Incoloy 800 (I800), 304 stainless steel, and 316 stainless steel] were selected for study in molten salt convection loops. The temperature range of the salt in convection loops was 350° to 600°C with a salt flow rage of about 0.5 cm sec^{-1}. The test results are noted below.[20]

- Below 600°C, corrosion is relatively benign although samples lose weight due to chromium depletion. At temperatures about 600°C, corrosion and depletion rates are several times faster. Further experiments are required to clarify the temperature limit for acceptable corrosion rates.

- There is no thermal gradient mass transport of chromium. Once the chromium is leached from the alloy, it remains in the salt and does not precipitate in the cold leg of the loop.

- Weldments behave similarly to the base metal.

- Preliminary results indicate similar corrosion rates between the open or breathing loops and those employing an inert cover gas.

The Martin Marietta Company was awarded a contract to provide (1) preliminary design and costing of a full-size storage tank for molten salt cooled collector/receiver at temperatures of 566°C (1050°F); (2) component development for internally insulated liner; (3) design, construction, and evaluation of a 6.9-MW$_t$ storage subsystem experiment that is scalable to full size; and (4) assessment of full-size storage subsystem market potential. A conceptual drawing of the molten salt storage tank is illustrated in Figure 14.

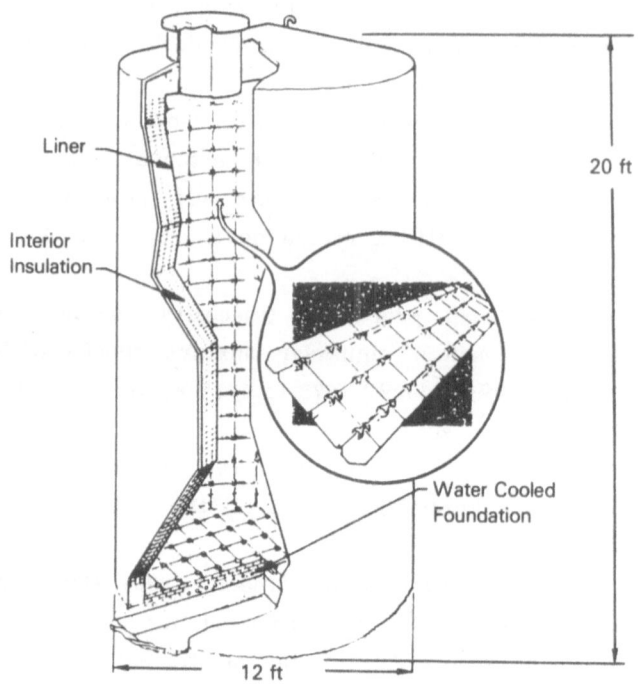

Liner

Interior
Insulation

20 ft

Water Cooled
Foundation

12 ft

81-156

Figure 14.   Molten Salt Storage Tank

Storage for Liquid Metal Cooled Sensible Heat Collectors/Receivers:
The goal of this element is to develop a second-generation storage
subsystem for an International Energy Agency (IEA) retrofit application
and possible repowering/industrial retrofit application.   The first
generation for advanced central receiver systems uses liquid sodium as the
storage   medium   with   externally   insulated   hot   and   cold   tanks.
Second-generation storage subsystem development would have emphasized
low-cost containment techniques such as internal insulation.

A potentially low-cost air/rock storage concept has been proposed for
use with liquid metal receivers.   Preliminary data by SNLL show little
evidence of rock fracture when thermally cycled (approximately 600 cycles)
from 316° to 593°C (600° to 1100°F) while under a mechanical load.[18]

Storage for Gas Cooled Sensible Heat Collectors/Receivers: The goal
of this element is to develop an energy storage subsystem capable of
operating at temperatures up to 816°C (1500°F).   Due to high-temperature

and high-pressure [3.4 MPa (500 psi)] operation, the conceptual design uses costly welded steel containment vessels. Development included verification of thermocline performance and low-cost storage media (e.g., rocks and internally insulated vessels). Testing of a refractory brick thermocline system test module was completed for a dish-mounted collector as part of the Small Power Systems Program. Test results confirmed the existence of a sharp thermocline under a range of air flow, inlet temperature, and pressure test conditions. The test module performed well after 100 cycles and at temperatures above 704°C (1300°F). At 927°C (1700°F), surface galling occurred on a valve after a few cycles.

Storage for Organic or Silicone Fluid Cooled Sensible Heat Collectors/Receivers: This element provides support for storage systems under development for mid-temperature [399°C (750°F)] solar thermal applications, such as irrigation, cogeneration, and repowering/industrial retrofit. First-generation organic fluid thermocline systems have been constructed for irrigation applications (Willard, New Mexico, and Coolidge, Arizona). Operational testing indicated that further development is warranted before the system characteristics are satisfactory. Second-generation storage development is to provide for continued development of diffuser designs that are used to separate the thermocline.

Dish-Mounted Latent Heat Buffer Storage: This element includes the development of a storage subsystem for small community systems or remote defense installations. The storage system under consideration uses a storage module mounted directly on a collector dish or in a counterweight position. The concept requires that the modules be small, lightweight, and configured to cast minimum shadow.

Ford Aerospace and Communications Corporation completed a conceptual design of a latent heat buffer storage system for a Dish/Organic Rankine system. The conceptual design is illustrated in Figures 15 and 16. Three phase-change materials having different melting points are used due to a large axial temperature gradient within the receiver. The phase change materials are $NaNO_3$-KNO_3 with a melting point of 220°C (428°F), KNO_3 with a melting point of 334°C (633°F), and K_2CO_3-Li_2CO_3-Na_2CO_3 with a melting point of 385°C (725°F). The storage capacity is 4 kW$_t$h which is sufficient to eliminate 70 to 80 percent of potential shutdowns due to cloud cover.

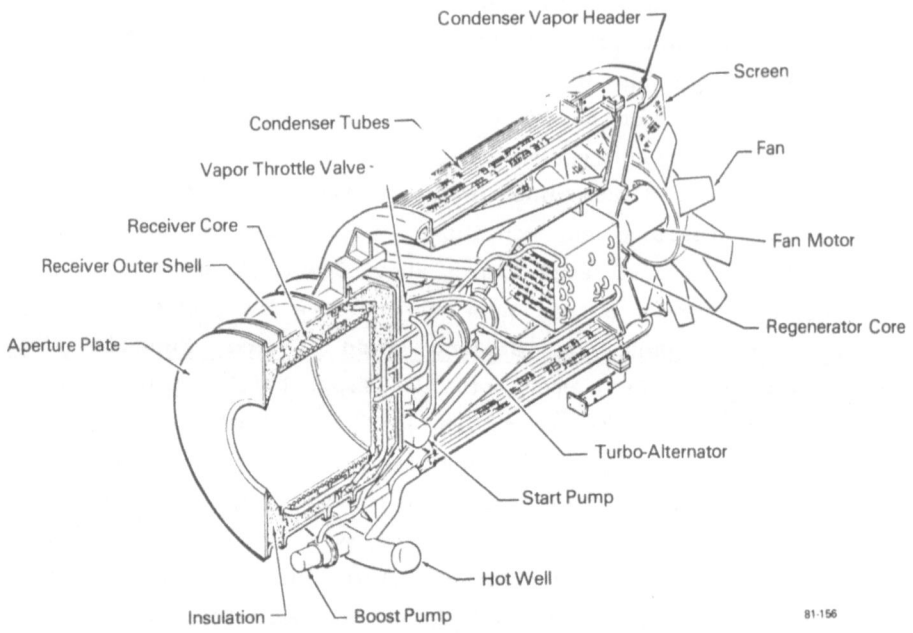

Figure 15. Dish/Organic Rankine Power Conversion Assembly

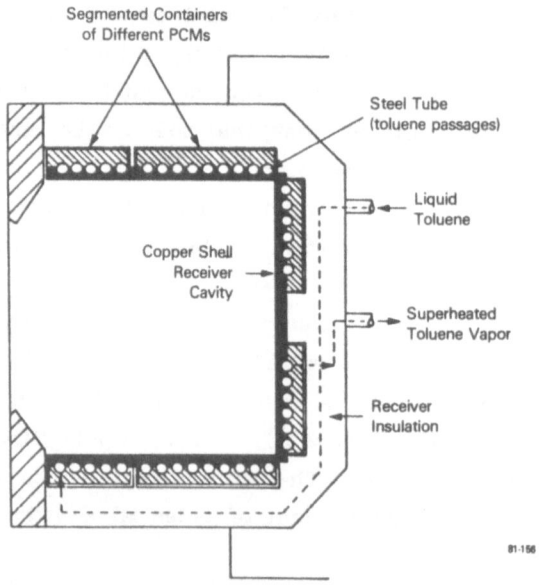

Figure 16. Integrated Dish Receiver/Storage Design

- 496 -

The Space Division of General Electric Company also completed a conceptual design of a latent heat buffer storage system for a Dish/Stirling system. The concept, illustrated in Figure 17, requires a heat pipe solar receiver to transfer energy to a storage phase change material [NaF-MgF$_2$ salt with a melting point of 826°C (1500°F)] and a secondary heat pipe to transfer energy from the phase change material to the engine. The storage capacity is 53 kW$_t$h which is sufficient to provide a power output of 25 kW$_e$ for 8 hours.

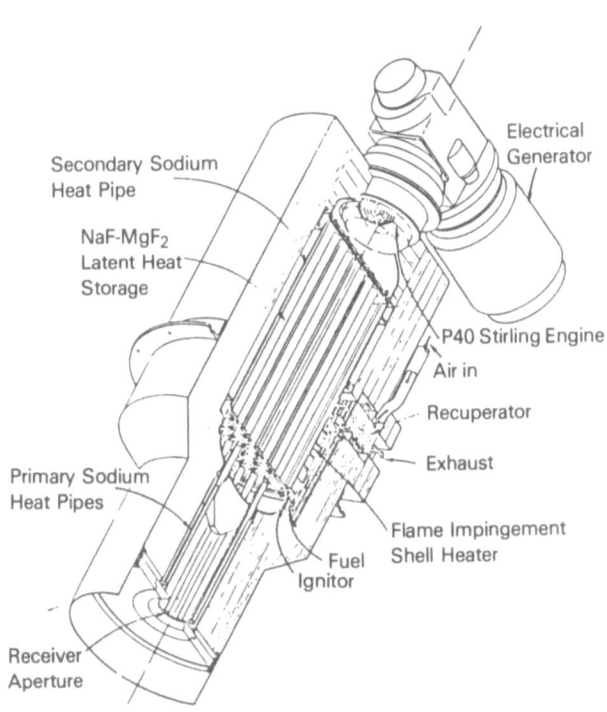

Secondary Sodium
Heat Pipe

NaF-MgF$_2$
Latent Heat
Storage

Primary Sodium
Heat Pipes

Receiver
Aperture

Electrical
Generator

P40 Stirling Engine

Air in

Recuperator

Exhaust

Flame Impingement
Shell Heater

Fuel
Ignitor

Figure 17. Dish/Stirling Latent Heat Storage System

4. PLANNED ACTIVITIES FOR 1982

The building heating and cooling subprogram activity will include development of phase-change materials for short-term storage components for active systems. For passive solar heat storage in building materials, the development of phase-change substances will continue. Design studies will be conducted on building heating systems that use waste heat recovery, storage, and transport technology.

The development of the sulfuric acid/water heat pump and the mixed hydride-hydrogen heat pump will be continued through the chemical heat pump subprogram.

In the seasonal thermal energy storage subprogram, water mobility studies, heat retention evaluations, and physical and chemical analysis of water and aquifer materials will be undertaken by Auburn University (Mobile, Alabama) and the University of Minnesota (St. Paul, Minnesota) at their experimental field test facilities. The studies by Auburn will be conducted at temperatures of up to 125°C (257°F). Similar experiments with high-temperature water will be conducted at the University of Minnesota. The aquifer experiments will be supplemented by numerical simulations. Hot water tests will be conducted at increasing temperatures to determine recovery effects and thermal efficiencies. Long-term injection experiments will be conducted and institutional problems addressed. Nonaquifer seasonal thermal energy storage concepts studies including engineering and economic assessments of promising concepts and selected research efforts on new concepts will also be investigated.

Technology support for solar thermal systems includes development of improved storage components, together with heat transfer and long distance heat transmission technology. Projects that will be supported include storage for liquid metal receivers, parabolic dish-Stirling engine systems, direct contact heat exchange processes, and thermochemical reactions for heat transport.

5. REFERENCES

1. "Distributed Thermal Energy Storage in the Residential Sector: Commercialization Readiness Assessment and Implementation Strategy," U.S. Department of Energy, Assistant Secretary for Conservation and Solar Energy, Office of Advanced Technologies, DOE/CS-0195, August 1980.

2. John Douglas, "Movement Toward Conservation," Electric Power Research Institute Journal, Vol. 6, No. 2, March 1981.

3. Joseph E. Rizzuto, "Design and Demonstration of a Storage-Assisted Air Conditioning System," Proceedings of the DOE Thermal and Chemical Storage Annual Contractors' Review Meeting, Conference 801055, McLean, Virginia, March 1981.

4. J.F. Martin, "ORNL Thermal Energy Storage Program Overview," Proceedings of the DOE Thermal and Chemical Storage Annual Contractors' Review Meeting, Conference 801055, McLean, Virginia, March 1981.

5. A.I. Michaels, "An Overview of the U.S.A. Program for the Development of Thermal Energy Storage for Solar Energy Applications," Thermal Storage of Solar Energy, (ed. C. den Ouden), Martinus Nijhoff Publishers, The Hague, 1981.

6. C.D. MacCracken, "Salt Hydrate Thermal Energy Storage System for Space Heating and Air Conditioning," Proceedings of the DOE Annual Active Solar Heating and Cooling Contractors' Review Meeting, ConferEnce 800340, Incline Village, Nevada, March 26-28, 1980.

7. R. Deal and A.D. Solomon, "The Simulation of Four Pure Conduction Paraffin-wax Freezing Experiments," ORNL/CSD-74, Union Carbide Corporation, January 1981.

8. Mark P. Ternes, "Crawl Space Assisted Heat Pumps," Proceedings of the DOE Thermal and Chemical Storage Annual Contractors' Review Meeting, Conference 801055, McLean, Virginia, March 1981.

9. C.E. Wyman, R.J. Copeland, R.G. Nix, and J.D. Wright, "Solar Energy Storage Program: FY 80 Annual Report," SERI/PR-631-1132, Solar Energy Research Institute, Golden, Colorado, April 1981.

10. E.C. Clark, D.K. Carlson, and O.M. Morgan, "The Sulfuric Acid/ Water Chemical Heat Pump/Energy Storage Program," Proceedings of the DOE Thermal and Chemical Storage Annual Contractors' Review Meeting, Conference-801055, McLean, Virginia, March 1981.

11. H.G. Lorsch and M.A. Baker, "Survey of Commercial Thermal Storage Installations in the United States and Canada," Annual DOE Technical and Economic Analysis Contractors' Review Meeting (proceedings in preparation by Argonne National Laboratory), Chicago, Illinois, April 22-23, 1981.

12. Lincoln B. Katter, "Applications of Thermal Energy Storage to Process Heat Recovery in the Aluminum Industry," Proceedings of the DOE Thermal and Chemical Storage Annual Contractors' Review Meeting, Conference 801055, McLean, Virginia, March 1981.

13. J. Carr, et al., "Applications of Thermal Energy Storage to Process Heat Storage and Recovery in the Pulp and Paper Industry," NASA-CR-159398, CONS-5082-1, September 1978.

14. J.E. Minor, "Seasonal Thermal Energy Storage Program Progress Report, January 1980 - December 1980," PNL-3746, Pacific Northwest Laboratory, Richland, Washington, March 1981.

15. P.L. Hendrickson, "Legal and Regulatory Issues Affecting the Aquifer Thermal Energy Storage Concept," PNL-3437, Pacific Northwest Laboratory, Richland, Washington, October 1980.

16. C.F. Meyer and W. Hausz, "Guidelines for Conceptual Design and Evaluation of Aquifer Thermal Energy Storage," PNL-3581, Pacific Northwest Laboratory, Richland, Washington, October 1980.

17. J.W. Mercer, C.R. Faust, W.J. Miller, and F.J. Pearson, Jr., "Review of Simulation Techniques for Aquifer Thermal Energy Storage (ATES)," PNL-3769, Pacific Northwest Laboratory, Richland, Washington, March 1981.

18. L.G. Radosevich, "Thermal Energy Storage for Solar Thermal Applications Program Progress Report (April 1980 - March 1981)," SAND81-8225, Sandia National Laboratories Livermore, Livermore, California, May 1981.

19. R.W. Carling, C.M. Kramer, R.W. Bradshaw, D.A. Nissen, S.H. Goods, R.W. Mar, J.W. Munford, M.M. Karnowsky, R.N. Biefeld, and N.J. Norem, "Molten Nitrate Salt Technology Development Status Report," SAND80-8052, Sandia National Laboratories Livermore, Livermore, California, March 1981.

20. R.W. Carling, "Molten Nitrate Salt Materials Studies," Proceedings of the DOE Thermal and Chemical Storage Annual Contractors' Review Meeting, Conference 801055, McLean, Virginia, March 1981.

NOTICE

INDEX